AF772780

The Petrochemical Handbook

The Petrochemical Handbook

Edited by **Andy Margo and**
Michael Dedini

NY RESEARCH PRESS
New York

Published by NY Research Press,
23 West, 55th Street, Suite 816,
New York, NY 10019, USA
www.nyresearchpress.com

The Petrochemical Handbook
Edited by Andy Margo and Michael Dedini

International Standard Book Number: 978-1-63238-440-9 (Hardback)

Printed in China.

Contents

Preface

There are thousands of types of physical materials that we work with every moment in our lives, and new ones are being discovered every day. Some are naturally occurring whole other are artificial and man- made. Petrochemicals fall under the man-made category. Petrochemicals are products that are derived chemically from petroleum. Typically extracted during the refining process of crude oil and natural gas, they are utilized in a variety of ways. There are so many day to day products that we use but never realize are petrochemical products, like plastic, soaps, pharmaceuticals and fertilizers. The synthetic fibers of a sweater to even the detergents we use are petrochemicals. Even these end products are being sidelined in the need to discover newer and better by products and end-use products from the refining process. Petrochemical research pertains to this arena. Newer methods of refining crude oil that lead to better and numerous petrochemical products as well as research that creates these products is what makes the foundation of this field. The research areas can be as diverse as the need to identify problems in the oil and petrochemical industries to monitoring scientific and technological development in petrochemicals as well as constantly creating a new breed of petrochemicals.

This book applies itself towards the collation of such research that brings all new technologies and methodologies under one cover. I am extremely thankful to those whose hard work has made this book a success as well as the readers who support us.

Editor

1

Highly active and selective catalyst for synthetic natural gas (SNG) production

Yu Huang · Haoyi Chen · Jixin Su · Tiancun Xiao

Abstract A novel nickel-based methanation catalyst has been prepared and tested for CO and CO_2 hydrogenation to synthetic natural gas under various conditions. The catalysts before and after reaction have been characterized using XRD, Laser Raman and IR spectroscopes. It is showed that the novel catalyst can give high methane selectivity and yield at a very broad ranges of temperature and H_2/CO ratios, although the selectivity to methane increase with the H_2/CO ratio, which, however, increases with the temperature rising from 200 to 350 °C, then gradually decrease. The characterization results showed that the there is surface carbons forming over the catalyst surface, and the nickel crystalline structure changes during the reaction. Despite this, the catalyst still gives the same performance, which suggested that the catalyst can operate in a very broad range conditions.

Keywords Nickel catalyst preparation · Co-precipitation · Methanation · Synthetic natural gas

Introduction

Biomass development is predicted to grow the fastest of all renewable energy in the next decade, although it is already being utilized for a variety of purposes, including biofuels, biopower (electricity), biomaterials, biochemicals ("green chemistry"), and biopellets. The different categories of biomass are all interrelated, and of all the sources of renewable energy, biomass can be most honestly labeled "home-grown." Among these applications, **S**ynthetic **N**atural **G**as (SNG) production from biomass has attracted increasing attention in recent years, due to the rising price of natural gas, the wish for less dependency from natural gas imports and the opportunity of reducing green house gases. Although solid dry biomass has [1–5] been used directly (e.g., wood and straw) for ages, converting them into SNG has is a more preferential process.

The advantages SNG are, besides the relatively high efficiency of its production, the already existed infrastructure such as pipelines and the efficient end use technologies such as compressed natural gas cars, combined heat and power plants, or combined cycle plant. Therefore, in recent years there has been a dramatic increase interest in the manufacture of a clean, high BTU gas energy source which will meet pipeline standards by synthetic means biomass.

SNG normally is produced through the gasified products, e.g., H_2 and CO-containing gas stream. While a number of metallic species are known to be active and selective methanation catalysts including, inter alia, nickel, ruthenium, cobalt and iron, their application to the manufacture of high BTU or pipeline gas has been less than satisfactory for several reasons. First, methanation reactions are strong exothermic reaction, which easily heats up catalyst to 800 °C in an adiabatic reactor. However, these catalyst systems generally cannot stand very high temperature, the reactor temperature needs to be limited to temperatures below 400 °C to avoid sintering and deactivation of the catalyst, so the highly exothermic nature of the methanation reaction itself provides severe operational

Y. Huang · H. Chen · T. Xiao
Guangzhou Boxenergy Technology Ltd, Guangzhou Hi-Tech Development Zone, Guangzhou, People's Republic of China

J. Su
Environmental Engineering College, Shandong University, No. 27 Shanda Nanlu, Jinan, People's Republic of China

difficulties in controlling catalyst temperature when CO concentration of the feed gas is in the range required for methane-rich gas manufacture. Further, the methanation reaction itself is considered to be a combination of several reactions including the primary reaction and secondary reactions (2) and (3)

$$3H_2 + CO \rightarrow CH_4 + H_2O. \quad (1)$$

$$2H_2 + 2CO \rightarrow CH_4 + CO_2 \quad (2)$$

$$4H_2 + CO_2 \rightarrow CH_4 + 2H_2O \quad (3)$$

The thermodynamic calculation results for this process are shown in Table 1.

These thermodynamic equilibriums of the reaction 1 shows that the equilibrium yield of methane is adversely effected at high temperatures, i.e., above 900 K, as can be seen, the delta G at 900 K is +1,742 J/mol, which does not favor the methanation reaction. In fact, it can be seen that reaction (2) is a combination of reaction (1) and the water gas shift reaction (4).

$$CO + H_2O \rightarrow CO_2 + H_2 \quad (4)$$

So far, there have been 3 processes proposed for the SNG production, one is high temperature processes [6–10], which requires very highly stable catalyst and several steps for heat exchange, so as to cool down the reaction stream to improve the catalyst performance and avoid the catalyst deactivation. Another is to the low-temperature methanation [11–13], which proceeds in a reaction configuring with heat exchanger so as to remove the reaction heat quickly, and may have low requirements on the catalyst stability, but requires the catalyst to have high activity and selectivity. So far, the methanation catalyst is normally nickel-based system, which is prepared using impregnation method. The catalyst thus has high dispersion which can also be used in fluidized bed reaction and slurry reactor for SNG production, where the heat transfer is very fast, but attrition is a more important parameter [1, 14–18], also the separation of the catalyst particles from the reactor is also an issue.

Recently, a novel reactor system has been developed by Boxenergy Tech Ltd, which can take the reaction heat away in a fixed bed reactor, thus to maintain a constant low isothermal reaction temperature. In such case, the catalyst with low temperature activity and selectivity is more important. Overview of the previous publications, there are few reports on the low-temperature methanation catalyst for SNG production. It has shown that the organic matrix combustion method can give a Fischer–Tropsch catalyst with super high activity [19, 20], thus in this work, a novel method, e.g., organics assisted catalyst preparation method has been developed, and tested under various conditions, a very nickel based robust methanation catalyst with high activity and selectivity has been developed.

Experimental

Catalyst preparation

The catalyst has been prepared using organic induced partial combustion method. It may be an evolution of the chelating and organic combustion method [21, 22]. In short, a specific amount of 20.8 g $Ni(NO_3)_2{\cdot}6H_2O$, 1.06 g $ZrO(NO_3)_2$, and 5 g citric acid are mixed with 20 ml of water, heating to 50 °C while stirring for 30 min to form a transparent solution. 10 g of AlOOH (beomite) is then added to the solution, and statically placed under ambient conditions for 4 h, dried at 80 °C for 2 h and calcined in static air at 500 °C for 4 h to give a black powder. The nickel content in the black powder is analyzed using atom absorption, and the content is 30.1 %. The resultant powder is pelletized using a mechanical pelletizing machine. When using for catalyst, the sample is crushed and sieved into 60–100 mesh particles.

Catalyst testing

Every time, 0.1 g of the crushed $NiZrO_x/Al_2O_3$ catalyst is loaded in a quartz tube as the reactor. The reactor is

Table 1 The thermodynamic calculations of the methanation reaction *1*

Reaction temperature (K)	Δ_H°(J/molCH$_4$)	Δ_G°(J/molCH$_4$)	Δ_{vol}(l)	Δ_S°(J/K)	Δ_{CP}(J/K)	K_{eq}
298.15	−250176.0	−150843.2	−7.3379E+01	−333.164	−4.642	2.6696E+26
300.00	−250184.6	−150226.8	−7.3834E+01	−333.192	−4.635	1.4327E+26
373.50	−250439.6	−125702.4	−9.1929E+01	−333.965	−1.580	3.7939E+17
600.00	−217924.0	−72548.9	−9.8469E+01	−242.292	−29.890	2.0691E+06
700.00	−220623.5	−48098.5	−1.1488E+02	−246.464	−24.193	3.8820E+03
800.00	−222782.8	−23298.6	−1.3129E+02	−249.355	−19.097	3.3205E+01
900.00	−224463.5	1742.5	−1.4770E+02	−251.340	−14.618	7.9226E−01

The above calculation is from: http://www.crct.polymtl.ca/fact/

inserted into a tubular furnace and connected to the gases. The catalyst is activated with 33.3 (vol) % H_2/N_2 flowing at 70 ml/min and heated at 2 °C/min to 350 °C and held at this temperature for 2 h. Then the reactor is cooled down to 200 °C in the flowing reduction atmosphere.

The gas is switched to a gas mixture of N_2:H_2:CO = 60:30:10 (volume) with a total following rate of 100 ml/min. when testing CO_2 hydrogenation, the mixture gas is set as N_2:H_2:CO_2 = 50:40:10. The other settings are all the same. The products flow through a cold trap where the produced vapour is condensed the flowing rate of the gas in and out were measured using a soap bubble meter, and the dried gaseous product concentration are analyzed using non-dispersive IR spectrometer.

The CO conversion and methane selectivity and yield are calculated as follows:

$$\text{CO conversion: } X_{CO}\,(\%) = \frac{V_{CO.in} - V_{CO.out}}{V_{CO.in}} \times 100$$

$$CH_4 \text{ selectivity:} S_{CH_4}\,(\%) = \frac{V_{CH_4.out}}{V_{CO.in} - V_{CO.out}} \times 100$$

$$CH_4 \text{ yield: } Y_{CH_4}\,(\%) = \frac{X_{CO}S_{CH_4}}{100} = \frac{V_{CH_4.out}}{V_{CO.in}} \times 100$$

where V_{CO}, V_{CH_4}in and out are the volume flowing rate under the ambient conditions calculated from the flowing rate and the content of CO and CH_4 in the inlet and outlet gas stream.

Catalyst characterization

XRD diffraction was carried out in a Philips PW1710 diffractometer equipped with Cu Kα radiation to detect the crystalline phase of nickel and alumina and coke after reaction. The morphology of the catalysts before and after reaction was observed in a JEOL-4000EX high-resolution electron microscope with an accelerating voltage of 400 kV. Raman spectra were recorded with a resolution of 2 cm^{-1} using a Yvon Jobin Labram spectrometer with an He^+ laser, running on a back scattered confocal arrangement.

Results and discussion

Normally when CO is hydrogenated, it is firstly converted into –CH_2– which either link together to form C_nH_m as liquid hydrocarbons. This is a well-known process for Fischer–Tropsch synthesis [23–26]. For SNG production, –CH_2– is expected to further hydrogenated into CH_4 rather than link to form the liquid. In such case, C_2^+ selectivity is more important. In SNG production, CH_4 selectivity as well as yield is more important [13, 15, 27, 28]. Therefore in the following test results, only CH_4 selectivity and yield are presented under various conditions.

Figure 1 gives the temperature effect on the CH_4 selectivity and yield of CO hydrogenation test under different conditions. It is seen that the main product is methane, no C_2^+ hydrocarbons were detected, the methanation reaction starts up from 200 °C, however, significant CH_4 yield appear from 220 °C, which may be due to the low CO conversion, although the methane selectivity is significantly high. The CO selectivity reaches 87 % at 220 °C and almost unchanged from 220 °C to the reaction temperature to 400 °C, while the yield of methane decreases gradually with the methanation temperature. This can be explained by the strong exothermicity of the methanation reaction. The higher methanation temperature may lead to some steam reforming reaction, which is strong endothermic reaction. It is, therefore, inferred that the methanation reaction should be kept at temperatures from 260 to 300 °C. However, as showed before, once methanation starts, the generated heat is huge, thus how to remove the reaction heat away quickly is an issue when applying the low-temperature methanation process [29].

In contrast to the CO methanation, CO_2 hydrogenation to methane showed different trends. As shown in Fig. 2, the CO_2 hydrogenation starts up at 200 °C, although the methane yield is significantly lower, only about 1.4 % over the $NiZrO_x/Al_2O_3$ catalyst. When the reaction temperature is raised to 250 °C, methane selectivity increases to 95 %, while the CH_4 yield is about 35 %. When the methanation temperature rises to 280 °C, methane selectivity reaches about 98 %, while the CH_4 yield is about 62 %, suggesting that the catalyst prepared in this way gave a high CH_4 selectivity, although its activity for CO_2 conversion is lower than that for CO hydrogenation. The CH_4 yield reaches the maximum at 340 °C, while the CH_4 selectivity is about 98 %, suggesting that the suitable operation conditions for CO_2 methanation is 340 °C when pressure is 1 bar. Also the methanation temperatures above 340 °C have less effect on the CH_4 selectivity, and CO_2 conversion cannot reach more than 90 %.

CO_2 methanation performance with the temperature can be explained by the reaction process steps. Generally, CO_2 hydrogenation has two steps

$$CO_2 + H_2 \rightarrow H_2O + CO \quad (5)$$

$$CO + 3H_2 \rightarrow CH_4 + H_2O. \quad (6)$$

Reaction 5 is endothermic, and reaction 6 is strong exothermic. The overall reaction heat is less than CO methanation, therefore its CH_4 yield increase in a broader temperature range than the CO only methanation.

The stability of the catalyst for CO methanation with time at 250 °C and 1 bar and GHSV of 60,000 h^{-1} is

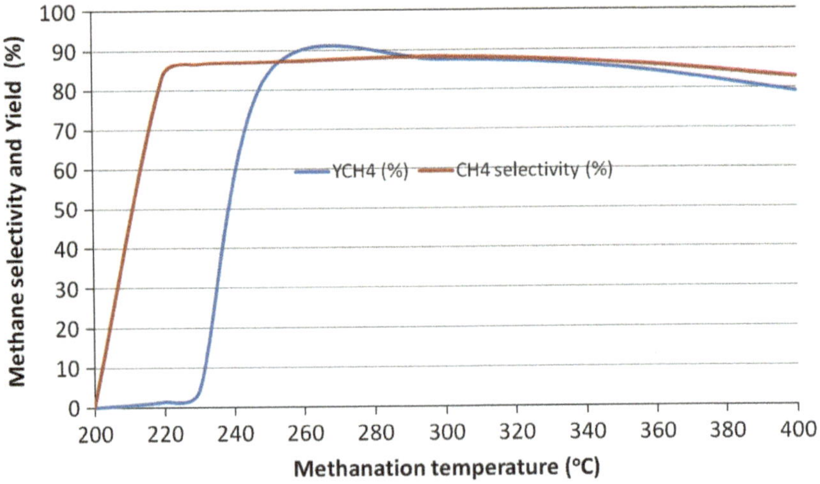

Fig. 1 Effect of temperature on the methanation reaction. Test conditions: GHSV = 60,000 h^{-1}, N_2:H_2:CO = 6:3:1

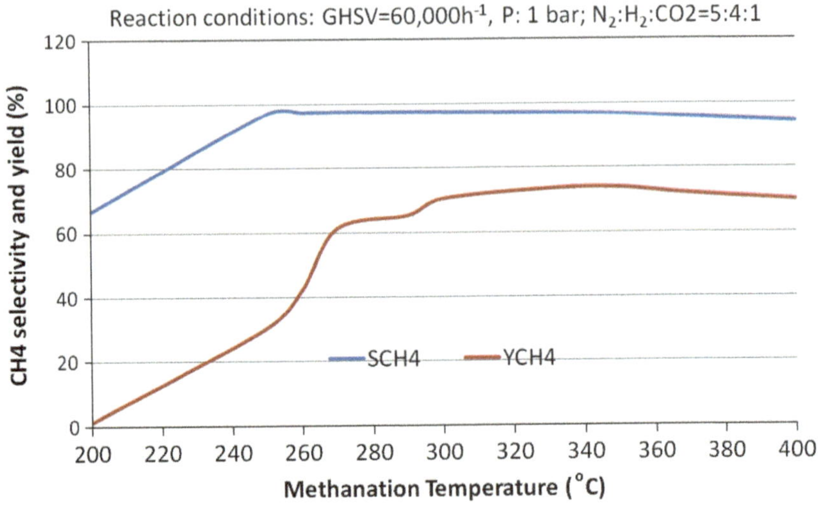

Fig. 2 Effect of reaction temperature on CO_2 methanation under the conditions of: GHSV = 60,000 h^{-1}, P: 1 bar; N_2:H_2:CO_2 = 5:4:1

shown in Fig. 3. The CO conversion under this condition at the start of run is nearly 100 %, although the selectivity to methane is about 95 %, and the CH_4 yield is 92.2 %. The CH_4 selectivity drops to nearly 94 % after 3.5 h running, but changes to 95 %, which drops again after 8 h to 94 %. This change might result from the analysis errors, because the gas products were detected by non-dispersive IR. Thus, it can be inferred that the CH_4 selectivity in fact does not change much during the 600 min time on stream. However, the CH_4 yield as shown by the blue curve in Fig. 3 changes from 92.1 % at the start of the reaction to 89.8 %, which suggests that the CO conversion may drops during the reaction. From the decline trend it can be found that the yield drops faster in the first 260 min, but slower after 430 min. This can be explained by that the fresh catalyst in the initial reaction stage may be more sensitive to the reactants, the reactant can induce the catalyst surface change or leads to carbon deposition, which may cover some active site, thus leading to the activity drop. With the time on stream, the active site tends to go to steady state, the carbon deposition rate is less or equivalent to the carbon hydrogenation rate, thus the yield tends to be stable. It is interesting to see that although the CH_4 yield drops slowly with the time on stream, which may be due to the loss of some active site or the nickel metal sintering, the methane selectivity over the supported nickel catalyst almost unchanged, suggesting that the catalyst active site change or nickel metal structure change does not alter the methanation reaction on the other active sites.

The effect of H_2/CO ratio on the products selectivity has also been studied over the $NiZrO_x/Al_2O_3$ catalyst. A methanation process normally involves in CO absorption over nickel site, which is then attacked by H_2 to give CH_4 and H_2O [30]. If there is not enough H_2 around the activated CO, it may convert into carbon and CO_2. So when H_2/CO ratio is 1.5, the selectivity to methane is about 80 %, and CO_2 about 20 %, no other hydrocarbons are produced. This suggests that the prepared nickel catalyst can work under low H_2/CO ratio gas mixture. With the H_2/CO ratios increasing, the CO_2 selectivity gradually drops,

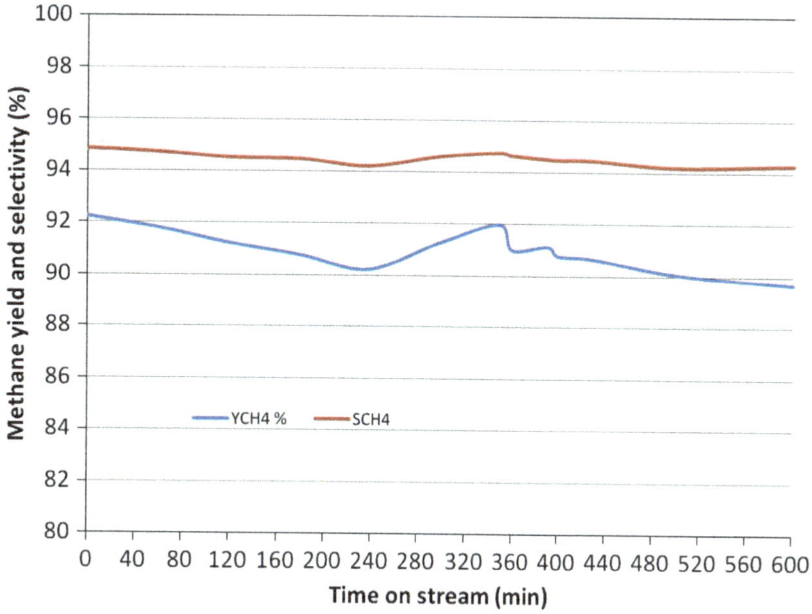

Fig. 3 Stability test of CO methanation for SNG production under the conditions of: Methanation reaction conditions: $H_2:CO:N_2 = 30:10:60$, P: 1 bar; reaction temperature: 250 °C, GHSV: 60,000 h^{-1}

while CH_4 selectivity increases, this is in agreement with the thermodynamic prediction (Fig. 4). When H_2/CO ratio reaches the theoretical value for methanation, the CH_4 selectivity is about 84 %, while there is still some CO_2 generated, which may be due to competitive reaction between the Boudart reaction and methanation [23, 31].

Overview of the above test results shows that the $NiZrO_x/Al_2O_3$ catalyst prepared in this way has high methane selectivity, little C_2^+ side products are generated. The catalyst showed high activity at low temperature and can operate in a wide range.

The catalyst before and after catalytic test have been characterized using XRD, and the results are shown in Fig. 5. For XRD pattern (the green curve in Fig. 5) of the as prepared catalyst only showed three broad diffraction peaks, which are assigned to the diffraction of gama-Al_2O_3 [32–34]. When the catalyst is tested for CO methanation in the mixture of $[(N_2 + CO_2)/H_2 = 4]$ at 350 °C for 5 h in stream, while the catalyst still had CH_4 selectivity more than 90 % and CO conversion more than 89 %, the XRD patterns of the alumina became sharper, compared to the fresh catalyst sample. This suggests that the catalyst may experience partly crystallization during the reaction. Also diffraction peaks at 38.7, 43.6, 58.5 and 62.5 appeared in this used sample. According to the literature, the peaks at 38.7 and 58.5 are due to the metallic nickel crystal with hcp

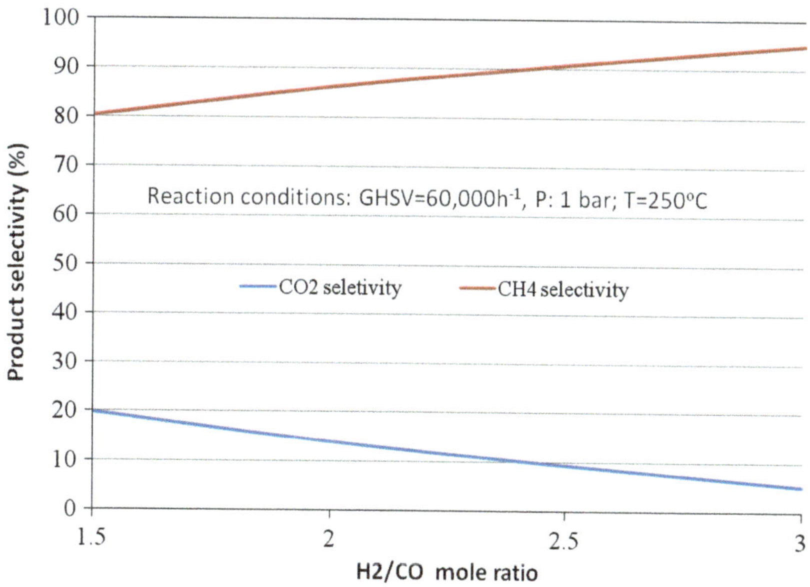

Fig. 4 Effcet of H_2/CO ratio on CO methanation. The test conditions are shown in the graph

structure, while the diffraction peaks at 43.6 and 62.5 are ascribed to the diffraction of metallic nickel with fcc structure. The peak of fcc form nickel is much stronger than that of hcp, suggesting that the main phase of nickel after CO_2 hydrogenation is fcc structure [35].

XRD patterns of the spent catalyst from the CO methanation reaction for 600 min is different from the catalyst unloaded from CO_2 methanation reaction. Probably the CO methanation reaction was controlled at 250 °C, the Al_2O_3 support diffraction peak is not as sharp as that from CO_2 system, suggesting that Al_2O_3 support still keeps most of its amorphous phases. Besides these peaks, there are small diffraction bands at 43.8°, 52.8° and 63.2° observed, which correspond to the diffraction of planes of 011, 200 and 220 of nickel crystal. Meanwhile, a small sharp diffraction peak at 27.6° is seen, which can be assigned to the deposited carbon [36, 37]. This suggests that there is carbon formed during the CO hydrogenation although the catalyst is still active.

The spent catalysts from different methanation reactions under various conditions have been measured using Laser Raman spectroscopy. In the fresh catalyst, almost no Raman band can be seen. The Raman spectrum of the spent catalyst after CO_2 methanation showed two distinct peaks at 1,404 and 1,882 cm^{-1}, and a shoulder peak at around 1,590 cm^{-1}. According to literatures, the Raman band at 1404 cm^{-1} can be assigned to the D band, and that at 1,590 cm^{-1} is due to the G band [38, 39]. The D band has arisen from structure defects or imperfection of graphite, whereas the G band is associated with a splitting of the E_{2g} stretching mode of graphite. In addition, a very weak D′ band is present at ca. 1,600 cm^{-1} as a shoulder of the G band, which has stemmed from the dangling band of disorder graphite. The broad bang at 1,882 cm^{-1} is unknown, and may need further study.

In the Raman spectrum of the spent catalyst from CO methanation, these peaks intensity increases significantly, suggesting that both D and G bands graphite increase significantly. This implies that over the spent catalyst from CO methanation more carbon is formed over the catalyst surface, which is in agreement with the XRD results (Fig. 6).

The TEM images of the nickel catalyst from CO methanation and CO_2 methanation are shown in Fig. 7. The image of the spent catalyst from CO_2 has smaller particles, the nickel particles size ranges from 20 to 60 nm. No whisker carbon is observed in the catalyst surface. The dark black fake may results from nickel particle or amorphous carbon as suggested by Laser Raman. However, in the TEM images of spent nickel from CO methanation, clearly the nickel particles and the surface carbon sizes are bigger, as suggested by XRD and Laser Raman. More carbon formed in the spent catalyst from CO methanation, However, no whisker carbon were seen over the catalyst, which might be the reason the catalyst is still active even after 600 min time on stream, because it is already pointed out that the whisker carbon normally account for the catalyst deactivation [36, 37].

Conclusion

Al_2O_3 supported Ni catalyst for CO_x methanation to produce synthetic natural gas has been prepared using organic decomposition method and tested under various conditions.

The catalyst prepared from this method catalyses CO methanation from 220 °C and CO_2 methanation starting

Fig. 5 XRD patterns of the methanation catalysts at various stages

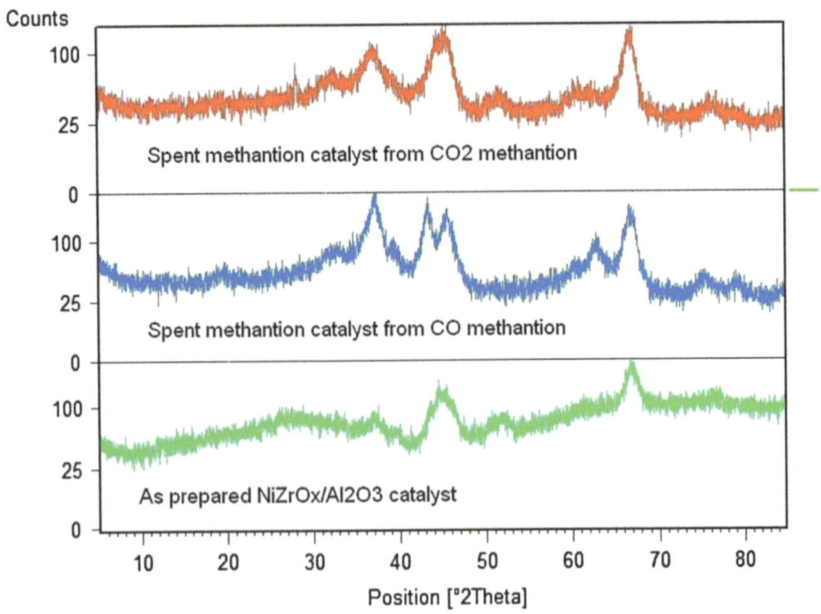

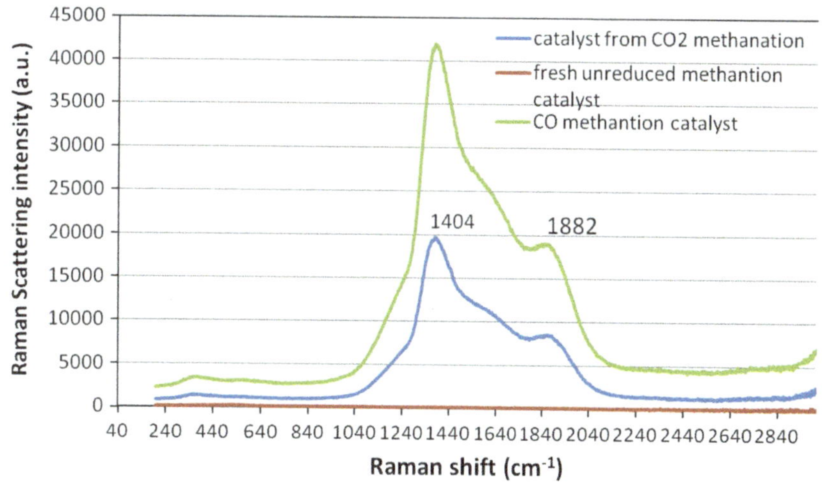

Fig. 6 Laser Raman spectra of the methanation catalyst at different stages

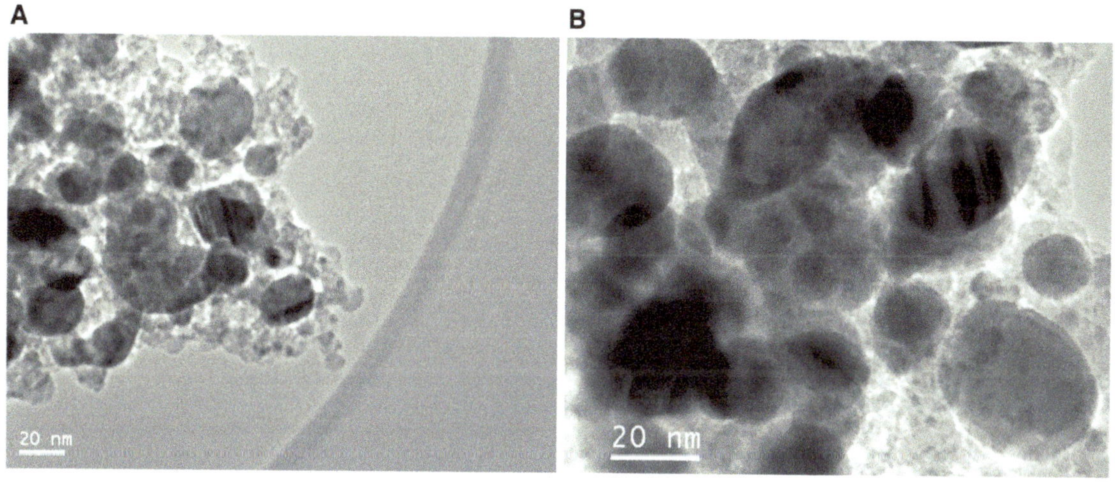

Fig. 7 TEM of the spent $NiZrO_x/Al_2O_3$ catalysts from **a** CO_2 methanation at 300 °C; **b** CO methanation at 250 °C

from 260 °C. It showed high CH_4 selectivity and yield, the catalyst can operate in a broad range of temperature.

The catalyst showed high stability and CH_4 selectivity for CO methanation, although more surface coke is detected over the catalyst surface.

The nickel crystallite becomes bigger and has different forms in the spent catalyst from CO_2 methanation reactions, while the one from CO methanation has different metallic form and more crystalline carbon.

Acknowledgments This work has been financially supported by Guangzhou Bairen (Hundred Talents) program. We would like to thank Dr. Zhenxing Liang for his discussion and professor Jixin Su for his activity test.

References

1. van der Meijden CM et al (2008) Production of Bio-CNG by gasification. In: Proceedings of 25th Annual International Pittsburgh Coal Conference, pp 133/1–133/8
2. Weiss AJ (1973) CRG [catalytic rich gas]-hydrogasification process for SNG [synthetic natural gas] production. Chem Eng Progr 69(5):84–90
3. Woodward C (1977) Methanation in substitute natural gas production. Energiespectrum 1(12):342–347
4. Wix C (2010) Process for the production of substitute natural gas. (Den.). Application, US, p 10
5. Lessard RR, Reitz RA (1981) Catalytic coal gasification: an emerging technology for SNG. Baytown Res. Dev. Div., Exxon Res. Eng. Co., Baytown, TX, USA, p 26
6. Hoehlein B et al (1984) Methane from synthesis gas and operation of high-temperature methanation. Nucl Eng Des 78(2):241–250
7. Luterbacher Jeremy S et al (2009) Hydrothermal gasification of waste biomass: process design and life cycle assessment. Environ Sci Technol 43(5):1578–1583

8. Rostrup-Nielsen JR, Pedersen K, Sehested J (2007) High temperature methanation. Appl Catal A 330:134–138
9. Udengaard NR, Olsen AN, Wix-Nielsen C (2006) High temperature methanation process-revisited. In: Proceedings of 25th annual international Pittsburgh coal conference, pp 25/1–25/5
10. Woodward C (1976) A high-temperature methanation catalyst for SNG applications. Am Chem Soc Div Fuel Chem Prep 21(4):22–29
11. Shinnar R, Fortuna G, Shapira D (1982) Thermodynamic and kinetic constraints of catalytic synthetic natural gas processes. Ind Eng Chem Process Des Dev 21(4):728–750
12. Yang Z et al (2010) Isothermal methanation for manufacture of substitute natural gas from coal or biomass. (Shanghai International Engineering Consulting Company, Peop. Rep. China). Application, CN, p 18
13. Kopyscinski J, Schildhauer TJ, Biollaz SMA (2010) Production of synthetic natural gas (SNG) from coal and dry biomass—a technology review from 1950 to 2009. Fuel 89(8):1763–1783
14. Bock HJ et al (1986) The Comflux-process for substitute natural gas (SNG). In: Proceedings of Eng. Found. Conf. Fluid, pp 489–496
15. Seemann MC, Schildhauer TJ, Biollaz SMA (2010) Fluidized bed methanation of wood-derived producer gas for the production of synthetic natural gas. Ind Eng Chem Res 49(15):7034–7038
16. van der Meijden CM et al (2009) Bioenergy II: scale-up of the MILENA biomass gasification process. Int J Chem React Eng 7
17. Blinn MB et al (1989) Advanced gasifier-desulfurizer process development for SNG (substitute natural gas) application. Final report August 1987–December 1988. KRW Energy Syst., Inc., Pittsburgh, PA, USA, p 210
18. Punwani DV, Arora JL, Tsaros CL (1978) SNG from peat by the PEATGAS Process. Inst. Gas Technol., Chicago, IL, USA, p 19
19. Xiao T, Chen H (2011) Methanation catalyst, its preparation process, and methanation reaction device having the same, Peop. Rep. China, p 11
20. Xiao T, Qian Y (2008) Promoted carbide-based Fischer-Tropsch catalyst, method for its preparation and uses thereof. Oxford Catalysts Limited, UK, p 35
21. Rezaei M et al (2006) Nanocrystalline zirconia as support for nickel catalyst in methane reforming with CO_2. Energy Fuels 20(3):923–929
22. Seo JG et al (2008) Preparation of Ni/Al_2O_3–ZrO_2 catalysts and their application to hydrogen production by steam reforming of LNG: effect of ZrO_2 content grafted on Al_2O_3. Catal Today 138(3–4):130–134
23. Boudart M, McDonald MA (1984) Structure sensitivity of hydrocarbon synthesis from carbon monoxide and hydrogen. J Phys Chem 88(11):2185–2195
24. Dalai AK, Davis BH (2008) Fischer-Tropsch synthesis: a review of water effects on the performances of unsupported and supported Co catalysts. Appl Catal A 348(1):1–15
25. Khodakov AY (2009) Fischer-Tropsch synthesis: relations between structure of cobalt catalysts and their catalytic performance. Catal Today 144(3–4):251–257
26. Stiegel GJ, Srivastava RD (1994) Natural gas conversion technologies. Chem Ind (London) 21:854–856
27. Bery RN (1973) SNG [substitute natural gas] from naphtha. Chem Eng (N. Y.) 80(13):90–91
28. Woodcock KE, Hill VL (1987) Coal gasification for synthetic natural gas (SNG) production. Energy (Oxford) 12(8–9):663–687
29. Tao P, Wang X (2007) Method for preparing synthetic natural gas (SNG) from coke oven gas (COG). (Southwest Research and Design Institute of Chemical Industry, Peop. Rep. China). Application, CN, p 9
30. Vogel F et al (2007) Synthetic natural gas from biomass by catalytic conversion in supercritical water. Green Chem 9(6):616–619
31. Ichikawa S, Poppa H, Boudart M (1985) Disproportionation of carbon monoxide on small particles of silica-supported palladium. J Catal 91(1):1–10
32. Hao Z et al (2009) Characterization of aerogel Ni/Al_2O_3 catalysts and investigation on their stability for CH_4-CO_2 reforming in a fluidized bed. Fuel Process Technol 90(1):113–121
33. Jun J et al (2008) Surface chemistry and catalytic activity of Ni/Al_2O_3 irradiated with high-energy electron beam. Appl Surf Sci 254(15):4557–4564
34. Xiao T-C et al (2004) Tungsten promoted Ni/Al_2O_3 catalysts for carbon dioxide reforming of methane to synthesis gas. Chem Res Chin Univ 20(4):470–477
35. Li H et al (2008) Fabrication and growth mechanism of Ni-filled carbon nanotubes by the catalytic method. J Alloys Compd 465(1–2):51–55
36. Koo KY et al (2008) Coke study on MgO-promoted Ni/Al_2O_3 catalyst in combined H_2O and CO_2 reforming of methane for gas to liquid (GTL) process. Appl Catal A 340(2):183–190
37. Martinez R et al (2004) CO_2 reforming of methane over coprecipitated Ni–Al catalysts modified with lanthanum. Appl Catal A 274(1–2):139–149
38. Reshetenko TV et al (2003) Catalytic filamentous carbon. Structural and textural properties. Carbon 41(8):1605–1615
39. Zhu X, Cheng D, Kuai P (2008) Catalytic decomposition of methane over Ni/Al_2O_3 catalysts: effect of plasma treatment on carbon formation. Energy Fuels 22(3):1480–1484

2

Theories and experimental investigations of the structural and thermotropic mesomorphic phase behaviors of metal carboxylates

Peter N. Nelson · Richard A. Taylor

Abstract Investigations on the phase behaviors and structural properties of mono-, di- and poly-valent metal carboxylates are reviewed with reference to developments in experimental and theoretical concepts surrounding their liquid crystalline properties. The main methods of structural investigation such as X-ray diffraction, infrared and ^{13}C-NMR spectroscopies are examined in detail on the basis of common synthetic routes leading to the isolation of pure compounds. A detailed review of the thermal behaviors of several metal carboxylates is presented along with proposed theories and molecular models for odd–even alternation, chain length effects, phase structures and mesophase formation. Theories explaining the effects of metal ion radii and chain unsaturation are also discussed. Proposed degradation mechanisms resulting in the formation of various products and kinetic studies are also considered. Though this review highlights a number of investigations on the structural and phase properties of the mostly widely studied carboxylates, the results presented here strongly indicate that there is room for further studies on some of these systems.

Keywords Metal carboxylates · Room temperature structures · Metal–carboxylate coordination · Odd–even alternation · Phase transition · Mesomorphism · Phase structures · Degradation mechanisms

P. N. Nelson
University of the West Indies, Mona Campus,
Kingston 7, Jamaica

R. A. Taylor
University of the West Indies, St. Augustine Campus,
St. Augustine, Trinidad and Tobago

Introduction

Overview: metal carboxylates as metallomesogens

Metal carboxylates (soaps) are of different types based on the metal ion radius, hardness, softness, valency and alkyl chain structure. As a result, they exhibit diverse physicochemical properties; hence, they find extensive commercial applications in protective coating agents, paints, ink driers, polymer stabilizers, catalysts, waterproofing agents, lubricants, fuel additives and fungicides [1, 2]. Some of these compounds are also used in photothermogravimetry [3], as low angle calibrants for neutron diffraction instruments [4, 5] and in the manufacturing of pharmaceuticals [6–9]. However, the most exciting aspect of metal carboxylate chemistry stems from their ability to undergo mesomorphic phase transitions (metal containing liquid crystal) of various textures. In fact, from a historical perspective, metal carboxylates were the first metal containing liquid crystal compounds (metallomesogens), reported in 1855 by Heintz, for magnesium(II) tetradecanoate [10]. Similar observations were reported by Vörlander [11] in 1911 for several alkali metal carboxylates. Subsequently, a number of other mesomorphic metal carboxylates were successfully prepared and shown to exhibit thermotropic lamellar and columnar mesophases in their phase sequences [10, 12, 13]. Furthermore, some metal carboxylates formed smectic type mesophases which are of considerable commercial importance since they are often applied in liquid crystal display devices (LCDs) where they offer considerable advantages over conventional nematic and twisted nematic display devices; that is, they off shorter (faster) switch times, larger field of view and higher complexity. They are also more environmentally friendly, cheap and easier to synthesize relative to currently used liquid crystal compounds [14–

16]. Hence, despite the advent of other forms of display mechanisms, such as light emitting diodes (LEDs), the technological relevance of liquid crystals remains significant [16]. However, the applicability of metal carboxylates in LCDs is limited since they form thermotropic liquid crystals (liquid crystals formed above ambient temperature). Therefore, considerable investigations have been carried out on their phase properties, resulting in several proposals [10–13] for the factors governing their phase transition behaviors. These studies also revealed that several metal carboxylates formed ionic melts which find application in ionic solvents and for the manufacture of portable energy storage devices such as lithium ion batteries which makes use of lithium salts [14–16]. The magnetic properties of cobalt(II) and manganese(II) *n*-alkanoates have been the center of investigations by Kambe and coworkers [17]. For instance, manganese(II) stearate was found to exhibit two-dimensional (2D) magnetic properties which was applicable in magnetic switches, hence, the possibility exists for their use in data storage devices. Additionally, metal carboxylates are generally insoluble in water but are usually soluble in various organic solvents in which they exhibit lyotropic mesomorphism, where the soap molecules exist as small aggregates of various shapes, classified as micelles [6, 18].

Theories explaining the structural and phase properties of these compounds are presently the center of a myriad of investigations. For example, a detailed theoretical study was carried out by the cancer research institute of Philadelphia on the three-dimensional (3D) binding of carboxylate moieties to different metals [19]. The relevance of the study stemmed from the fact that such groups were abundant in the active sites of several biological enzyme systems as part of glutamine and aspartate side chains that are active metal binders. However, despite the many uses of these compounds, most of the basic theoretical aspects of their physical and chemical properties remained largely underdeveloped up until the last two decades, where the majority of investigations were more meticulously carried out and focused on zinc(II), lead(II), copper(II), lithium(I), sodium(I), potassium(I), silver(I) and lanthanide(III) carboxylates. Such studies revealed a step-wise melting mechanism for these compounds, as indicated by the presence of several pre-melting transitions in their phase sequences [20, 31]. Unfortunately, though highly detailed thermodynamic explanations were proposed for some compounds, the majority of these studies resulted in no molecular and lattice structure–phase behavior correlation models. However, for lead(II) [21–27], copper(II) [28, 29], zinc(II) [30–32] and silver(I) [33–36] carboxylates where both crystal–crystal and mesomorphic transitions were observed in their phase sequences, a number of thermodynamic phase models were proposed. Several studies have also been carried on the thermal degradation of lead(II) [37, 38] and copper(II) [39, 40] carboxylates where their decomposition mechanisms, activation parameters and significant kinetic factors adduced. Theoretical investigations on the phase behavior of metal carboxylates have also been carried out by a number of authors via molecular modeling calculations such a density functional theory (DFT) [41, 42]. Considering the commercial importance of metal carboxylates, understanding their molecular structures and phase behaviors should be of considerable interest. Therefore, in this review attempts will be made to give a general overview of investigations on the structures and mesomorphic behaviors of the more widely studied metal carboxylates and the contributions of such reports to the current body of knowledge on the phase properties of metallomesogens.

Approach to synthesis

The synthesis of pure metal carboxylates presents a real but not insurmountable challenge. This involves simple reactions in which the acidic hydrogen of the alkanoic acid is replaced by a metal [20]. Such changes can be accomplished in several ways: the action of base (hydroxide or oxide) on fatty acids; double decomposition of the soluble salt of a fatty acid and the salt of a mineral acid or by saponification of glycerides or other esters of fatty acids with a base or metallic oxide [20]. These reactions hold true generally for mono- and di-valent soaps such as potassium, sodium, lithium, and lead carboxylates among others. On the other hand, preparation of trivalent and higher valence soaps appears to be particularly difficult. For example, there have been many attempts to utilize conventional methods, but these did not result in the formation of pure anhydrous compounds [20]. In many cases, soap hydrates as well as basic $[M(OH)_x(RO_2)]$, and or acidic $[(RCO_2)_nM(RO_2H)]$, compounds were obtained. For example, studies involving the synthesis of aluminium(III) carboxylates resulted in the formation of both acidic and basic products [43–45]. After many attempts to produce the anhydrous tri-soap, it was demonstrated that it could be prepared only under anhydrous conditions [20, 46, 47]. Similarly, for chromium(III) stearate, the normal products have been shown to be a mixture of basic soaps and excess of acid [20, 48]. Various studies showed that basic chromium(III) carboxylates were easily prepared if hydroxyl species or water were present [49–51]. However, it has been shown that tri-soaps of chromium(III) could be prepared, with difficulty, only under anhydrous conditions [52]. In contrast, the hydrated trivalent lanthanide(III) soaps were readily prepared even in aqueous solution [20, 53, 54].

In light of these and other similar observations, the general methods for preparation of the carboxylates of the poly-valent metal ions include [20]: precipitation from aqueous solutions of sodium or potassium carboxylates by the addition of solutions of appropriate salts of the poly-valent metal; ligand exchange between the metal chloride or acetate and the long chain fatty acid in an appropriate anhydrous solvent, fusion of metal oxides, hydroxides or carbonates with the fatty acids and direct reaction of the metal and the fatty acid. The precipitation route is particularly simple, and whilst earliest studies involved, in general, reactions in aqueous solutions, it is usually more prudent to use alcohols or alcohol–water mixtures as solvents. The use of aprotic solvents, however, seemed to be more appropriate in these precipitation reactions as they generally produced anhydrous compounds. Upon precipitation, soaps were purified by recrystallization from organic solvents, such as ethanol, benzene or toluene or from neat fatty acids.

Molecular and lattice structure of metal carboxylates

Overview

As indicated before, the predominant liquid crystal phase exhibited by most metal carboxylates is the smectic phase, particularly types *A* and *C*; SmA and SmC [20]. This is because metal carboxylates possess layered structures in the solid state which is maintained on heating. The smectic phase is an example of a calamitic liquid crystal phase where in its structure the molecules possess more order than in the nematic (N) phase and are also arranged in layers which are preferentially aligned, on average, in a particular direction (called the director, *n*) [55]. Of the various types of smectic phases, two are more predominant, especially for metal carboxylates: the SmA, where the director is parallel to the layers' normal and the molecules are arranged such that the average tilt angle is zero; the SmC, where the director and the molecules are tilted at an angle, θ, to the normal (Fig. 1) [56]. There are also the SmB, SmD, SmE, SmF and SmG phases. In each case, there is considerable fluidity and easy diffusion between the layers is possible though their viscosities are much higher than nematic phases [55, 57]. In general, these phases, arise from rod-like shaped molecules in which the axial part of the molecule is larger than the radial part.

The rod-like molecular shape of some metal carboxylates can be said to derive from the basic structure of carboxylic acids, where they exist as dimers through interaction between hydrogen bonding which allows for the formation of a small core relative to the size of the peripheral part of hydrocarbon chains. This type of structure allows for a lattice in which the molecules have lateral intermolecular van der Waals interactions, responsible for stabilizing the layered phase structure, leading to a smectic phase [58].

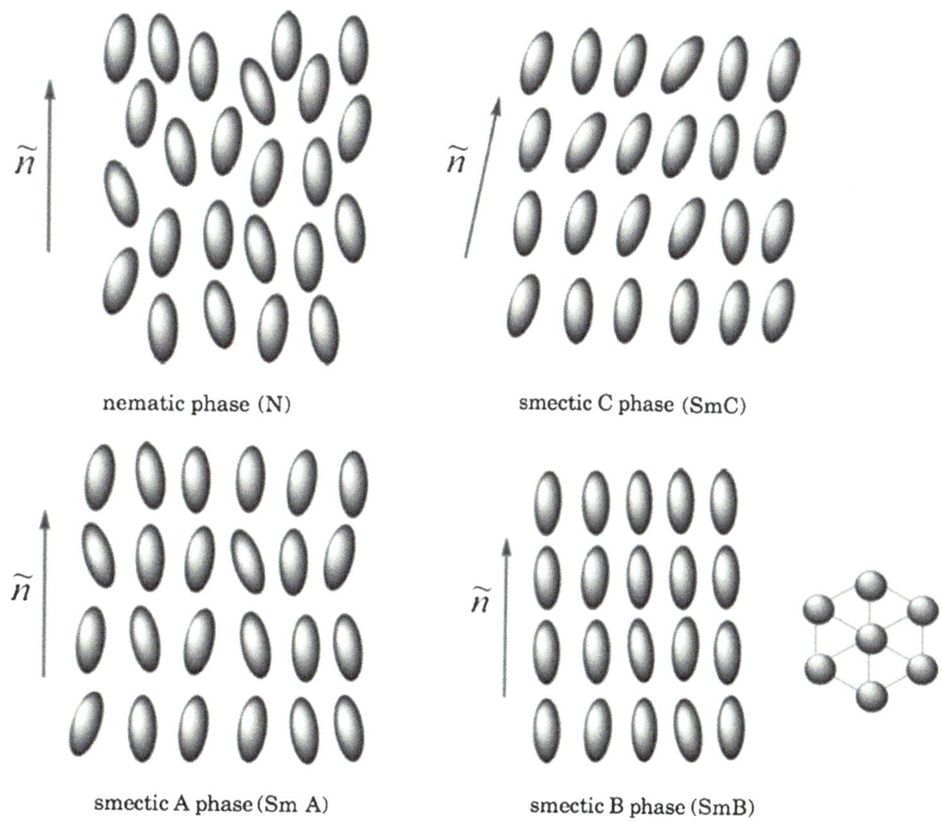

Fig. 1 Prevalent molecular arrangements of calamatic phases [58]

Incorporation of a metal, interacting through the carboxylate groups, essentially maintains the rod-like structure with a small core and flexible peripheral parts leading to layered packing. The type of packing structure is dependent on (1) the nature of the metal and associated coordination geometry; (2) the type of carboxylate bonding mode and (3) the type of substituents on the carboxylate, viz. hydrocarbon chain length, branched hydrocarbon chain, etc. [18, 31, 58]. Invariably, the elongated molecules packed in a layered lattice must possess two characteristics: (1) positional order and (2) orientational order, which are responsible for the formation of a particular type of mesophase [55, 56]. Positional order is controlled by the basic molecular structure and shape, particularly from the size of the core relative to the molecular length and orientational order, which is strongly dependent on the extent of chain tilting. Certainly, changing the chain length is an important approach to inducing mesomorphism and it is widely understood that the longer the peripheral hydrocarbon chain length relative to the core, the more likely will a smectic phase be exhibited [20]. Not only do the metal ions offer rigidity at the center, which is important for positional order, they also maintain molecular polarizability along the molecular axis, giving rise to important intermolecular interactions responsible for mesophase formation [58].

Metal–carboxylate coordination and solid-state lattice structure

As can be inferred from the foregoing, the structural chemistry of metal carboxylates is quite fascinating because carboxyl groups can coordinate in different ways to the metal ion resulting in the formation of mono-, di- and polymeric forms of these compounds [59]. It has been shown in numerous reports that metal carboxylate coordination was central to their molecular packing and lattice structures. Therefore, various techniques have been employed to elucidate the structures of the room temperature solids [10, 20], though, for many long chain adducts few reports have been published on their molecular and lattice arrangement in the crystal system. This challenge arises because many compounds do not produce crystals of suitable morphology for single crystal X-ray analysis [12, 20, 60–62], that is, the crystals are usually twined or exist as thin needles that are very fragile and do not diffract well. This is due to their low molecular symmetry and flexibility of the polymethylene chains. Nevertheless, single crystal and powder X-ray diffraction (PXRD) studies in conjunction with infrared (IR), Raman, high-resolution solid-state nuclear magnetic resonance and UV/Vis absorption spectroscopies as well as but to a lesser extent, magnetic susceptibility and photoluminescence studies have allowed some advances in molecular structure determination [10, 18, 61, 63]. Notwithstanding the challenges in structural elucidation, a very useful approach that can allow for fairly precise postulations about the packing structure involves the use of composite data from PXRD, IR and molecular calculations. In recent times this approach, used by several authors [1–7, 20], has proven to be quite reliable since structures of metal carboxylates have fairly typical IR spectra (Fig. 2) and X-ray diffraction patterns arising from the nature of the carboxylate coordination, layered packing and chain conformation.

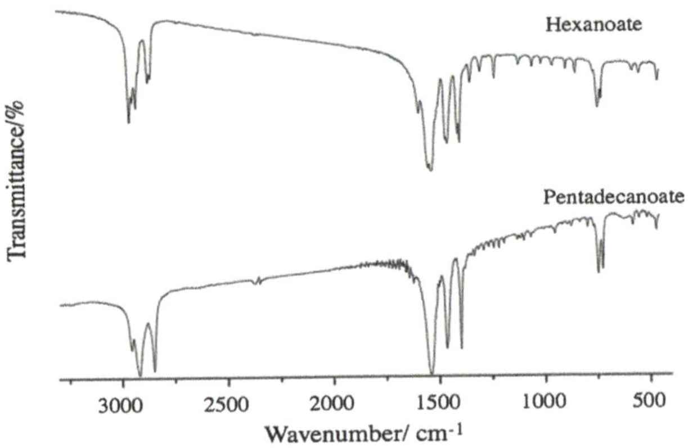

Fig. 2 Typical infrared spectra for metal carboxylates, exemplified by zinc carboxylates [31]

Infrared spectroscopy

The use of vibrational spectroscopy in the study metal–carboxylate chemistry provides invaluable insights where metal–metal interactions, chain conformation and packing are concerned; hence, it is extensively used by a number of authors on various metal carboxylate systems [18, 45, 60, 64]. Typical IR spectra for metal carboxylates, exemplified by zinc(II) hexanoate and pentadecanoate (Fig. 2), were characterized by strong absorptions in the region of ca. 1,700–1,200 cm^{-1} and weak to medium bands between ca. 1,400 and 500 cm^{-1} [31, 46, 60]. The absence of hydroxyl absorption bands in the region of 3,500–3,300 cm^{-1} confirmed that the compounds were anhydrous as well as absence of a carbonyl absorption band in the region of 1,730 cm^{-1} and its replacement by absorptions in the region of 1,500, 1,400, 950, 580 and 540 cm^{-1} indicates that there was complete resonance in the C–O bonds of the carbonyl group as a result of coordination with the metal [18, 31].

The C–O bands were assigned as: asymmetric stretch, ν_{as}(COO) ~1,400–1,530 cm^{-1}; symmetric stretch, ν_s(COO) ~1,410–1,390 cm^{-1}; deformation, ν_d(C–C)COO ~940–960 cm^{-1}; bending, ν_δ(COO) ~745 cm^{-1}; out of plane twisting, ν_τ(COO) ~580 cm^{-1} and rocking, ν_ρ(COO)

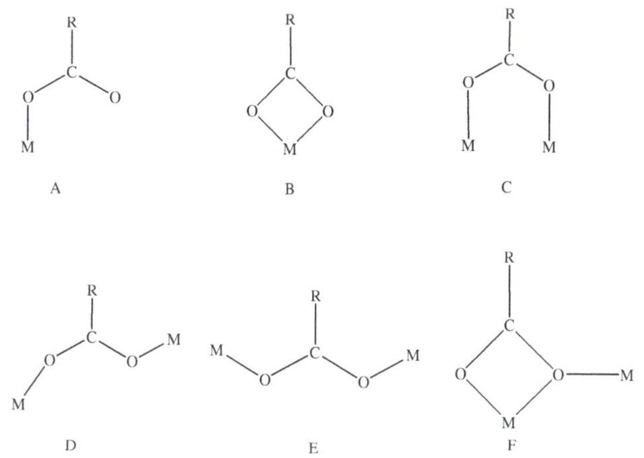

Fig. 3 Possible coordination modes for the carboxylate ions [29, 60]

~550 cm^{-1}. Splitting of C O bands correlated with a lowering of COO symmetry and the relative strengths of the metal–carboxylate bond due to head group and side chain lateral van der Waals interactions [20, 31, 60]. For a homologous series, near invariance of the ρ(COO) absorption at around 550 cm^{-1} gives evidence of isostructures. Additionally, for triclinic or hexagonal chain packing, a single band for $\rho(CH_2)$, at around 720 cm^{-1}, is expected, however, splitting of this band is indicative of orthorhombic or monoclinic packing [11, 65]. The nature of chain packing can be inferred from $\rho(CH_2)$ and $\delta(CH_2)$ absorption bands with splitting indicating horizontal packing and if not, parallel packing is assumed [11]. Further, the progression of regularly spaced absorptions, in the region of 1,347–1,232 cm^{-1}, arising from the wagging vibrations of methylene groups, are typical for aliphatic chains and indicate *all-trans* conformation of hydrocarbon chains [65].

In order to understand the importance of IR spectroscopy in obtaining structural information, it is useful to explain the points outlined in the foregoing with respect to coordination. To that end, Nakamoto and co-workers [66] stated that, site group or factor analysis, based on knowledge of the crystal structure, is desirable when taking into consideration the symmetry of the ligand, from the spectra obtained, in the crystalline state. For the most part, it was anticipated that the effects of coordination on ligand absorption were much stronger than the effects of the overall crystal field, hence, rigorous site group analysis may be ignored. Furthermore, by coordination, all the fundamentals (bands) were more or less shifted according to their modes of vibration. As the metal–ligand bonds become stronger, shifts to lower wavenumbers become greater. When coordination lowers the symmetry of the ligand, forbidden vibrations of the free ions are permitted, and degenerate vibrations became energetically distinct [64, 66]. Therefore, vibrational energy shifts, the magnitude of splitting and the intensity of the newly permitted bands are useful indicators of the effect of coordination. In addition, it is generally expected that emerging bands of the metal complex as well as those of the free ion, increase in intensity on formation of the metal complex. Such changes are affected by the type of metal carboxylate coordination mode(s), of which there are various possibilities (Fig. 3): chelating-type bidentate; *Z*, *Z*-type bridging bidentate; *Z*, *E*-type bridging bidentate; *E*, *E*-type bridging bidentate and bridging tridentate [66, 67].

Evaluating the mode of coordination in metal carboxylate systems involves use of the magnitude of separation, $\Delta\nu$ cm^{-1}, between ν_a(COO) (carboxyl asymmetric) and $\nu_s(COO^-)$ (carboxyl symmetric) stretching vibration bands [60, 63, 64, 66]. However, according to Edwards and co-workers [68] the use of $\Delta\nu$ can be misleading, especially in cases of variable coordination modes, in some carboxylates, which make it difficult to assign, unambiguously, the correct coordination mode/s. Nevertheless, IR studies on a series of metal acetates, showed that for chelating bidentate coordination $\Delta\nu$ was in the vicinity of 100 cm^{-1} or less but for bridging bidentate bonding the value was ca. 150 cm^{-1} [20, 60, 61, 68]. Additionally, studies on some metal haloacetates indicated that [20] a higher value of $\nu_{as}(-COO^-)$ relative to the values for the ionic soaps (Na^+ and K^+) was indicative of bridging coordination whilst lower values confirmed chelating coordination [60, 64]. Since monodentate coordination removes the equivalence of the C–O bonds resulting in a higher value for $\nu_{as}(COO^-)$ and a lower value for $\nu_s(COO^-)$, $\Delta\nu_{mon}$ will be much greater than $\Delta\nu_{ion}$ (ca. 138 cm^{-1}) [60]. Moreover, when COO^- is unidentate, one of the C–O bonds should have enhanced double-bond character and should give rise to a high-wavenumber band [69] in the 1,590–1,650 cm^{-1} region of the spectrum. This approach has been utilized by many authors [31, 57, 61] and has resulted in fairly accurate descriptions of the metal–carboxylate coordinations for several compounds.

Clearly, for these compounds, molecular packing was a function of the type of carboxylate coordination; hence, various packing arrangements such as monolayers, bilayers, interdigitated bilayers were observed [20]. Other effects such as alternating orientation of the metal and hydrocarbon planes were observed. Typically, the hydrocarbon chains in the lamellar are packed in the fully extended *all-trans* conformation. This was proven by a number of authors [20, 27, 31] on the basis of IR data as outlined in the forgoing.

Structural information from powder X-ray diffraction

Typical room temperature X-ray powder diffraction patterns, for metal carboxylates (Fig. 4) are characterized by

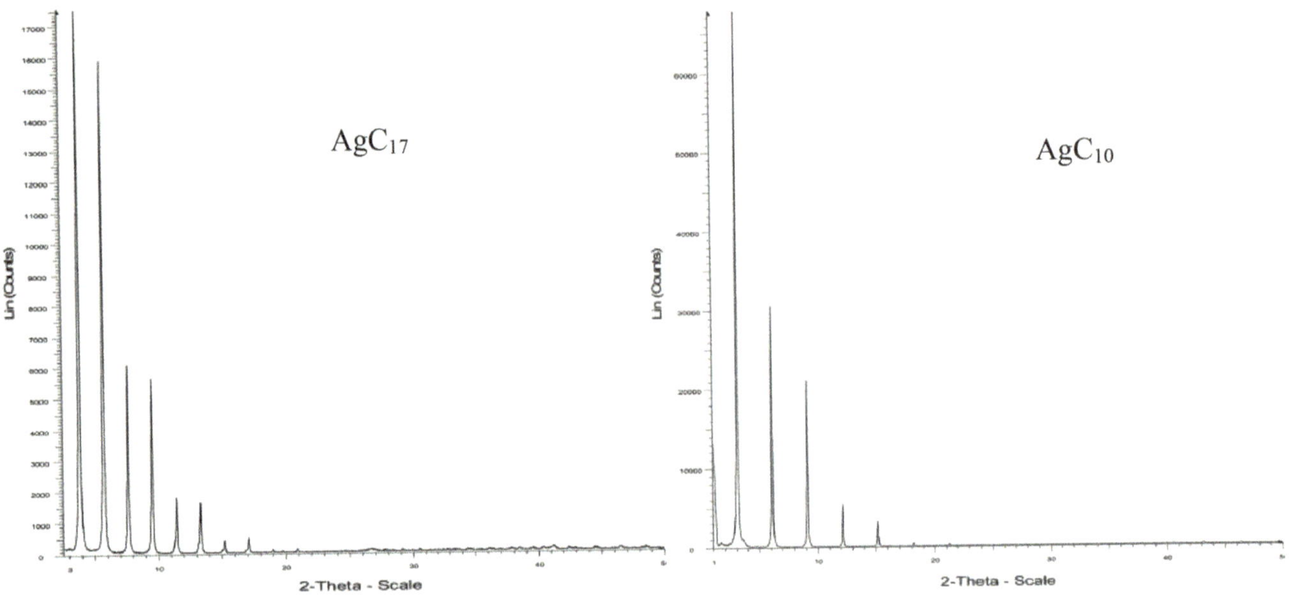

Fig. 4 Powder X-ray diffractograms of silver(I) carboxylates [70]

evenly spaced reflections from successive d planes in the order of: 1:1/2:1/3:…1/n characteristic of a lamellar structure [30, 31, 61–64]. Generally, the shorter chain compounds have evenly spaced reflections at low angles (2θ) but for the longer chain adducts a progression of evenly spaced low-intensity reflections are usually observed at, $20° \leq 2\theta \leq 25°$ which becomes irregularly spaced at higher angles due to side chain packing interactions of the methylene hydrogen [61–63]. Indexation and refinement of the diffraction patterns are usually carried out to determine the unit cell parameters and *hkl* miller indices (typically, 00*l*, 0*k*0, *h*00) from which the d values were obtained [20, 31]. Such parameters were useful in presenting a picture of the molecular and lattice packing of these compounds. Further calculations carried out by several authors [18, 27, 31] have shown that comparison of the d-spacing, d_{exp}, which represents the distance between successive metal planes, and the calculated molecular length, d_L [$d_L = d_{C\text{–}H} + (n - 1)d_{C\text{–}C} \sin 55° + d_{C\text{–}O} + r_M^{x+}$], was indicative of whether the compounds were packed in a monolayer, bilayer or interdigitated bilayer arrangement within the lamellar.

Monovalent alkanoates

The phase behaviors of monovalent metal alkanoates have been the subject of several investigations for a number of decades, especially those of lithium, sodium, potassium, thallium and silver. In the case of lithium carboxylates, some of the most comprehensive studies have been carried out by Shoeb et al. [71], Ferloni and Westrum [72], and Martinez Casado et al. [27], prior to which only the crystal structure of some hydrated lithium alkanoates (lithium formate monohydrate and lithium acetate dihydrate) have been determined [27].

Sodium(I) and potassium(I) carboxylates

Despite being extensively used as surfactants in detergents, the solid-state structures of sodium and potassium carboxylates have not been investigated to any appreciable extent. This is possibly due to their inability to form crystals of suitable morphology for X-ray single crystal diffraction, especially for the long chain homologues. Nonetheless, according to Kovalenko et al. [42], molecular modeling calculations, based on the isolated propanoate ion, were in agreement with experimental data. These results showed that the most probable coordination mode, for sodium carboxylates, was via chelating bidentate or monodentate bonding [42]. However, no molecular models were presented for these compounds, hence, very little is known about their three-dimensional lattice structures, intermolecular interactions and exact metal–carboxylate bonding. Nonetheless, fair amount of infrared analysis was carried out on these compounds, which gave a clear picture of the structure of the hydrocarbon chains. For example, Bui [73] in his studies of some short chain sodium n-alkanoates, with $n_c = 1\text{–}12$, inclusive, showed that the intensities of the methyl and methylene group symmetric and asymmetric vibration bands were related to the number of methylene groups in the alkyl chain in agreement with previous reports on other metal carboxylate systems [30, 60–64]. In this study, it was suggested that the longer chain sodium compounds had greater ionic character relative to the shorter chain adducts. This suggestion is quite contradictory since it is well established in the literature [15–26] that shorter chain compounds were of higher melting points, indicating greater ionic charter. Furthermore, since $\Delta\nu = 217$–$227\ cm^{-1}$ [73] the clear inference was for ionic metal–carboxylate coordination, irrespective of chain length and not covalent as suggested. This assertion was supported by the presence of a UV absorption in the region of 186–187 nm corresponding the π–π^* transition of C=O.

For potassium carboxylates, no accurate crystal structures have been reported that shows clearly, the type of metal–ligand bonding and their lattice packing. According to Lewis and Lomer [74] and Lomer [75] these compounds showed extensive chain length-dependent polymorphism; that is, compounds with 4–10 carbon atoms, inclusive, when crystallized at ambient temperature from alcohol formed a distinct phase, classified as *A* whilst those with $n_c = 12$–18, inclusive, formed phase *B*. Furthermore, with the exception of the butanoate and hexanoate, all compounds transitioned to a third phase; *C*, on heating [75]. X-ray data, in agreement with data from Piper [41], showed that the compounds crystallized in a monoclinic crystal system with *P*21/*a* space group for the *A* polymorph; however, for polymorph *B*, the compounds were triclinic with *P*1 symmetry [75]. Though the structure for potassium decanoate *A* has been investigated by Lomer [75], no three-dimensional molecular models were proposed to explain its' molecular and lattice packing. Nonetheless, these results showed that the molecules were tilted at ca. 57.9° with respect to the metal basal plane within $1.123\ g\ cm^{-3}$ unit cells. Interestingly, UV absorption measurement carried out by Bui [73] showed two main transitions associated with the strong π–π^* and low probability n–π^* transitions of the C=O group. These transitions, observed in the range of 186–189 nm, were used to support the IR data in proposing that metal–carboxylate bonding was either via ionic or bridging coordination, though no detailed explanations were presented for such conclusions.

Lithium(I) carboxylates

Shoeb et al. [71] in their studies on a series of anhydrous lithium soaps ($n_c = LiC_{8 \ldots 18}$) suggested, on the basis of infrared data, that the compounds were isostructural with ionic lithium–carboxylate interactions. However, the ionic character of the metal–carboxylate bonding increases with decreasing chain length as indicated by the observed decrease in melting point with increased chain length. Furthermore, PXRD data indicated that there was a change in the tilt angle of the hydrocarbon chains relative to the lithium basal planes with chain length [71]. However, this study was limited since there was no description of the metal carboxylate coordination, type of molecular packing and/or the effect of chain length on their lattice structures. Recently, more extensive structural studies were carried out by White et al. [76] and it was concluded, based on room temperature PXRD and IR data, that the ionic molecules were packed as lamellar bilayer structures in a triclinic crystal system composed of fully extended *all-trans* hydrocarbon chains. Interestingly, single crystal structures reported by Casado et al. [77] for lithium(I) propanoate and pentanoate showed carboxylate groups bound in a chelating bidentate mode to tetrahedral lithium(I) ions. These structures showed that the tilted hydrocarbon chains were fully extended and arranged as non-overlapping lamellar bilayer structures. Unit cell parameters presented by White and Ellis [113] for some long chain lithium(I) carboxylates, adduced from PXRD data, showed that the layers within the triclinic unit cells were chain length dependent. However, according to Casado et al. [77], the tilt angle of the alkyl chains, of LiC_2 and LiC_3, relative to the ionic layer was, on average, 60. The crystal structures (Fig. 5) showed that metal carboxylate bonding in the basal was such that a two-dimensional lithium polymeric structure was formed (coordination polymer). Frequency differences between asymmetric and symmetric COO stretching ($\Delta\nu = \nu_{as} - \nu_s$), $\Delta\nu \approx 125\ cm^{-1}$ vibrations confirmed bridging carboxylate coordination as indicated by PXRD. However, since $\Delta\nu \sim 144\ cm^{-1}$ for the longer chain adducts, as reported by White and Ellis [113], the clear inference was for change in head group coordination with chain length.

Along with infrared data, recent studies by Bui [73] suggested, based on UV absorption data, that lithium carboxylates ($n_c = 1$–12, inclusive) were of a greater covalent character relative to other alkaline metal compounds. The spectra of these compounds were characterized by a single π–π^* transition in the range of 192–188 nm. Surely, the wavelengths of these absorptions were related to the type of metal carboxylate interaction, however, no such analyses were presented in this study. Interesting, the Raman and infrared spectra were, for the most part, identical. However, the Raman spectra did not show clear splitting of the carboxyl asymmetric vibration. This indicated significant mechanical coupling with another vibrational mode, possibly the stretching of the C–C bond adjacent to the COO.

Silver(I) carboxylates

Perhaps the most similar metal–carboxylate analogues to the dimeric alkanoic acids are silver(I) carboxylates. For example, investigations by Bremmer et al. [78] showed that the hydrogen bonded protons of the carboxylic acid dimers were replaced by silver ions to form a *Z*, *Z*-type bridging bidentate coordination structure. However, since the coordination spheres of the Ag^+ were not satisfied by the two carboxylates groups, additional ligands, typically neutral

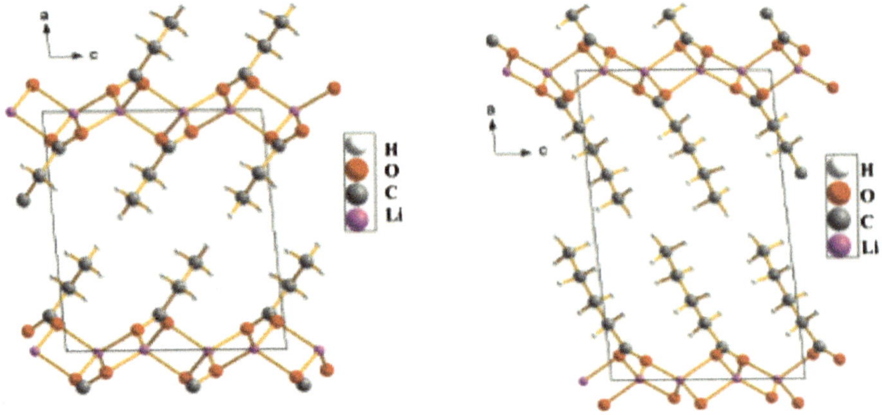

Fig. 5 Single crystal structures for short chain lithium(I) *n*-alkanoates [77]

ones, can become coordinated to the metal centers, leading to polymeric structures serving as secondary building units for metal–organic frameworks. Not only are silver(I) carboxylates useful in this regard but the use of the behenate, stearate and/or a mixture in thermography and photothermography has attracted significant interest in their thermochemical properties [78]. Though these compounds were known to exist as lamellar structures, the exact 3D arrangement of the silver–carboxylate coordination and hydrocarbon packing for longer chain compounds were unknown [18, 79]. This was because they did not produce suitable crystals for X-ray single crystal analysis. Nonetheless, investigation by Tolochko et al. [80], on the structure of silver stearate, indicated based on EXAFS and PXRD, that the compounds were arranged as dimeric units, similar to that proposed by Bremmer et al. [78]. This data also showed that the Ag-to-Ag distance in the dimers was ca. 2.90 Å which was similar to the bond distance in metallic silver, indicating possible metal–metal interactions. A more extensive investigation on a homologous series of silver(I) carboxylates, AgC_3–AgC_{24}, inclusive, by Binnemans et al. [79] indicated that the fully extended hydrocarbon chains which were slightly tilted with respect to the metal plane, for these compounds, were arranged to form lamellar bilayer structures. Furthermore, the results showed that the inter-lamellar distance increased linearly with chain length, though the chains were proposed to be slightly interdigitated. Together with data from Tolochko et al. [80], Binnemans et al. [79] concluded that the structures for the series were as shown in Fig. 6, where the silver ions were bridged by carboxylate ligands to form eight-membered rings, held together by extensive head group intermolecular interactions to form infinite 2D sheets. Interestingly, since in the aforementioned eight-membered rings, the Ag–Ag distances were equivalent to those in metallic silver, coulombic interactions were expected to be significant. Indeed, such postulates were supported by DFT calculations of the electronic structure for the dimeric silver(I) acetate structure, reported by Olson et al. [81]. Furthermore, in this study, vibrational frequencies for the Ag–Ag bond were confirmed by experimental and theoretical Raman studies.

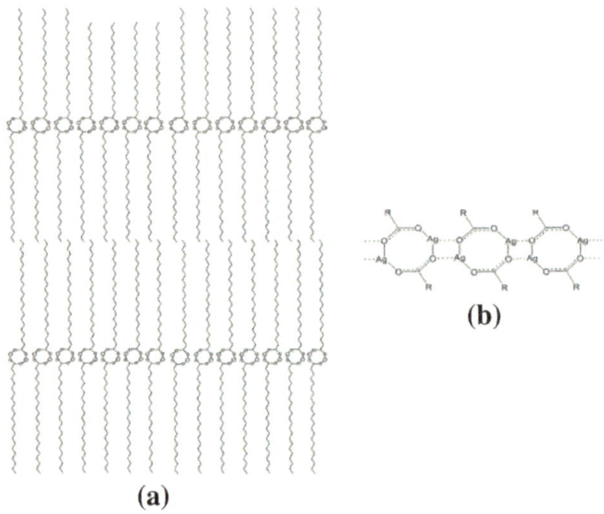

Fig. 6 Lamellar (**a**) and head group (**b**) structures of Ag(I) *n*-alkanoates, proposed by Binnemans et al. [79]

In addition to their coordination structures, PXRD data, presented by Nelson and Ellis [70], confirmed that all silver carboxylates with n_c 8–20, inclusive, were crystallized in a monoclinic crystal system with the fully extended *all-trans* hydrocarbon chains tilted ca. 75° with respect to the metal basal plane within the lamellar. Importantly, methyl groups from opposite layers in the bilayer do not overlap but were closely packed, resulting in methyl–methyl carbon–hydrogen van der Waals interactions [70]. This assertion was supported by molecular calculations and the presence of minimal reflections in the $20° \leq 2\theta \leq 30°$ region of the powder diffractograms, associated with side chain van der Waals interactions. Extensive head group intermolecular interactions resulting in the formation of polymeric molecular layers, was also confirmed by infrared and solid-state ^{13}C-NMR spectroscopies. This study culminated in a detailed structural model adequate in explaining the observed odd–even alternation in certain features that will be discussed in a following section [70].

The structures of other monovalent metal alkanoates, such as rubidium and caesium, have been investigated, though not in significant detail relative to other such compounds [73]. These salts, synthesized via fusion of the carboxylic acids with the metal carbonates, in anhydrous methanol, showed significant ionic character with decreasing chain length. Such assertions were made based on infrared data, which showed Δν in the range of 231–249 cm^{-1}. Based on these observations ionic or bridging coordination was proposed. For caesium and rubidium carboxylates the UV absorption spectra showed a strong π–π^* transition at ca. 200 nm and a weak n–π^* transition at longer wavelength. Though no mention of possible structural similarities were mentioned by Bui, these data strongly indicated similar structural features for these compounds [73].

Divalent alkanoates

Though many metal carboxylato complexes were composed of ligands coordinated in a unidentate mode, there are frequent exceptions; for example, in some copper(II) carboxylate, metal ions were bridged by bidentate carboxylate ligands to form a robust cage-like 3D structures [42]. For lead(II) formate, adjacent lead ions were bridged by formate ligands to form an eight-coordinated 3D polymeric structure as was similarly, for a number of other short chain lead(II) carboxylates; some of which, showed preference for the eight- or seven-coordinate structure [44]. This was supported by Ellis and co-workers [20] who also proposed the same structure for the longer chain adducts.

Copper(II) carboxylates

Of the bivalent carboxylates, perhaps the most extensively studied and well understood structures are those of copper(II); hence, there have been little or no ambiguities about their structure relative to other metal carboxylate systems. This is partly due to the fairly easily predictable square planar coordination geometries, typical for copper complexes, as well as evidence adduced from single crystal X-ray diffraction studies, carried for a number of copper(II) carboxylates [20, 82–84]. These structures showed binuclear one-dimensional (1D) pseudo-polymeric chain or ladder structures, where each copper ion was surrounded by five oxygen atoms together with a single Cu–Cu bond shown in Fig. 7, for the octanoate [83]. Further, the square planar copper ions formed weak axial intermolecular interactions with one of the neighboring molecules resulting in columns arranged parallel to each other, to form columns as shown in Fig. 7b.

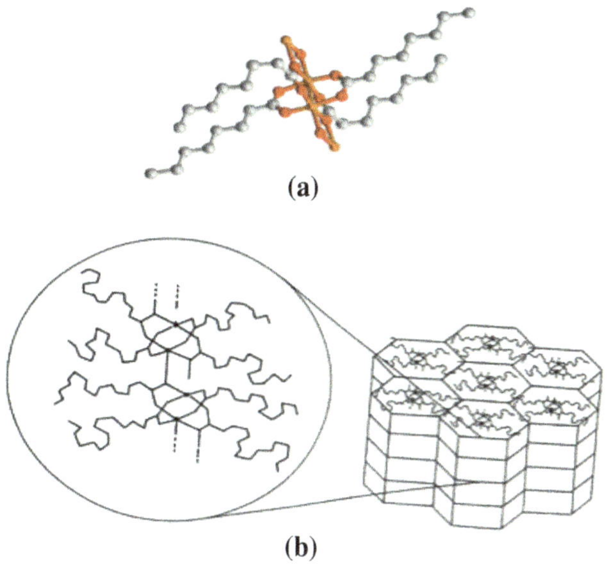

Fig. 7 Crystal structure of copper(II) octanoate (**a**) [83] and molecular for the columnar structures of copper(II) alkanoates (**b**)

Structural information for long chain copper(II) carboxylates which do not produce suitable crystals for X-ray analysis was obtained from electronic and infrared absorption spectroscopies [85]. Accordingly, the electronic absorption of the long chain adducts showed a band at 680 nm ascribed to the *d–d* transition of the copper(II) ion whilst the intense absorption at 300 nm was attributed to the ligand-to-metal charge transfer transition. The shoulder observed at ~370 nm indicated a metal-to-metal transition as a result of the metal–metal bond as was confirmed by magnetic susceptibility measurements [16, 58]. Additionally, IR data for all the compounds showed strong absorptions for the asymmetric and symmetric COO^- stretching at ca. 1,590 and 1,410 cm^{-1}, respectively, which gave $\Delta\nu = 170\ cm^{-1}$ confirming the bridging bidentate coordination as was predicted from PXRD data [18].

Zinc(II) carboxylates

Zinc(II) carboxylates have been synthesized in various polymorphic forms [20, 61, 62, 86–88], by a number of authors including Taylor [31]. Unfortunately some of these studies featured zinc(II) *n*-alkanoates synthesized in low purities and yields [20, 89, 90]; hence, discrepancies in the proposals for their structures and mesomorphic properties were evident. Additionally, though the structures for some compounds have been investigated the majority of studies were focused on the chain length homologues such as the acetate, butanoate, hexanoate and octanoate [87, 90–95]. Unfortunately, like most metal carboxylates, these compounds do not readily produce single crystals, consequently there had been no comprehensive proposals for the molecular structures and lattice packing for longer chain homologues. Though it was proposed that bidentate carboxylate coordination was most probable [18, 88, 90–96], an accurate picture of the molecular and lattice arrangement was not presented.

Investigations by Ishioka et al. [95, 96] involving the use of XAFS and IR spectroscopy, for ZnC_{18} and PXRD analysis involving ab initio Rietveld structure solutions, for ZnC_{11}, ZnC_{12}, and ZnC_{14}, by Mesbah et al. [97] showed that the structures were arranged in layers with tetrahedral zinc ions, bridged by oxygen atoms of carboxylate groups coordinated in a bidentate mode. The calculations presented by Mesbah [97] were very useful, since proposed structures showed two crystallographically independent chains, for ZnC_{11}; that is, hydrocarbon chains in one of the layers within the lamellar were packed parallel to the zinc plane whilst fully extended and those in the other layer were tilted at an angle of 3.96° to the metal planes. However, for ZnC_{12} and ZnC_{14}, the chains in the layer were crystallographic equivalent. Hence, as expected the tetrahedral oxygen environment for the zinc ions was different from those for ZnC_{11} and ZnC_{12}. For ZnC_{11}, there were alternating ZnO_4 tetrahedrons along the *a*-axis whilst for ZnC_{12} and ZnC_{14}, this was not observed. Furthermore, the structure for ZnC_{11} was similar to single crystal data for the shorter chain adducts [97]. Based on these results it was concluded that for a homologous series there was a change in structure with chain length. For example, with all short chain compounds, with an odd number of carbons, an alternating bilayer arrangement was adopted whilst for the longer chain adducts, $n_c > 10$, with an even number of carbons, the opposite was proposed [31].

Structural investigations carried out by Taylor and coworkers [31, 61, 62] on a homologous series of anhydrous zinc(II) butanoate to eicosanoate, inclusive, made use of several techniques such as IR spectroscopy, X-ray single crystal, PXRD and conoscopy in conjunction with density measurements and molecular calculations. These results confirmed that at room temperature the compounds were isostructural, where metal–carboxylate coordination was concerned; that is, metal–ligand bonding was via asymmetric *Z*, *E*-type (syn-anti) bridging bidentate mode. The hydrocarbon chains were all crystallized in the fully extended *all-trans* conformation and were tilted at an average angle of 60° with respect to the zinc basal plane within the lamellar. Interestingly, similar to Mesbah [97], the X-ray data indicated that the hydrocarbon chains for short chain adducts were packed differently from the long chain homologues. For example, a tail-to-tail bilayer arrangement was observed from the butanoate to the octanoate, inclusive, whilst an interdigitated bilayer arrangement was proposed for the longer chain adducts [31]. Though most of these observations were in agreement with Mesbah [97], no interdigitation of the hydrocarbon chains was proposed. Single crystal analysis of zinc(II) butanoate, pentanoate and hexanoate confirmed both the carboxylate coordination and lamellar bilayer arrangement of these short chain homologues. These results also showed that the compounds crystallized in the monoclinic crystal system with *P*2(1)/*c*, *P*1 (*P*21/*a*) and *P*1 (*P*21/c1) Bravais lattices, respectively [31, 88, 90–97]. The short chain homologues were composed of alternating bilayers between alternating zinc basal planes arising from crystallographically distinct zinc ions. Interestingly, powder data in agreement with the single crystal results indicated that the short chain length homologues had two molecules per unit cell (*Z*) but for long chain adducts $Z = 1$ [31]. For the butanoate, pentanoate and hexanoate, the complete structures consisted of 3D polymeric layered lattices in which component sheets were stacked in the direction of the longest unit cell side, similar to the structure proposed for the other homologues [96]. Interestingly, plots of interlayer (lamellar) spacing or unit cell long axis versus chain length were linear; however, there was an abrupt discontinuity between ZnC_8 and ZnC_9 as a result of changes in their molecular packing [31].

More accurate molecular calculations by Nelson et al. [98] considered zinc basal plane thickness and intermolecular distances between the hydrocarbon chains with in the lamellar; hence, it was shown (Fig. 8) that the hydrocarbon chains started to overlap for zinc compounds with $n_c \geq 9$; the degree of which increases with increasing chain length. Furthermore, measured densities and melting points, indicated differences in molecular packing for odd and even chain length compounds, attributed to geometries necessary for higher packing efficiencies and greater lattice stability. A more detailed discussion of the odd–even behavior including a more accurate description of the extent of packing was proposed by Nelson et al. [98] and will be discussed in a following section.

Lead(II) carboxylates

Though much of the early investigations on the structures of lead(II) carboxylates were focussed on the short chain length homologues, extensive studies have been carried out on the even chain adducts of various chain lengths. PXRD and solid-state NMR data confirmed lamellar structures, where the tilted hydrocarbon chains were crystallized in the fully extended *all-trans* configuration [20, 63]. However, as a consequence of the paucity of accurate structural data, for these compounds, both mono- and bi-layer arrangements were proposed for the lamellar packing of Pb(II) decanoate molecules [20]. Nonetheless, investigations by Harrison and Steel [99], on the crystal structure of anhydrous Pb(II) formate as well as the acetate trihydrate

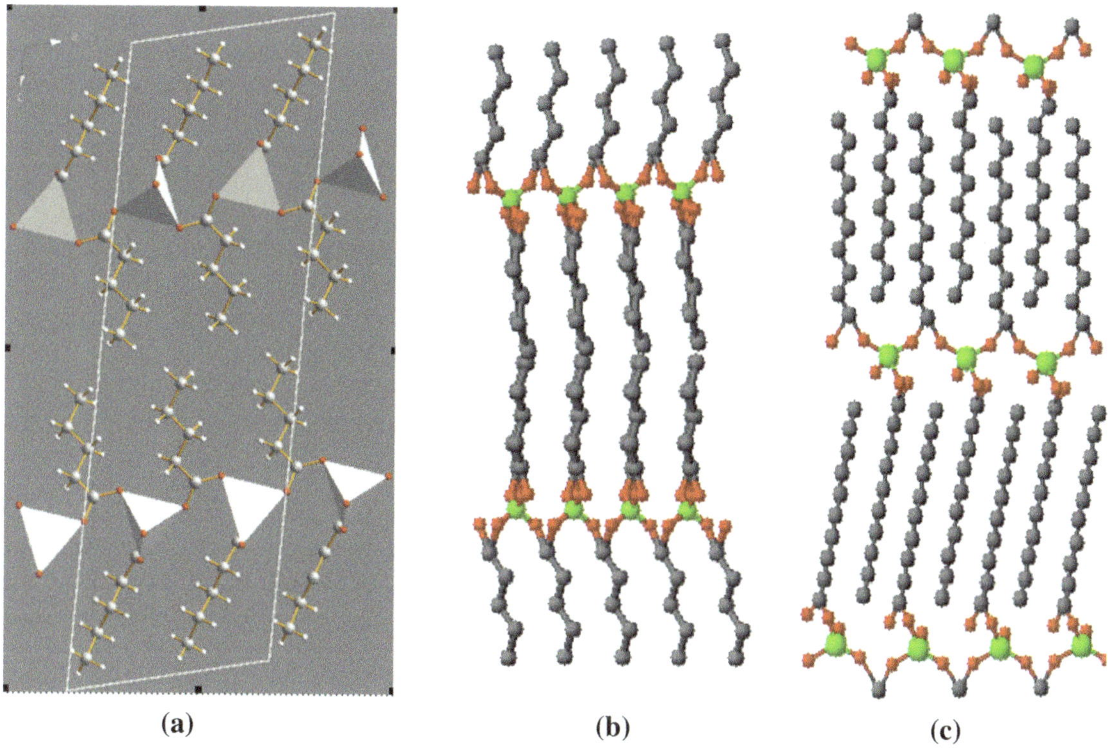

Fig. 8 Crystal structure of zinc(II) pentanoate (**a**) and molecular models for lamellar bilayer arrangement for short (**b**) and long (**c**) chain zinc(II) *n*-alkanoates [31, 62]

showed that the formate comprised a 3D polymeric network in which adjacent lead atoms were bridged by formate groups to form distorted eight-coordinated lead octahedrons. Interestingly, the data showed that the formate groups were dissimilar; that is, in some formate groups the C–O bonds (1.21 and 1.26 Å) were not equivalent. Though the high coordination number of the lead ions was surprising, other structures for the acetate trihydrate, pentafluorobenzoate and ethylene tetraacetate also showed similar coordination numbers and modes [20, 63].

Extensive IR, PXRD and microscopy studies for a homologous series of anhydrous Pb(II) carboxylates by Ellis and co-workers [20, 63], confirmed that head group bonding does not change with chain length in agreement with Mesubi [100]. Analysis of the IR data reported by Ellis et al. [63], showed that there was a non-equivalence of the CO bonds, hence, the Pb–O bonds were not identical in length, similar to the formate, where ν_a stretches were more intense than ν_s; an effect which increases with chain length. Interestingly, $\Delta\nu$ values decrease with decreasing chain length suggesting unsymmetrical chelating bidentate coordination with possible carboxylate bridging, for the shorter chain compounds ($PbC_{8...12}$). Furthermore, differences in the PXRD data resulted in the proposal, that there were differences in the molecular and lattice packing between short and long chain compounds; that is, for $n_c < 12$, a bilayer arrangement of the hydrocarbon chains existed but for $n_c > 12$, a monolayer arrangement was proposed. For example, marked differences in the high angle region, $20° \leq 2\theta \leq 25°$, of the powder diffractograms, associated with reflections from side chains packing interactions were observed [63]. However, despite the immense number of investigations carried out on these compounds, no detailed molecular models were proposed for them.

Lanthanide(III) alkanoates

Lanthanide(III) ions are known to take high coordination numbers such as 7–10 with the most popular coordination geometries being octa-coordinate square antiprism and the ennea-coordinate tricapped trigonal prism geometries [67]. As a consequence, lanthanide(III) carboxylates, exhibited diverse dimeric or linear polymeric structures, where the unit complexes were bridged by carboxylate ligands to form 3 or 2D structures as observed for some formates [18]. Furthermore, some structures were composed of many types of inter-metal ionic bridges (metal–metal bonds), which were different in some complexes; that is, they appeared as alternating with the polymeric chains [67, 101]. Some compounds showed simple bridges, where each pair of unit complex were linked by a single bidentate carboxylate group through both an *E, E*- and a *Z, E*-type bridging bidentate coordination. For example, trimeric scandium(III) formate was reported as octahedral, with three pairs of *Z, Z*-type carboxylate ions bridging three metal ions to form a 2D polymeric complex [67]. Similarly, scandium(III) chloroacetate was octahedral with metal ions bridged by three *Z, Z*-type carboxylate ions, respectively, to form a linear trimer [67]. Based on these results it is clear that some of the structures were quite complex and required a great deal of analytical sophistication for elucidation of their structures.

Though lanthanide carboxylates are widely used their anhydrous structures are not well known; that is, the majority of compounds isolated were hydrated. For example, lanthanide acetate was isolated as $Ln(CH_3COO)_3{\cdot}xH_2O$ ($x = 1, 1.5, 3, 4$) [58]. Such complexes were easily obtained as crystals of suitable morphology for X-ray single crystal analysis from slow evaporation of aqueous solutions; hence, their coordination structures and molecular packing were well known. However, the structures for anhydrous acetates and longer chain compounds have not been extensively studied, up until the last 15 years when they were isolated as anhydrous powders. Hence, the majority of structural investigations on these adducts were carried out via PXRD and IR spectroscopy, which indicated that the compounds were crystallized as lamellar bilayer structures composed of 100 % *all-trans* methylene group conformers in the hydrocarbon chains, packed in either a triclinic or hexagonal lattice. For these compounds, the use of IR data was unsuccessful in distinguish between triclinic or hexagonal lattice. Nonetheless, IR analysis indicated that a symmetrical chelating bidentate coordination was present, since $\Delta\nu = 120–133\ cm^{-1}$ and the COO bands were shifted to lower frequencies relative to the ionic soaps [20].

Investigations by Binnemans et al. [59] on a series of neodymium(III) carboxylates ($n_c = 3–20$, inclusive), indicated the presence of mesomorphic transitions in their phase sequence which showed significant structure dependence. The structure of the butyrate monohydrate was obtained from single crystal data but for the longer chain length homologues, PXRD and IR data were used to propose structural models. It was shown that, for this series, the structures consisted of ionic layers of neodymium ions, separated by bilayers of carboxylate anions anchored via *Z, E*-type bridging tridentate COO groups to the metal planes. Furthermore, the lamellar was composed of two crystallographically distinct neodymium ions, both having coordination number 9, allowing for a monocapped square antiprism geometry. In fact, the crystal structure for the butyrate (Fig. 9) showed Nd1 ions surrounded by two water molecules and five COO groups, four of which were bridging tridentate and the other bridging bidentate [59]. The Nd2 ions were coordinated by four bridging tridentate, one bridging bidentate and one chelating bidentate COO group. Based on these observations, the use of only IR data would be misleading since $\Delta\nu$ of 125 cm^{-1} suggested only chelating bidentate COO coordination. The assumption was made that half the water molecules can be lost without a change in the carboxyl coordination mode of Nd1, however, the experimental results showed that there was a change in coordination number from 9 to 8 accommodated by the bridging bidentate COO group, transforming to the bridging tridentate mode [59].

PXRD and IR investigations, by Hinz et al. [102], on a homologous series of praseodymium carboxylates gave similar results to that adduced from single crystal analysis, for lanthanum(III) butyrate monohydrate, previously reported [59]; hence, it was concluded that the structures for long chain lanthanide carboxylates were typical: a phenomenon possibly related to the lanthanide contraction effect. Binnemans et al. [57] isolated several lanthanide dodecanoates (Ln = Y, La, Ce–Lu; except Pm), as hemi- or mono-hydrates, however, all compounds were obtained in the anhydrous forms on heating and cooling from the melt. These compounds showed structures arranged in a lamellar bilayer arrangement, with alkyl chains in the *all-trans* conformation anchored to the metal basal plane, with respect to which they were tilted. Calculated and experimental values for the interlamallar spacing from PXRD did not decrease in magnitudes comparable to the decrease in metal ion radius and was suggested to be attributed to random deviations in chain flexibility (which affected the accuracy of the calculations, since 100 % *all-trans* conformation was assumed) and or experimental error [57]. It was proposed that lattice stability derives from the interplay between van der Waals forces, ascribed to the polymethylene chains and the electrostatic interactions associated with the Ln–OOC ionic layers. For the dodecanoates, it was assumed that if ionic radius decreases, then the bilayer was likely to contract resulting in increased interactions between the end methyl groups, thereby preventing the lattice from contracting too much. Hence, it was suggested that instead of a contraction of the overall lattice, there was an increase in the separation between the layers or tilting of the hydrocarbon chains which offsets any decrease in lanthanide ionic size [57, 58]. Therefore, the bilayer structure remained constant as evidenced from PXRD data. However, at increased temperatures, the repulsive electrostatic interactions which change with ionic radius became more prominent in controlling the lattice order since the van der Waals forces from the alkyl chains decreased significantly. Certainly, it would have been useful to calculate the thickness of the ionic layer as ion size changes. The structures of these compounds were assumed to be similar to that of Nd carboxylates based on observed similarities in their IR spectra and PXD patterns [57, 58].

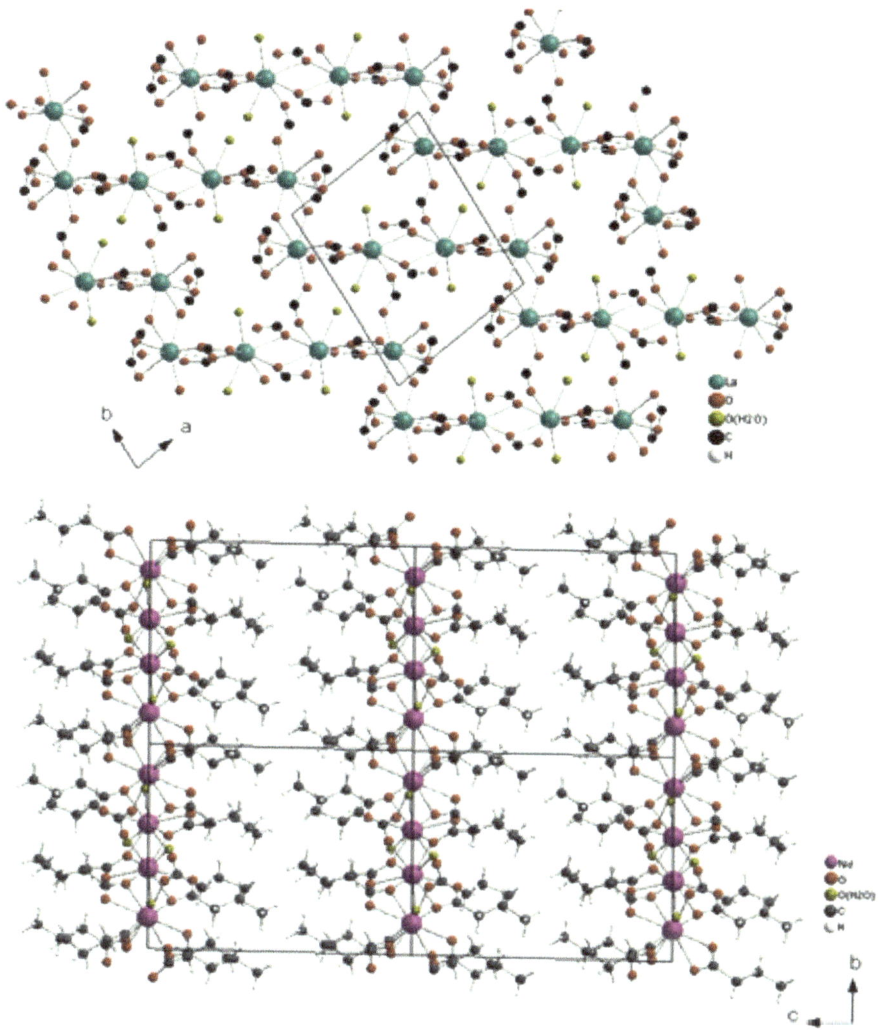

Fig. 9 Crystal structures of neodymium(III) butyrate monohydrate [59]

Odd–even alternation

One of the challenges arising from a lack of crystal structures, for an extended homologous series, is the difficulty associated with modeling structural features responsible for odd–even alternation/variation in melting point, density and viscosity data, observed for several metal carboxylates [31, 61, 64, 70]. Similar to *n*-alkanes, *n*-alkanoic acids and other *n*-alkane derivatives, several metal carboxylates display this type of behavior, which was widely understood to be attributed to a difference in the packing of the hydrocarbon chains in the lattice. Though the phenomenon of odd–even alternation is well documented [20, 30, 31, 58], no plausible explanations were proposed to explain this effect until recently. Boese and co-workers [103–105], proposed the first comprehensive geometric model for odd–even variation in alkyl chain derivatives by studying the melting behaviors of *n*-alkanes and α,ω-alkanedithiols. This model was proposed based on X-ray data and involved representing the packed alkane chains in appropriate quadrilaterals which allowed for the revelation of void spaces in the packing structures [103–105]. The result was that the planes of the hydrocarbon skeletons, for the even chain alkanes, depicted parallelograms whilst the odd chain alkanes depicted trapezoids (Fig. 10). Based on these models, it was clear that in order to allow for the most efficient packing of geometric units, the approaching methyl groups adopted a staggered conformation resulting in columnar packing of the hydrocarbon units which were shifted with respect to each other. For even chain alkanes, intermolecular contact between the parallelograms at all sides resulted in fewer void spaces, whereas for the odd chain adducts only some parts of the molecules were efficiently packed; hence, more void spaces were present. Therefore, even chain alkanes had higher densities and melting points than their odd chain counterparts [103]. A similar approach was used to explain the melting point alternation in α,ω-alkanedithiols [103–105].

For metal carboxylates, there has been no definitive proposal, adequately explaining odd–even alternation in

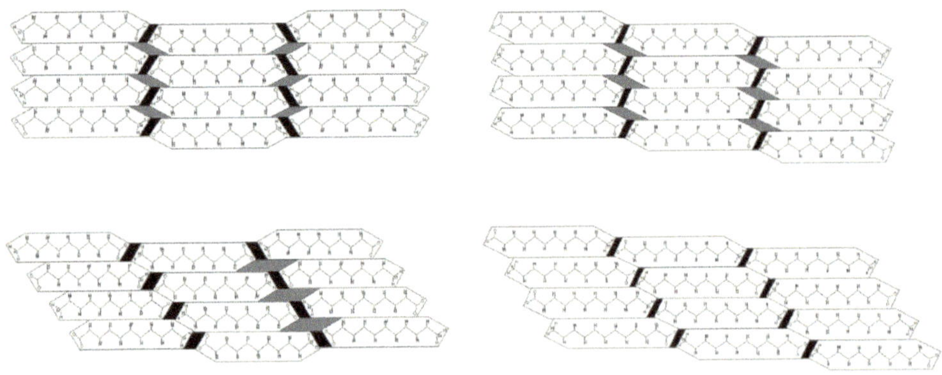

Fig. 10 Geometric models for the packing of *n*-alkane chains [103–105]

melting point, density, IR and ^{13}C-NMR, until recent studies by Taylor [31] and Nelson and Ellis [70] for the zinc(II) and silver(I) carboxylates, respectively. In these studies IR, ^{13}C-NMR spectroscopic and PXRD data were used to provide evidence for the various lattice interactions. Molecular calculations were also used to provide evidence for the spacing between interacting groups within the molecular lattice. For example, long chain zinc(II) *n*-alkanoates, $n_c = 9$–20, inclusive, showed odd–even variation in the methyl group, $\nu_{as}(CH_3)$ vibration and the solid-state ^{13}C-NMR chemical shift (δ) [31]. According to Taylor [31], the *all-trans* conformation of the hydrocarbon chains and chain interdigitation resulted in the CH_3 groups of even chain homologues being in close proximity to the zinc ions as opposed to the odd chain homologues whose natural zig-zag structure resulted in the CH_3 groups pointing along the zinc basal plane and was therefore, at a greater distance from the zinc ion. This model was correlated with observed odd–even variation in IR, ^{13}C-NMR, density and melting point [31]. However, further analysis by Nelson et al. [98] lead to the proposal of more accurate molecular models for zinc(II) carboxylates. In these models short chain length compounds, $n_c < 9$, did not show overlapping of the hydrocarbon chains, hence, methyl–methyl carbon–carbon interactions between the layers within the lamellar did not influence their molecular packing; therefore, odd–even alternation was not observed. However, for longer chain compounds, $n_c > 9$, the chains formed overlapping instead of interdigitated bilayers. The extent of lamellar overlap increased with increasing chain length and as such, odd–even alternation observed in the $\nu_a(CH_3)$ values were due to methyl–methylene interactions of chains from opposite layers within the lamellar [98]. This close approach of the methyl and methylene group carbons resulted in London dispersion forces that cause both carbons to become partial positively polarized; hence, partial negative charges were induced on the hydrogen atoms of the methyl and methylene groups, resulting in attraction between the hydrogen atoms of the alkyl chains from opposing layers. This resulted in reduced methyl C–H bond strength and frequencies as the vertical distance between the hydrocarbon planes within the lamellar decreased. Hence, these effects were greater for compounds that were more efficiently packed carboxylates [98].

In the case of silver(I) *n*-alkanoates [70], odd–even alternation was observed in $\nu_{as}CH_3$ absorptions, in addition to the $\nu_{as}COO$ frequencies, which was never reported for any other metal–carboxylate system; that is, $\nu_{as}COO$ values were consistently higher for the even chain compounds indicating greater metal–carboxylate interactions. For the methyl groups, stretching frequencies for even chain adducts occurred at higher frequencies, than for odd chain compounds, as a result of the relative difference in the closeness of approach of the hydrocarbon chains from different layers in the bilayer. This resulted in different extents of methyl–methyl and methyl–methylene and van der Waals interactions between the chains [70]. The chemical shifts for even chain length compounds were also consistently higher than for odd chain adducts; that is, they were more deshielded. By calculating the distance between hydrocarbon tails within a lamellar, $D_{bt} = (d_{exp}/\sin r) - 2d_L$, Nelson and Ellis [70] showed that methyl–methyl carbon–hydrogen interactions between tail groups from opposite layers in the bilayer were clearly possible. Such interactions resulted in hydrogen atoms from closely packed tail groups becoming partial positively polarized by nearby methyl carbons from opposite layers within the lamellar, hence, the methyl carbons were partially negatively polarized; an effect which was a function of the vertical distance between tail group carbons. Consequently, tail group C–H bonds lengthened and weakened, as evidenced from the lower asymmetric methyl stretching frequencies and greater electron densities on methyl carbons for odd chain homologues. Odd–even alternation was also reported for the chemical shift, of the carboxyl carbon which is a first in metal carboxylate chemistry [70]. Based on this observation it was proposed that the carboxyl carbon for odd chain compounds had a greater electron density as a consequence of higher resonance in C–O bonds of carboxyl groups. However, though they did not discuss the basis for this difference in COO resonance for odd and even chain compounds, it may be assumed that the extent of metal–ligand interactions for odd and even chains may have influenced polarization along the chains, the way in which they were packed and also head group interactions. Perhaps the symmetry of the head groups was different for odd and even chain adducts [70].

Thermotropic phase transition behavior

Overview: mesophase structures

Metal carboxylates constitute a large class of metallomesogens, several of which have been reported on by various authors [20] and it was first observed by Luzzati and co-workers [116] in the1960s that alkali, alkaline earth and transition metal soaps exhibited non-linear liquid crystallinity. This investigation provided the foundation for modern understanding of self-assembly of amphiphilic systems and importantly, non-lamellar mesomorphism. Furthermore, formalized analysis of self-assembled systems in terms of molecular shape of constituent amphiphiles, is of fundamental importance in all fields involving molecular interfaces, from the study of synaptic endocytosis in neural transmission to the understanding of primitive cellular membranes. The study of the thermal behavior of these compounds allows for these applications in various fields of modern science.

Several phase structures including lamellae, discs, rods and ribbons in various 1D, 2D and 3D arrangements have been proposed for several metal carboxylates systems [20, 58, 106]. The phase behaviors of these carboxylates depend on both the metal and carboxylate chain length, but the assignment of many of the mesophases is often controversial and certain of the proposed structures must be treated with caution. The transitions are normally first order, although second order transitions have been suggested for several compounds such as cadmium(II) and barium(II) stearates [20, 58]. For many metal carboxylates, not only are the phase sequences dependent on alkyl chain length, but they are also dependent on lattice arrangement and metal–carboxylate interaction.

Monovalent carboxylates

Their interesting thermochemical properties and possible commercial applications have resulted in several decades of studies on the phase properties of metal carboxylate systems. For example, studies on the physicochemical properties of these compounds have shown that when sufficiently heated, the room temperature crystalline solid transforms into the isotropic melt via several intermediate steps, some of which were liquid crystalline for some metal carboxylates [107–110]. However, assignment and accurate characterization of these transitions have been controversial, hence, some of the proposed phase structures must be treated with caution. According to Skoda [111] the phase sequences for even chain lithium(I) carboxylates with $n_c = 14–18$, inclusive, consisted of three transition ranges. These were classified as the polymorphic transition range at ca. 383 K adjacent to the higher temperature crystalline modification phases which transitioned into a mesomorphic ribbon phase at ca. 473 K followed by the isotropic melt in the region of ca. 503 K. A second order transition was also observed at 353 K for these compounds. Similar studies were carried out by Ferloni and Westrum [72] via DSC analysis, on lithium carboxylates with $n_c = 1–20$, inclusive. Their phase sequence showed several crystal–crystal pre-melting transitions associated with changes in the hydrocarbon chain tilting and lamellar packing. The melts were associated with molten hydrocarbon chains anchored to unperturbed polar head groups [72]. According to Gallot and Skoulios [112], the phase sequences for lithium carboxylates with $n_c > 11$ showed several pre-melting transitions, most of which were plastic crystal or lamellar phases, composed of ribbon-like structures, assembled in a rectangular centered mesh. Based on these results and POM studies, the following stepwise fusion mechanism was proposed: [112]

Lamellar crystal I ↔ lamellar crystal II ↔ plastic crystal ↔ melt.

Though several meta-stable phases associated with changes in van der Waals interactions in the lattice were observed for these compounds, the majority of transitions were reversible [113]. Furthermore, this was supported by White and Ellis [113] who showed that lithium(I) soaps with $n_c = 8–18$, inclusive, exhibited enantiotropic phase behaviors. Their phase sequences were composed of only solid–solid transitions; the number of which increased with increasing chain length as a result of greater alkyl chain flexibility. For example, only one intermediate phase was observed for $LiC_{10\cdots12}$ [112, 113], when heated from the room temperature lamellar crystal to the melt:

Lamellar crystal I ↔ lamellar crystal II ↔ isotropic liquid and two for $LiC_{14\cdots18}$:

Lamellar crystal I ↔ lamellar crystal II ↔ plastic crystal ↔ isotropic liquid.

Since, super-cooling was observed for all transitions in the phase sequences, for all lithium carboxylates, it was concluded that reversion of the alkyl chains to the original lamellar arrangement was kinetically controlled [113]. This phenomenon, which was more pronounced for longer chain adducts, due to greater chain flexibility, confirmed that the observed transitions were solid–solid rather than liquid crystalline [114]. Calculated molar enthalpies ($\Delta_{fus}H_m$) and entropies ($\Delta_{fus}S_m$) of fusion, for lithium carboxylates, reported by several authors [111], showed reasonable agreement indicating that the compounds in these studies were structurally similar and were of high purity. Since plots of $\Delta_{fus}H_m$ and $\Delta_{fus}S_m$ versus n_c showed no appreciable change with increasing chain length, for compounds with $n_c < 13$, similar fusion mechanisms were proposed for these adducts [113, 114]. However, for longer chain adducts a linear relationship between these parameters and chain length was observed, confirming strong chain length dependence and different fusion mechanisms for short and long chain lithium carboxylates [113, 114]. Furthermore, since the plot of total molar enthalpy ($\Delta_{total}H_m$) versus n_c was linear with a slope less than the contribution per-methylene group (3.8 kJ mol^{-1}) [115], for the fusion of *n*-alkanes, incomplete alkyl chain fusion was proposed for these compounds in the melt [113]. Phase behaviors for the shorter chain adducts, investigated by DSC and electrical conductivity measurements, revealed a single pre-melting transition at 549 K followed by the melt at 604 K for lithium propanoate [77]. For the pentanoate, fusion was observed at 576 K subsequent to two pre-melting transitions at ca. 206 and 325 K. Since the pre-melting transitions showed super-cooling of up to 5 K, it was concluded that the observed transitions were crystal–crystal rather than liquid crystalline [115]. Based on electrical conductivity data, the basis of these transitions was assigned to increased metal and carboxylate ion mobility as the compounds become increasingly molten with increased temperature [115]. Such conclusions were supported by variable temperature infrared data, which indicated increased methyl and methylene group rotation to yield conformational disordered states (*con-dis*) of various energies with increased temperature. This was further supported by the fact that the ΔH values, for the pre-melting transitions, were in the energy range for rotation of methylene groups [115]. Interestingly, there was a general consensus among authors where assignment of the phases for lithium(I) carboxylates were concerned.

Other alkali metal alkanoates have also been the subject of extensive investigations. For instance, Luzzati et al. [116] concluded, based on X-ray powder diffraction analysis, that the phase textures for anhydrous sodium carboxylates were composed of ribbon-shaped polar head group aggregates. These were dispersed in a sea of liquid hydrocarbon moieties that were arranged in a two-dimensional oblique or rectangular lattice [117]. In earlier studies [118–120] these phases were classified as sub-waxy, waxy, super-waxy and subnet, based on their lattice parameters. Interestingly, some of these phases were optically isotropic whilst others were birefringent [116]. According to Corkery [65], sodium(I) carboxylates exhibited both thermotropic and lyotropic mesomorphic transition behaviors. Their phase textures appeared as birefringent oily streaks for moist samples whilst for dry samples they were weakly birefringent at ambient temperature. In the region of ca. 393–403 K, a streaky feathery fan-like texture which deformed under a shear stress was detected by POM studies [65]. However, since this phase changed to spherulites when aged, the observed mesophases might have been due to the presence of impurities; hence, these phase textures must be treated with caution. Similarly, Sokolov [121] reported a fusion temperature of 571 K, for sodium(I) propanoate, in addition to four uncharacterized solid–solid transitions at ca. 560, 490, 468 and 350 K detected by a visual-polythermal technique. However, Baum et al. [122] obtained a fusion temperature of ca. 563 K via hot-stage polarizing microscopy for this same compound. Subsequent reports by Ubbelohde et al. [123] indicated a mesophase at 567 K which was above the fusion temperature of 566 K ($\Delta_{fus}H_m$ = 8.8 kJ mol^{-1}) adduced from DSC measurements. For this same compound, DSC analyses carried out by Ferloni et al. [124] showed a fusion temperature of 561 ± 1 K along with two uncharacterized solid–solid transitions, at ca. 498 and 475 K. Other reports have also been published on this compound; however, there was hardly any agreement where the transition temperatures, number and textures were concerned [125–127]. This is probably due to polymorphism, which is common for these compounds, and or the presence of impurities, since no C, H analyses results were presented in any of the aforementioned reports. Similar inconsistencies have been reported for long chain sodium *n*-alkanoates [128]. For instance, it was concluded, based on X-ray diffraction data, that sodium octadecanoate, eicosanoate and docosanoate formed a mesomorphic phase at the clearing point (after the melt). According to these reports a smectic I mesophase which transitioned into a smectic II phase was observed for the octadecanoate; however, for the longer chain adducts only a smectic II mesophase was formed at the clearing point [128]. These results are quite contradictory to the majority of results presented for metal carboxylates systems. For example, the assertion that there were post-melting transitions in the phase sequences for these compounds indicated inaccurate phase characterization. Such inconsistencies confirmed the need for more intensive investigations into the phase behaviors of sodium carboxylates from a theoretical and experimental perspective.

A study of the thermodynamic and transport parameters of some molten ionic salts showed variable thermal stabilities even in the absence of air, due to the great diversity of degradation pathways [129]. However, the presence of excess fatty acid retarded the decomposition rate by absorbing hydroxyl ions accumulated in the salts during synthesis. Melting volumes were studied and shown to follow a linear regression equation, $\rho = \rho^* - @(T - T^*)$, where ρ and ρ^* as well as T and T^* were associated with the solid and liquid phases, respectively. The densities were measured using specially designed equipment. Interestingly, after a single heating–cooling cycle increased crystal volumes were observed for sodium butyrate, indicating the development of rifts in the crystals. Furthermore, based on DSC data it was concluded that in some compounds one or more of the observed pre-melting transitions were liquid crystalline. However, since no C, H-analysis or microscopy data were presented to prove the purities of these samples and to characterize the phases, the results presented should be accepted with caution. Electrical conductance measurements, carried out in a conventional Pyrex dipping cell, showed a general increase with increasing temperature. However, for sodium butyrate there were three points of steep increases, indicating significant changes in the mobility of the current carriers. For example, the steepest increase in conductance for sodium isobutyrate was observed at the melt. For the proposed liquid crystalline phase no significant change in conductance was observed, indicating that they were not real mesophases. This was confirmed by the high viscosities ($\sim$100 P in most cases) observed for them [129]. Interestingly, the viscosities of most compounds decrease to $\sim$0.2 P at the clearing points, indicating significant changes in structure as they transition to the purely liquid phase. It was postulated that such changes might be associated with positional randomization of the anions and or cations as well as configurational changes due to thermal motion of various parts (–CH_2–, –COO– and C–C) of the anion or cation and or the aggregation of ions [129].

Though several investigations have been carried out on the thermal behaviors of potassium carboxylates, no C, H-analysis results were presented for these compounds. Hence, the proposed phase sequences and corresponding thermochemical data must be viewed with caution. For instance, the phase sequences for potassium butanoate [124] showed five solid–solid transitions, at ca. 133, 464, 552, 626 and 677 K, inclusive of a liquid crystalline phase. For potassium propanoate two solid–solid transitions were observed, at ca. 255 and 353 K, prior to the isotropic melt at 638 K [130]. Dilatometric analysis of potassium acetate showed three transitions between room temperature and the melt at 577 K [131]. Variable temperature crystallographic analyses showed that for KC_1 these transitions were associated with changes in crystal packing from orthorhombic to monoclinic: phase I (428 K) to phase II (349 K) [131]. These results indicated that the phase behaviors of these compounds were controlled by changes in alkyl chain packing with increased temperature. Investigations on the palmitates (hexadecanoate) of alkali metals from Li to Cs, inclusive, carried out by Vold and Vold [120], showed a total of three transitions at ca. 366, 462 and 484 K for lithium palmitate but for the potassium adduct, four transitions were observed at ca. 340, 403, 447 and 527 K. Dilatometric traces, collected between 273 and 423 K, showed two pre-melting transitions, at ca. 242 and 388 K for Rb and 234 and 368 K for Cs carboxylates. However, DSC traces for samples of these compounds, prepared by a different method, showed temperature values that were different from those adduced from dilatometry. This indicated possible contamination of the samples analyzed via dilatometry or that the second samples were different polymorphs of the compounds [120]. DSC traces for lithium palmitate showed four transitions, inclusive of the melt, whilst five transitions were observed for potassium and four for rubidium and caesium palmitates [120]. The transition at lowest temperature (374–340 K), for all compounds, was classified as an *inter-crystal* transition based on X-ray single crystal and powder diffraction analyses. These results were also used to classify the second transition, in the range of 400–470 K, as a *waxy* transition and subsequent phases, as sub-neat and neat in accord with Bolduan and coworkers [132]. Increased fusion temperatures observed on going from Li to Cs indicated that increased metal ion radius resulted in a more rigid carboxylate network [120]. Unfortunately, despite the myriad of investigations carried out on these compounds the underlying theories of their phase properties remains largely underdeveloped. Nonetheless, a study of the enthalpy of formation of the alkali metal salts was reported by Piedade and coworkers [133]. This investigation made use of some basic concepts of classical thermodynamics. Therefore, the molar enthalpies of formation, $\Delta_f H_m$, for Li, Na, K, Rb and Cs acetates were calculated to be 714.4, 711, 722, 723 and 726 kJ mol^{-1}, respectively [133]. These results indicated that the metal ion radius of metal carboxylate systems affected their packing energies; hence, the significant increase, $\Delta_f H_m$, observed on going from Na to K indicated significant changes in lattice packing at this point due to changes in the metal ion radius. Indeed, calculated lattice energies using the Kapustinskii [134, 135] approximation: $\Delta_{lat}U_0 = 1.079 \times 10^5(vZ_+Z_-/(r_+ + r_-))$, where v is the number of ions and Z_+ and Z_- were the charges on them, showed decreasing values with metal ionic radius from 911 for CH_3COOLi to 668 kJ mol^{-1} for CH_3COOCs in agreement with Jenkins and Roobottom [136]. Recent studies by Bui [73], on some short chain monovalent carboxylates (MC_1–C_{12}, where M = Li, Na, K, Rb, Cs), confirmed the presence of several solid–solid pre-melting transitions in their phase sequences; the number of which increased with chain length. The calculated transition enthalpies for lithium compounds were in the range of 36–330 J/g. The onset of decomposition occurred in the range of 600–780 K. Interestingly, decomposition occurred at higher temperatures for the longer chain adducts, as did the fusion. This is quite contradictory to the results published by several authors; [20, 58, 113, 115, 130] that is, melting temperature decreases with increased chain length as the more weaker van der Waals forces becomes prevalent relative to head group electrostatic forces. Such discrepancies in the report by Bui [73] might be due to impurities in the samples as no elemental analysis data were presented to validate the purity of the compounds. For sodium carboxylate, the onset of decomposition occurred in the range of 670–750 K with weight losses in the range of 14–76 %. Interestingly, the percentage weight loss increased with increasing chain length. Unfortunately, no explanations were proposed for this observation except to say that the likely paralytic intermediate was $Na_2C_2O_4$. Furthermore, though a number of pre-melting phases were observed in the phase sequences of these compounds, the number of such transitions was not in agreement with the literature. This is possibly due to polymorphic differences between the samples reported by Bui [73] and those in the literature or the presence of impurities in his samples [117–124]. Interestingly, though significant super-cooling was observed for the observed pre-melting phases, some of them were classified as liquid crystalline. Surly, this interpretation is incorrect, as liquid crystalline phases do not under cool. Furthermore, though an ionic mesogenicity rule was used to show that these compounds could form liquid crystal phases, metal carboxylates are not totally ionic; hence, the Buis' phase postulates were inaccurate. Furthermore, though odd–even alternation was present in the melting point data, no mention of this feature was made by Bui [73] or explanations presented.

TGA measurement for potassium, rubidium and caesium carboxylates showed weight losses in the regions of 17–70 % ($T = 860$–750 K), 10–58 % ($T = 834$–760) and 12–50 % ($T = 737$–751), respectively [73]. These mass losses were associated with the loss of metal oxalate and an organic moiety such as a ketone. DSC data showed a greater number of transitions for K, Rb and Cs salts, compared to Li and Na compounds, indicating that some of the transitions observed might be associated with changes in the metal basal plane structure. For example, a total of seven transitions were observed for Rb octanoate and Cs heptanoate. However, for the smaller K ion, six transitions were observed for the decanoate.

The results of several investigations, carried out by various methods, on the phase behaviours of thallium(I) carboxylates (TlC_1.....TlC_n, $n = 1, 2, 3...$) have shown that they go through several pre-melting phases when heated from the room temperature solid to the melt, some which were observed at sub-ambient temperatures whilst the majority were super-ambient phases of various textures and morphologies, depending on the chain length [137–141]. For example, the phase sequence of thallium(I) hexanoate showed two first order transitions at ca. 204 and 208 K [139–141] in addition to several solid–solid phases and a single mesophase [142, 143]. For the decanoate [144] four solid–solid transitions were observed at ca. 233, 388, 306 and 327 K, followed by a mesophase at 405 K with ΔH and ΔS of 5.67 kJ mol^{-1} and 13.97 J mol^{-1} K^{-1}, respectively, whilst the melt was observed at 484 K. Phase sequences for the longer chain thallium(I) *n*-alkanoates [138] from the tridecanoate to heptadecanoate showed several transitions, some of which were classified as mesomorphic based on their appearances. For instance, the phase sequence for the tridecanoate showed a single solid–solid transition at 365 K followed by the fusion at 468 K. For TlC_{15}, a single solid–solid phase was observed at 353 K followed by a mesophase at 455 K and the melt at 488 K. Similarly, for TlC_{17} a single solid–solid phase followed by a mesophase and the melt were observed at 302, 421 and 502 K, respectively, whilst for TlC_{16} two solid–solid phases were observed followed by a mesophase and the fusion at ca. 398, 415, 425 and 500 K, respectively. Odd–even alternation in heat capacity C_p (at constant pressure) versus n_c was observed; however, no model was proposed to explain this behavior. POM studies by Peltzl and Sackmann [145] revealed that the optical properties of thallium(I) carboxylates were chain length and temperature dependent. This was attributed to the rotation of CH_2 groups about carbon–carbon bonds resulting in the formation of gauche conformers in the alkyl chains [138]. Thallium(I) carboxylate were also found to exhibit lyotropic mesomorphic phase behavior when combined with the parent acid [146–148]. Unfortunately, no detailed investigations have been carried out to adequately explain this phenomenon.

The thermophysics of dimeric silver(I) carboxylates $(RCO_2)_2M_2$, is of commercial importance since they are applied in photothermogravimetry and other imaging techniques as a source of silver [35]. However, their interesting phase properties represent their most significant possible industrial application since they exhibit phase transitions of various textures when heated from the room temperature crystal to the melt. Some of their pre-melting phases have been described as liquid crystalline as well as solid–solid phases [79]. Mesophase to mesophase transitions have also been reported for several adducts. Furthermore, according to several authors [149–152], these compounds melted in the range of 383–393 K with decomposition, to yield silver nanoparticles for both short and long chain adducts. Investigations by Binnemans and coworkers [79], on a homologous series of silver

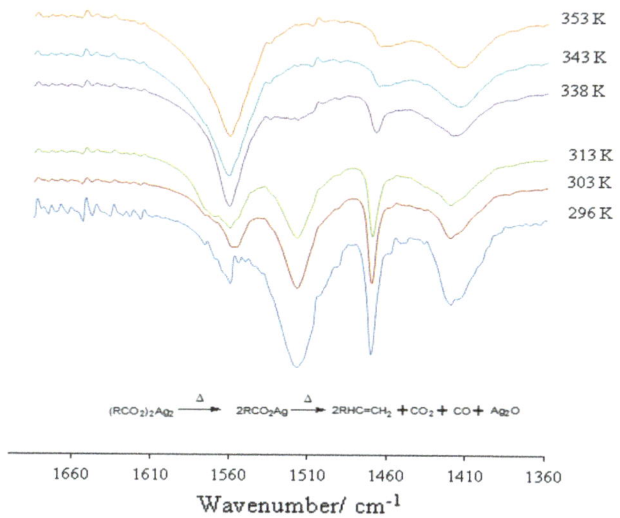

Fig. 11 Changes in infrared spectra for silver eicosanoate with increased temperature [70]

n-alkanoates, $n_c = 4$–26, inclusive, showed at least two smectic type mesophases in the phase sequences of several compounds. However, recent studies, though not refuting claims for the formation of mesomorphic phases, did not confirm more than one such phase for a single compound [70]. Furthermore, though there was a general consensus among authors, where the textures of the pre-melting phases were concerned, there were some discrepancies surrounding the number of transitions. For example, according to Binnemans [79], five pre-melting transitions at 376, 387, 400, 447 and 484 K (mesophase) followed by the melt at 508 K were observed for silver(I) *n*-dodecanoate. However, investigations by Nelson [70] showed that the phase sequence for this compound was composed of three pre-melting transitions at 374, 388 and 485 K followed by the melt at 510 K. Similar discrepancies were observed in the ΔH and ΔS values as well as the number of transitions in the phase sequences for the other homologues [70, 76]. Decomposition of these compounds at the melt was well documented; [20] hence, several proposals for their decomposition products and mechanism have been reported. For example, based on variable temperature infrared measurements (Fig. 11), Nelson [70] proposed the formation of an alkene which resulted from the decarboxylation of the dissociated dimeric silver carboxylates; that is, a mixture of carbon dioxide and carbon monoxide were released along with the metal oxide in accord with Binnemans [79].

Iweg and Hecht [153] posited that the decomposition of silver(I) butanoate produced butanoic acid, carbon dioxide, carbon and the metal by way of several elementary reactions. Similar results were reported for silver acetate, however, molecular hydrogen was also proposed as a decomposition product [154]. For the longer chain adducts, metallic silver, carbon dioxide and an alkyl radical, which subsequently combined with another such radical, were proposed [149] in agreement with Akanni [20].

Divalent carboxylates

Phase transition behaviors for long and short chain lead(II) carboxylates in both the solid (neat) and solution state have been investigated extensively [20–26]. For example, though, sub-ambient solid-state DSC analyses down to 180 K, revealed no additional transitions, super-ambient scans revealed a single intermediate transition for compounds with $n_c < 10$ whilst for $n_c = 10$ two such transitions were observed in the ranges of 337–382 and 343–379 K. The phase at ca. 355 K, showed extensive super-cooling and was, therefore, characterized as crystal–crystal rather than liquid crystalline [23]. However, the birefringence and texture of the phase centered at 361 K indicated a SmC mesophase. Increased volume observed on transition into this phase [25], was indicative of significant disorder. However, solid-state [201] Pb-NMR measurements [25, 155] suggested that little or no perturbation in metal–carboxylate bonding and by extension the entire metal basal plane, occurred on going from the solid to the melt. These results indicated that the phase transitions, especially the melt, were associated with disordered alkyl chains anchored to unperturbed metal basal plane aggregates. This assertion was confirmed by transverse ^{1}H-NMR relaxation studies which revealed that for lead(II) decanoate the sample was partly amorphous at ambient temperature, however, the percentage amorphous content increased with increasing temperature, without reaching the polar head group, as a result of progressive activation of conformational disorder in the polymethylene chains.

Though the formation of a SmA mesophase was first proposed for PbC_{10} [21–27], it has been agreed that only a single SmC phase with fan-like texture was formed for this compound. For the shorter chain adducts, $n_c = 2$–5, inclusive, the formation of ionic melts and various glass phases have been proposed [20, 27]. For instance, on heating lead(II) pentanoate, from the room temperature crystal to the melt, two endotherms were observed at 328 and 356 K ascribed to a solid–solid phase transition followed by the isotropic melt. Interestingly, slow cooling of the sample, at 5 K min^{-1}, showed only the crystallization transition whilst reheating showed a new transition at 322.9 K, classified as a glass phase [27]. These results indicated significant changes in the morphology of the sample on heating. However, these changes were temporary since aged samples (stored for 60 days at ambient temperature) showed total reversion to their original thermal behaviors. Phase sequences for lead(II) acetate and propanoate were almost identical; that is, only a single transition corresponding to the fusion was observed for both compounds [20]. Rapid cooling (20 K min^{-1}) of these samples resulted in the formation of a glass phase. However, on second heating two glass transitions were observed prior to the fusion [115]. The thermal behavior of PbC_4 was similar to that for PbC_2 and PbC_3, however, there were some noteworthy differences. For example, on heating, only a single endotherm ascribed to the fusion was observed for this compound. Interestingly, on cooling, the thermograms showed significant dependence on the time elapsed before reheating; that is, when the thermogram was recorded immediately after cooling, a single glass transition was observed but when aged, for a time in excess of 10 h., at room temperature, a new solid–solid transition was observed. These results indicated that the phase behaviors of lead(II) *n*-alkanoates were kinetically controlled and chain length dependent. Though very little investigations have been carried out on the unsaturated analogues, Adeosun and coworkers [156] showed that in contrast to lead(II) *n*-octadecanoate the mono-unsaturated adduct melted, in a single step, at a lower temperature. However, the alkyldiynoates which have been investigated for possible polymerization to poly(diacetylene) showed fairly ordered SmC phases at ambient temperature [157]. Such results indicated that the formation of room temperature liquid crystals were most likely for carboxylate with unsaturated hydrocarbon chains.

Thermal degradation mechanisms and kinetics for lead(II) *n*-alkanoates investigated by Ellis [37, 38] and Baraldi [158] via infrared emission spectroscopy, DTA and TGA showed that decomposition of the compounds, in the range of 465–640 K, occurred via a two step process which resulted in an overall mass drop of 59 % [38]. The major decomposition products, for PbC_{10}, characterized by infrared spectroscopy, NMR measurements and elemental analyses, were nanodecan-10-one and lead(II) oxide along with small amounts of CO_2 and CO gases [38]. However, according to Baraldi [158] decomposition of lead(II) formate resulted in the formation of the intermediate metal carbonate which subsequently decomposed to yield a mixture of the raw metal and metal oxide. These results also indicated that thermal degradation took place at a higher temperature in air than under inert atmosphere [158]. The decomposition kinetics for PbC_{10} investigated via thermogravimetric-mass spectrometry in conjunction with gas–liquid chromatography [158] indicated that the rate of decomposition was controlled by a diffusion mechanism at a contracting interface or by the remaining amount of reactant. Based on the popular Freeman and Carroll [159] model it was shown that the decomposition of PbC_{10} followed zero order kinetics with activation energy of 42.05 ± 2.3 kJ mol^{-1} in accord with Rasheed and Bhode [160]. Furthermore, it was proposed that the decomposition of the PbC_{10} took place via a free radical mechanism. Though it was expected that the thermal degradation mechanism for this compound would have been identical to that for the longer chain adducts, the observed differences for the dodecanoate, tetradecanoate and octadecanoate indicated slightly different degradation pathways [37]. For example, decomposition of PbC_{12} and PbC_{14} took place via two rapid steps in the range of 690–730 K, terminating after a total mass drop 62 and 65 %, respectively. However, for the octadecanoate, decomposition occurred via a single step resulting in an overall mass drop of 71 %. Based on these results, it was concluded that one of the major decomposition products for all compounds was lead(II) oxide [37]. The kinetic data for the decomposition of these compounds were analyzed through application of the modified Freeman–Carroll model [161], expressed as: $\Delta\log(dw/dt) = n\,\Delta\log W_r - (\Delta E/2.303R)\,\Delta(1/T)$ where dw/dt was the rate of reaction, **n** was the order of reaction, **ΔE** was the activation energy, **R** was the gas constant and **T** the absolute temperature. These results indicated that, similar to the degradation mechanism for PbC_{10}, the decomposition process for PbC_{12}, PbC_{13} and PbC_{18} were zero order and were controlled by a diffusion mechanism with overall activation energies of 629, 103 and 139.31 kJ mol^{-1}, respectively [37]. Since the activation energies were different, it was concluded that there was a systematic relationship between chain length and the energy requirements for degradation of these compounds.

The monotropic mesomorphic transitions for copper(II) carboxylates of various chain lengths, have been characterized by POM, X-ray diffraction [65, 162–164], DTA, DSC [85], viscosity measurement [165], dilatometry [166] and variable temperature infrared spectroscopy [167], among other methods [168]. According to Giroud-Godquin et al. [164], the phase sequence for copper(II) dodecanoate consisted of a single discotic meophase at 380 K ($\Delta H = 17.1$ kcal $mol^{-1)}$ which was stable up to 523 K. However, phase sequences for the longer chain adducts, $n_c = 10$–18, inclusive, showed a single lamellar crystal to columnar mesophase transition in the region of 383–393 K, followed by the isotropic melt at ~473 K, after which the samples decomposed [168]. EXAFS studies revealed that the metal basal plane remained virtually unchanged both in the lamellar crystal and mesomorphic phase [169]. However, variable temperature infrared analyses showed an increase in the stretching frequencies of the CH_2 group as a result of disordering of the alkyl chains in the mesophase [167] in agreement with dilatometric and ^{1}H-NMR studies [170]. These results confirmed that the mesophase was composed of partially disordered alkyl chains anchored to unperturbed metal basal plane aggregates. Interestingly, high temperature X-ray powder diffraction for two

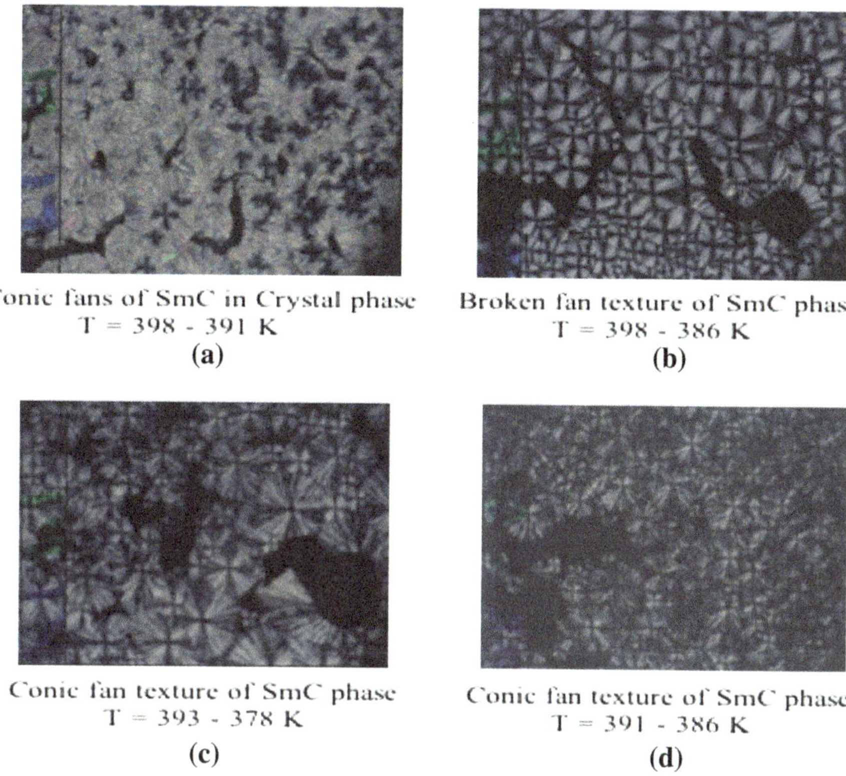

Fig. 12 SmC phase textures for **a** ZnC_{17}, **b** ZnC_{18}, **c** ZnC_{19} and ZnC_{20} [31]

polymorphs of copper(II) decanoate indicated that for these compounds polymorphism does not play a direct role in their liquid crystalline properties [171–173]. For instance, since the *d*-spacing for both samples in the disordered columnar mesophase were ca. 15.6 Å and showed an identical increase of 1.2 %, their liquid crystal phases were concluded to be of the same structure.

Though most investigations on copper(II) carboxylates were focused on straight chain compounds, studies on the copper complex of $(n\text{-}C_9H_{12})_2CHCH_2CO_2H$ revealed the formation of a columnar mesophase at 348 K [28]. Furthermore, copper(II) complexes where the terminal group (methyl or tail group) had been replaced by various groups such as a benzene ring, 1, 2-ditholane or 4-cyclohexyl phenyl moieties have also been investigated [29]. For copper(II) 1, 2-ditholane and 4-cyclohexyl phenyl substituted adducts, only a crystal to melting transition was observed. However, both copper(II) phenyl-propanoate and 13-henicosenoate showed a single mesomorphic transition at 428 and 336 K, respectively. Based on these results it was concluded that the tail groups played an important role in the phase behaviors of these compounds [28].

Though the thermal degradation of copper(II) carboxylates is to this day largely under investigated. Studies by Van My and coworkers [174] showed that the thermal degradation of short chain copper(II) carboxylates produced metallic copper under inert atmosphere whilst the oxide was produced in air [39, 40] along with the parent alkanoic acid and or the α-olefin in some cases [175]. However, no comprehensive mechanism or kinetic models were proposed for their degradation behaviors.

The structures and phase textures for zinc(II) *n*-alkanoates, when heated from the room temperature lamellar solid to the isotropic melt, have been investigated by several authors [30, 31, 58]. For example, recent studies by Ellis and Taylor [30, 31] on a homologous series of anhydrous zinc(II) carboxylates, $[(CH_3(CH_2)_nCO_2)_4Zn_2]$, from the butanoate to eicosanoate, inclusive, revealed a single SmC transition in the phase sequences for the heptadecanoate to the eicosanoate, inclusive. The broken fan textures were observed under plane-polarized light, in the range of ca. 398–378 K (Fig. 12), on cooling from the isotropic melt, in agreement with Roberts and Friberg [32]. For the other homologues only solid–solid transitions were observed, in accord with Konkoly-Thege et al. [176].

These transitions were associated with disordering of the alkyl chains due to methylene group rotation as a result of reduced van der Waals interactions caused by increased temperature in the lattice [177] and not the presence of impurities as previously suggested [178]. Furthermore, increased $\Delta_{Tot}H_m$ and $\Delta_{Tot}S_m$ values with increasing chain length were proposed to be a result of increased van der Waals interactions [32, 100]. Odd–even alternation observed in plots of $\Delta_{Tot}H_m$ and $\Delta_{Tot}S_m$ versus n_c was attributed to different packing arrangements and intermolecular interactions in the lattice [30, 31]. Using XAFS data from Ishioka et al. [95]. Taylor [31] proposed that melting of the compounds proceeded via dissociation equilibrium:

$$Zn_2(O_2CR)_4 \rightleftarrows Zn_2(O_2CR)_3^+ + RCO_2^- \rightleftarrows Zn(O_2CR)_2 + 2RCO_2^- + Zn^{2+}$$

This suggested that the isotropic liquid was a mixture of dissociated dimer and monomers as well as relatively small amounts of carboxylate and zinc ions.

Studies of unsaturated zinc(II) carboxylates such as the *cis* and *trans* isomers of zinc(II) 9-octadecenoate and 9, 12-octadecadenoates via DTA revealed a single transition associated with the fusion but at a lower temperature than for the corresponding saturated adducts. For other *cis–trans* isomers, the *cis* adduct was of lower melting point as a result of more reduced van der Waals interactions in the lattice [20]. These results also showed that the presence of an additional double bond resulted in further lowering of the fusion temperature for these compounds. Recent studies by Nelson et al. [98] on two geometric isomeric unsaturated zinc(II) carboxylates (zinc(II) 9-undecynoate and 10-undecynoate), consisting of a single triple bond in the alkyl chains, showed significantly different phase properties. For instance, though the compounds were of identical chain lengths, a SmC type mesophase was observed of zinc(II) 10-undecynoate at 344 K followed by the fusion at 350 K whilst only the fusion, at 429 K, was observed for the other isomer. Furthermore, ageing (60 days) of these compounds resulted in a 50 % increase in their transition enthalpies. Such observations along with extensive supercooling of the crystallization transition were due to slow molecular equilibration as the hydrocarbon chains rearrange to their original room temperature lamellar order [98]. Unfortunately, reports on the thermal degradation of zinc(II) carboxylates are sparse; however, studies by Dollimore and Tonge [179] revealed that the major products of their decomposition were H_2O, CO, CO_2 and H_2 along with the metal oxide. Other investigations have revealed the formation of a mixture of ketones, CO_2, $ZnCO_3$ and ZnO [180, 181].

The thermal behaviors of mercury(II) carboxylates have been investigated by X-ray diffraction [182, 183], DTA [107, 184], DSC analyses and POM [20]. Though it was initially suggested that even chain mercury(II) carboxylates [20] from the octanoate to octadecanoate, inclusive, showed smectic type mesomorphic transitions in their phase sequences, more recent studies by Ellis [183] revealed only solid–solid pre-melting transitions. Based on these studies it was concluded that the melt consisted of disordered alkyl chains. Unfortunately, though these compounds exhibited step-wise melting, not many investigations have been carried out on them. Hence, much of the theoretical and experimental aspects of their phase properties remain largely uninvestigated to the best of our knowledge. Nonetheless, thermal degradation studies on these compounds revealed the formation of a mixture composed of the carboxylic acid, and the corresponding α-olefin as a result of decarboxylation to release carbon dioxide and the raw metal [185].

Phase sequences for even chain cadmium(II) carboxylates of various chain lengths, investigated by X-ray diffraction [186], dilatometry [188] and DTA [108, 185, 187], showed a mesophase which consisted of infinite cylindrical micelles at ca. 373 K previous to the melt at ca. 503 K. However, these results were questioned by Konkoly-Thege et al. [176] on the basis of DTA studies which indicated that the results of previous investigations were due to impurities in the samples. Nonetheless, it was suggested that cadmium(II) carboxylates, from the dodecanoate to octadecanoate, formed a single mesophase ~10–15 K below the highly viscous melt, observed in the region of 373–393 K, which was composed of infinite cylindrical aggregates similar to those proposed for the mesophase [20]. For the decanoate a single SmA mesophase was reported by Mirnaya and coworkers [188]. Since, calculated total molar enthalpies and entropies of fusion were much less than for the corresponding lead(II) and zinc(II) *n*-alkanoates [20], incomplete alkyl fusion was concluded for the melt. However, increased conformational disorder in the alkyl chains with increasing temperature was indicated by variable temperature infrared measurements, DSC and Raman spectroscopy [176, 178, 179]. Such results confirmed that in the melt the hydrocarbon chains became molten without perturbation of the metal basal plane in accord with suggestions by Luzzati [116]. Interestingly, the hydration numbers of these compounds were found to be an influential factor where their phase behaviors were concerned.

Though no degradation or kinetic models have been proposed for these compounds it was shown by Yakerson [189] that the decomposition of $Cd(CH_3COO)_2{\cdot}2H_2O$ was a multi-step process; that is, both moles of water were lost at ~423 K which was followed immediately by endothermic decomposition at a reasonable rate to yield the metal oxide and carbon dioxide by way of the meta-stable metal carbonate intermediate [190].

Polyvalent carboxylates

Lanthanide carboxylates

The thermochemical properties of various di- and trivalent lanthanide alkanoates [191, 192] have been investigated by DSC, [60, 193] TGA [60], X-ray powder diffraction [194] and POM [60]. The thermal behaviors of cerium(III) carboxylates [60] with $n_c = 8$–12, inclusive, were shown to be markedly different from those with $n_c > 12$; that is, the melting point for the shorter chain adducts were higher than for the long chain homologues. Such results indicated that cerium(III) carboxylates were significantly ionic at short chain lengths. The phase sequence for cerium(III) alkanoates [193] with $n_c \geq 10$, showed a single lamellar crystal to mesophase transition in the vicinity of 351 K followed by the fusion which was quite gradual [60]. However, since super-cooling was observed for all transitions, for these compounds, it was concluded that the pre-melting phases previously classified as liquid crystalline were not well formed; hence, they were classified as disordered smectic mesophases based on POM studies [60]. However, according to Binnemans and coworkers [193] the phase sequence for cerium(III) decanoate was composed of a single SmA mesophase at 379 K followed by the isotropic melt at ca. 411 K. Enthalpy values calculated for the first heating cycle showed a 10–15 % decrease during second heating; hence, incomplete fusion was concluded in support of the pseudo-isotropic melt, indicated by POM studies [60]. Total molar enthalpies ($\Delta_{Tot}H_m$) for conversion from the lamellar crystal phase to the isotropic melt showed a linear relationship with chain which confirming that the main theoretical contributors to this parameter were: [195] $\Delta_{Tot}H_m = \Delta U_{conf} + \Delta U_{vdW} + \Delta U_o + p\Delta V$, where ΔU_{conf} and ΔU_{vdW} were energy variations associated with intra-molecular conformational disorder such as *gauche-trans* isomerism and van der Waals interactions, respectively. Other energy contributions, such as head group electrostatic interactions, were represented by ΔU_o. Since changes in molecular volume were negligible, $p\Delta V \approx 0$. This expression was used by Marques and coworkers [60] to explain the phase behaviors of some even chain Ce(III) carboxylates. Interestingly, though the decomposition of these compounds occurred at relatively low temperatures, in the region of 523–573 K, no comprehensive investigations on the kinetics of this process have been reported.

Partially hydrated praseodymium(III) *n*-alkanoates, $Pr(C_nH_{2n+1}COO)_3]. xH_2O$ ($x = 0.5–1$), showed at least a single SmA-type mesomorphic transition, for all compounds with $n_c = 5–20$, inclusive. These conclusions were based on POM and variable temperature X-ray studies [102]. However, since super-cooling was reported for these phases, such conclusions must be viewed with caution. Furthermore, according to Binnemans and coworkers [102] the melting temperatures for these compounds increase with increasing chain length, contrary to the results of previous studies. The effect of chain length on phase behavior is highlighted by the formation of two mesophases, for all compounds from the pentanoate to the octadecanoate whilst for the longer chain adducts, $n_c = 9–20$, inclusive, only a single liquid crystalline phase was observed in the region of 358–403 K. The low temperature mesophase in the region of 385–363 K, for all compounds, showed no appreciable change in *d*-spacing and was classified as a lamellar mesophase in which molecular rotation was restricted due to the nature of the metal–carboxylate coordination geometry [102]. Such an observation represented a sharp contrast to the SmA phase, where reduction in *d*-spacing was quite marked.

Neodymium(III) *n*-alkanoates form the butyrate to eicosanoate, inclusive, isolated as the hemi- and monohydrates, showed two mesophases for compounds with $n_c < 15$, inclusive. However, for the longer chain adducts only a single endotherm ascribed to the fusion was observed [59, 196]. Similar to Pr(III) [102], the high viscosity low temperature mesophase, in the region of 350–363 K, was labeled as an "M" phase whilst the other pre-melting transition, in the region of 365–374 K, was classified as SmA mesophase followed by the melt in the range of 388–439 K [59]. Since on cooling, the observed transitions were birefringent only under increased pressure, it was concluded that the molecules were aligned homeotropically (with alkyl chains perpendicular to the cover slip). Furthermore, X-ray powder diffraction studies on the mesomorphic transitions for these compounds showed a decrease in the *d*-spacing which was indicative of *gauche–trans* isomerism in the alkyl chains and possibly chain kinking. Based on these results it was concluded that the low temperature solid to mesophase transition was caused by partial melting of the hydrocarbon chains without changes in their tilt angle whilst for the SmA phase they were totally molten. Such descriptions of the phase structures for these compounds were nearly opposite to those for Li(I) [201–204], Ag(I) [70, 78, 79], Pb(II) [20] and Zn(II) [20, 30, 31, 65] among other metal carboxylate systems, adduced from X-ray diffraction, variable temperature ^{1}H and ^{13}C-NMR and variable temperature IR data. For all neodymium(III) carboxylates the melts were due to destruction of the ionic layers and the formation of an ionic salt melt [196]. However, since no electrical conductance data were presented to validate such claims, these results must be treated with caution.

Europium $(C_{n-1}H_{2n-1}CO_2)_3Eu$, $n = 14, 16, 18, 20$, investigated by Li and coworkers [194] showed mesomorphic-like phase behavior in the temperature range of 373–398 K. However, observed super-cooling indicated that these transitions were solid–solid rather than liquid crystalline in agreement with Binnemans and coworkers [59]. Though several transitions were observed for these compounds on heating, the cooling traces showed fewer phases. This indicated that some of the phases observed on heating were metastable or that the mechanism of cooling caused them to coalesce [59, 196]. Furthermore, X-ray data indicated that heating resulted in gradual melting of the alkyl chains to yield states consisting of varying degrees of *gauche* conformers [193]. Therefore, the observed phases were mainly controlled by alkyl chain structure and packing. Unfortunately, despite the importance of these heavy metals, very little investigations have been carried out on the thermal behaviors of their unsaturated and branched chain alkanoate salts. Furthermore, detailed searches of the literature revealed no data for the thermal degradation kinetics and or mechanism for these compounds. Nonetheless, Binnemans and coworkers [196], reported that the decanoates for metals from lanthanum(III) to lutetium(III), inclusive, between one and two pre-melting transitions were observed. DSC thermograms for compounds with Ln = Sm–Lu, inclusive and Y, showed a single peak corresponding to the melt whilst for La^{III}, Ce^{III}, Pr^{III} and Nd^{III} an additional endotherm was observed, ascribed to a SmA mesophase. Super-cooling observed for these phases indicated the absence of mesomorphic transitions in their phase sequence [59]. Interestingly, the melting point decreased from La to Lu indicating that the phase properties of these compounds were dependent on the properties of the metal ion.

Carboxylates with metal–metal bonds

Subsequent to the discovery that copper(II) alkanoates were dimeric, several other di-metal carboxylates were prepared. Of these, the first were rhodium(II) carboxylates, which also had a binuclear structure similar to copper(II), but with a metal–metal double bond [197]. Giroud-Godquin et al. [168] showed that like copper(II) alkanoates, the rhodium(II) complexes formed columnar mesophases on heating from the room temperature solid to the melt. These transitions occurred between 338 and 428 K, corresponding to a transition from the crystalline phase to a mesomorphic state followed by decomposition above 493 K. It was also demonstrated that a reverse transition (mesophase to crystal) occurred at slightly lower temperature which was attributed to slow nucleation and crystal growth kinetics. Not surprisingly, as with many carboxylates, the change in phase on heating was accompanied by disordering of the aliphatic chains in the lamellar at room temperature, as evidenced from PXRD data, however, the columns are maintained by the rigid order of the Rh cores. Evidence from Raman spectroscopy showed that the Rh=Rh bond was sensitive to increased temperature but EXAFS measurements showed that the core structure is maintained in the mesophases. ^{13}C-NMR spectral data suggested that the di-rhodium core underwent a fast axial bond exchange in the mesophase [197].

Di-ruthenium carboxylates ($Ru^{II}Ru^{II}$) containing Ru=Ru double bond as well as the mixed-valent ($Ru^{II}Ru^{III}$) species formed columnar mesophases [197]. It was suggested, based on PXRD and EXAFS data, that the $Ru^{II}Ru^{III}$ complexes had the same molecular organization as the copper alkanoates. Not surprisingly, branching and unsaturation of the carboxylates lead to a decrease in the melting points and the presence of two types of mesophases: hexagonal columnar and lamellar columnar. In the case of the $Ru^{II}Ru^{III}$ carboxylates, magnetic susceptibility measurement as a function of temperature showed a sharp change in magnetic moment at the solid to mesophase transition, suggesting that there is some distortion of the core, similar to the di-rhodium compounds. These $Ru^{II}Ru^{III}$ species which are cations, counterbalanced by X^- anions (X^- includes Cl^- and $RCOO^-$), to give $Ru_2(RCOO)_4X$, carboxylates, are such that the anion had a strong influence on the mesomorphism; those compounds with $n_c = 5$–9, 12, 16 and $X^- = Cl^-$, are not mesomorphic whereas, those with $X^- = RCOO^-$, exhibit a columnar mesophase on heating. It was shown that packing between the chloride and carboxylate ions created void spaces between the dimers and as such, the use of bulky anions or a combination of bulky carboxylates with a small anion, effected a mesophase as they filled the interdimer space [20]. The molecular model proposed for the mesophase is shown in Fig. 13 in the case for the propanoate, $Ru_2(prop)_5$; (a), (b) show the polymeric dimer anion chain, and (c) the columnar hexagonal structure of the mesophase, built up by four polymeric chains.

Extensive studies by Chisholm and coworkers [199] showed that carboxylates containing Mo–Mo and Cr–Cr quadruple bonds also exhibited columnar phases. The molybdenum alkanoates were mesomorphic from $n_c = 4$–10 and the melting points were fairly constant in the region of 425 and 381 K, for the series, except for $n_c = 4$ and 5. The clearing temperatures on the other hand decreased linearly with chain length, with no mesophases occurring above the decanoate. Similar phase behavior was observed for the di-chromium complexes, where they show crystal to liquid crystal transition but decompose at about 473 K. For the tungsten carboxylates, no mesophases were observed. These differences in phase transitions were surprising, especially considering that their structures were similar, even to those of the copper, rhodium and ruthenium di-carboxylates. It was proposed that since the phase behavior of these are almost identical to those of copper(II) alkanoates, similar dimeric interactions, axial metal–oxygen intermolecular interactions were preserved and responsible for the formation and stabilization of these phases. In fact, analysis of structural parameters of the compounds show that metal–oxygen intermolecular bond lengths were in the order of 2.33, 2.65 and 2.67 Å, for Cr, Mo, and W, respectively. Furthermore, it was also proposed using data from PXRD and ^{13}C-NMR that the axial bonds are kinetically labile, being readily broken and

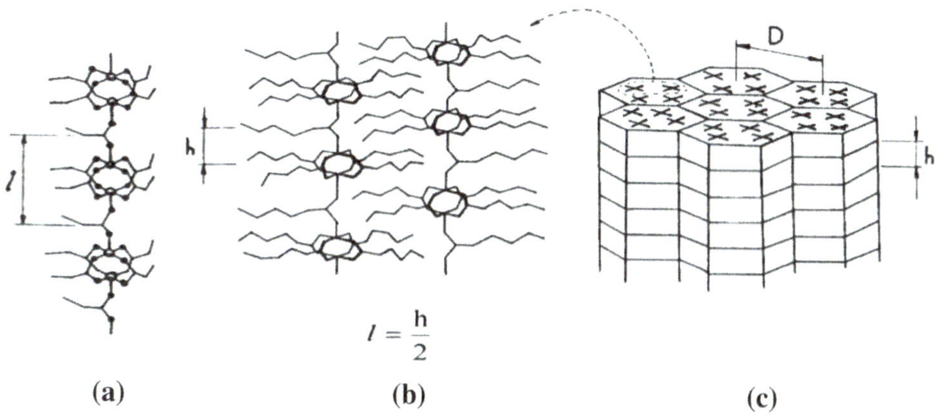

Fig. 13 Schematic representation for the mesophase of mixed-valent di-ruthenium pentacarboxylates [198]

formed, allowing for a dynamic behavior, as depicted through the columnar packing [199].

Other metal carboxylates

Though most investigations have been focused on the saturated carboxylates of Li(I), Na(I), K(I), Tl(I), Ag(I), Pb(II), Zn(II) and Cu(II), other systems have been studied but to a lesser extent. For example, the phase sequences for barium(II) and calcium(II) octadecanoates (stearate) showed more than one second order transition in addition to several first order phase changes [200]. Other alkaline earth metal carboxylates have also been investigated by various methods [201–206] and the results reviewed elsewhere [207]. These investigations revealed several intermediate transitions in their phase sequences, when heated from the room temperature crystal to the isotropic melts, some of which were classified as mesomorphic phases based on X-ray diffraction studies. They were classified as lamellae, discs, rods and ribbons, arranged in 1D, 2D and 3D networks [20]. However, based on X-ray diffraction studies in conjunction with other results, it was suggested that the observed transitions, for alkaline earth carboxylates, did not represent liquid crystal phases in a classical sense [208]. Further, Raman spectral and rheological measurements on calcium stearate as a function of temperature showed marked changes in yield stress, elastic moduli and viscosity, at ca. 398 K, confirming transition to a mesophase [209]. These results also showed that fusion temperatures increase from Mg to Ba, as expected. For aluminium(III) carboxylates, though several studies showed at least two transitions at ca. 373 and 438 K, in most cases the samples were either basic or hydrated [210]. Hence, these results must be treated with caution. Manganese(II) carboxylates, from the octanoate to hexadecanoate, showed a single intermediate phase, characterized as mesomorphic based on electrical conductivity and DTA measurements [103, 184, 211]. This phase was composed of infinite cylindrical micelles similar to Cd(II) carboxylates. However, for Mn(II) octadecanoate, two such transitions were observed. Though several endotherms were reportedly observed in the phase sequences of iron(III) dodecanoate and octadecanoate, in the region of 330–378 K, detailed and timely searches of the literature revealed no proposals for their phase structures [103, 212]. For cobalt(II) stearate, magnetic susceptibility and DTA measurements showed that the transitions observed at 372 and 381 K were associated with a solid–solid followed by a solid to plastic mesophase transition, respectively [17]. However, on cooling, an additional mesomorphic transition was observed in the region of 317–327 K. Such changes were due to perturbation in head group bonding. Phase sequences for rhodium(II) [164, 213, 214] and ruthenium [164, 215] carboxylates were similar to Cu(II) *n*-alkanoates, forming columnar mesophases consisting of infinite cylindrical micelles of binuclear metal carboxylate aggregates.

Binary metal carboxylate systems

Though the majority of investigations on the structures and phase behaviors of metal carboxylates were focused on pure compounds, in recent times, several studies have been carried out by various authors on the phase properties of binary carboxylate systems [20, 37, 38, 48, 55–59]. For example, studies by Spegt and Skoulios [206] showed that at high temperatures, a mixture of calcium and strontium octadecanoates was homogeneously arranged in a hexagonal phase. However, X-ray diffraction studies revealed that this phase was heterogeneous for a mixture of magnesium and calcium octadecanoates, corresponding to the phase behaviors of the individual carboxylates [201, 202]. Similarly, significant studies have been carried out by Hagino [216] on the phase properties of some binary even chain sodium carboxylate system, from the dodecanoate to octadecanoate, inclusive. However, though the phase diagrams for these systems have been presented, the composition and distribution of the components of the mixture were not discussed. Phase diagrams for binary systems of lead(II), zinc(II), cadmium(II), manganese(II) and

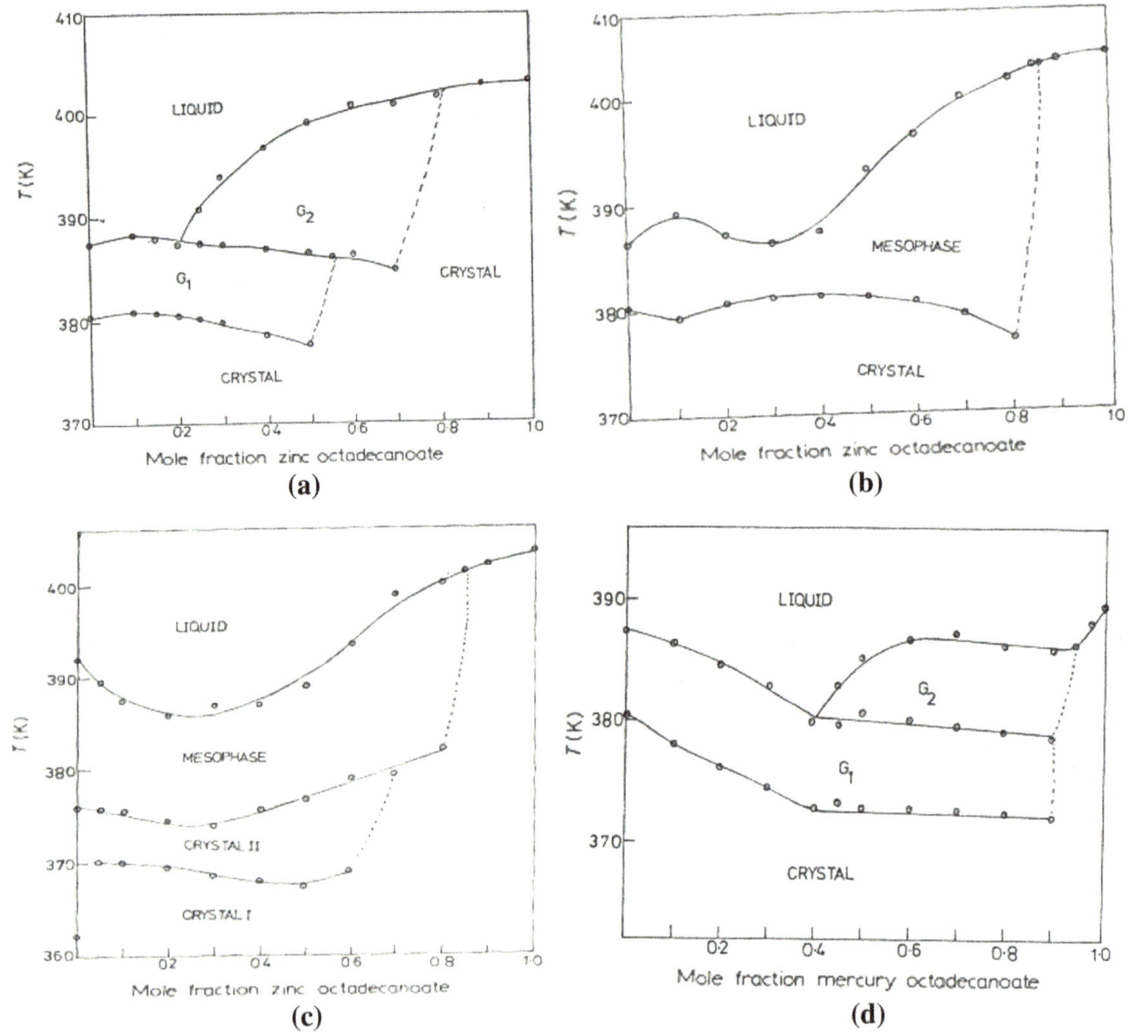

Fig. 14 Phase diagrams for various binary metal carboxylate systems [217]

mercury(II) octadecanoates have been presented by Adeosun and Ellis [217]. These investigations revealed that, though lead(II) octadecanoate showed an SmG phase when heated and the zinc(II) adduct melted directly, the addition of <0.2 mol fraction (x) of zinc octadecanoate to the lead(II) sample did not result in the formation of a new phase (Fig. 14a). However, when $x > 0.2$, a new phase was formed which was structurally different from the G phase and was labeled G_2. However, when the dopant was changed to cadmium octadecanoate a new mesophase formed when $x > 70$ %. No such change was observed when the dopant was manganese octadecanoate [217].

Interestingly, when cadmium(II) octadecanoate was doped with the corresponding zinc(II) carboxylate, no new phase changes were observed up to 0.8 mol fraction of the dopant; however, at higher mole fractions there was a direct transition to the melt (Fig. 14b) [217]. The simple phase sequence of manganese octadecanoate, crystal to mesophase to melt, showed an additional phase on the low temperature side of the DTA curve, upon addition of the corresponding zinc adduct (Fig. 14c), which was resistant to a shear stress. Doping of lead(II) octadecanoate with the corresponding Hg(II) adduct resulted in the formation of a new laminar G phase when $0.4 < x < 0.9$ as shown in Fig. 14d. However, at higher mole fractions, only the fusion was observed, in contrast to an earlier study [210] which showed a single pre-melting phase. Such an observation was due to pre-melting effects which occurred during thorough mixing of the samples. Mercury(II) octadecanoate have been mixed with other octadecanoates of various metals such as Mn(II), Cd(II) and Zn(II). Interestingly, the phase sequences for all systems were different indicating different interactions of the components in the mixture [217]. However, the solid-state structures of these mixtures were not investigated; hence, the exact interactions and packing dynamics of the components in lattice are unknown.

Phase behaviors of some binary lanthanide(III) carboxylate systems, $La_xLn_{1-x}(C_{11}H_{23}COO)_3$, where lanthanum dodecanoate was doped or used as a dopont with other lanthanide (Ln) dodecanoates, such as Nd, Eu, Ho, Tb and Yb, have been investigated by Binnemans and coworkers [218]. These results indicated that for systems of Eu and La, where the mole fraction of La was greater than or equal to 0.04, a SmA mesophase was observed at temperatures that showed no relationship with the percentage of the non-mesomorphic europium and lanthanum dodecanoates. This was due to the stabilizing effects of the larger La ion which also increased the melting and clearing temperatures of the substrate. For the Nd–Eu systems, decreased Nd mesophase stability was observed with increased Eu concentration. This effect was not predicted by the additivity law $T_{cl} = T_1 + x(T_2 - T_1)$ [219] where T_{cl} represents the clearing point of the mixture, T_1 and T_2 were the clearing temperature for the individual components and x was the mole fraction of the dopant. This was due to competition between the metal basal plane and alkyl chain fusion; that is, electrostatic and van der Waals forces. Phase diagrams for $La_xTb_{1-x}(C_{11}Ho_{23}COO)_3$, $La_xHo_{1-x}(C_{11}H_{23}COO)_3$ and $La_xYb_{1-x}(C_{11}H_{23}COO)_3$ systems showed that addition of the mesomorphic La(III) dodecanoate to the non-mesomorphic dodecanoates of Tb(III), Ho(III) and Yb(III) resulted in the formation of a liquid crystalline phase in each case. However, the fraction of La(III) dodecanoate required for mesophase formation was much greater for La–Yb than for the Eu–La system [218]. Based on these results it was concluded that meophase formation in doped lanthanide(III) carboxylate systems was strongly dependent on metal ion radius which provided a stabilizing effect for mesophase.

Binary mixtures of lithium(I) and thallium(I) carboxylates underwent both eutectic and peritectic reactions especially for the Li(I) and Tl(I) propanoates(C_3) where a new compound, $Li_2Tl(C_3)_3$, was formed [220]. This new compound showed a single solid–solid transition at ca. 368 K [$\Delta H = 4.9$ kJ mol^{-1} of $Li_2Tl(C_3)_3$ or 1.6 kJ mol^{-1} of mixture] followed by the melt at 512 K. For the $Li_2Tl(C_5)_3$ system formed from Li(I) and Tl(I) pentanoates (C_5) a single solid–solid phase transition and an incongruent melt, similar to the peritectic fusion of $Li_2Tl(C_3)_3$, was observed. Interestingly, these results also showed that doping of TlC_5 with $x$$LiC_5$, where $x = 0 - 0.56$, resulted in an extension of the mesophase stability from 35 K in the pure TlC_5 to 110 K in the eutectic mixture of LiC_5:TlC_5, 0.19:0.81 ratio. Based on POM studies, this transition was identified as a SmA mesophase (Fig. 15) consisting of the typical focal conic domains of a neat phase [220].

Phase diagrams for binary systems of sodium and barium(II) *n*-butanoates have been presented by Ferloni et al. [221]. These results indicated that barium butanoate exhibited a stepwise melting behavior, transitioning through a single pre-melting phase at ca. 519 K before melting at ca. 589 K. However, addition of sodium(I) butanoate (NaC_4)

Fig. 15 Smectic fan-like texture for the SmA phase of Li2Tl(C5)3 [220]

resulted in significant changes in the phase sequence on heating and cooling. For example, on heating, a eutectic point was observed at ca. 425 K where the mole fraction of NaC_4 was 0.62. Interestingly, the phase sequences for the different mixtures showed a SmA mesophase at $0.58 < x < 1.00$, identified by POM studies. Two metatectic points were observed at 445 and 480 K where $x = 0.47$ and 0.20, respectively. These were associated with interactions of a solid phase, an unspecified mesophase (M) and an isotropic liquid phase [220]. Unfortunately, no structural models were presented to account for the interactions between the components of these mixtures. Nonetheless, studies on various other binary carboxylate systems [222] have been reported as well as metal carboxylate systems doped with other compounds [223–226]. However, no studies on the catalytic properties, degradation kinetics or mechanism have been reported for these systems. This indicates that there is significant room for further investigations in this area and possible extension in ternary and quaternary systems.

Conclusion and outlook

Over the last century, several investigations on the chemistry of metal carboxylates have been carried out by a number of authors. These studies were carried out manly via DSC, TGA, EXAFS, FT-IR, spin decoupled solid-state ^{13}C- and ^{1}H-NMR spectroscopy, dilatometry, POM, etc. Such studies were focused on the structure and thermal behaviors of mono-, di- and poly-valent metal carboxylates of saturated, unsaturated, branched and straight chain carboxylic acids. Interestingly, the majority of results showed that the type of metal carboxylate bonding observed in these compounds were dependent on metal ion properties such a radius, hardness, softness and valence. The structure of the carboxylate ion also played an important role in the symmetry of the metal–ligand coordination sphere. Hence, though sodium and silver were of the same ionic radius, when bound by carboxylate groups of the same structure, their observed bonding types were different. Similarly, the head group symmetry and phase behaviors of zinc(II) 9-undecynoate was different from that of zinc(II) 10-undecynoate. This indicated that metal carboxylate bonding was not controlled only by metal ion radius. The coordination structure for zinc(II), Pb(II) and Cu(II) decanoates were all different despite being composed of metal ions with the same oxidation state. This showed that metal–carboxylate coordination is not directly related to the charge on the metal ion. Odd–even alternation observed for long chain zinc(II), lithium(I) and silver(I) carboxylates were due to differences in molecular packing between, the more asymmetric odd chain adducts and their even chain counterparts. Since this behavior was also observed in the methyl group chemical shift and unusually the carboxyl chemical shift, for silver(I) compounds, differences in electron density distribution was indicated for odd and even and short and long chain compounds. Unfortunately, theories providing insights into the three-dimensional structures of these compounds from PXRD data are sparse. Such theories would be highly useful in understanding the structures of these compounds since they do not readily produce crystals of suitable morphology for single crystal diffraction. Furthermore, since it was established that the phase behaviors of these compounds were dependent on molecular packing which was a function of metal–carboxylate bonding, structural theories, for these compounds, could allow for the designing of new ionic liquids.

Furthermore, despite the numerous pre-melting transitions observed, for several compounds, complete and quantitative theoretical understanding of the structural features responsible for these phases remain underdeveloped for most of them. Such theories would allow for phase manipulation and possibly energy storage in the structures of these compounds. Being able to control their phase behaviors could also allow for their application in dye sensitized solar cells [227, 228] as electrolytes or even as ordered ionic solvent systems for controlling reaction conditions on a molecular scale. Therefore, future work in this field should place significant emphasis on the theory of the phase and structural properties of these compounds.

Conflict of interest The authors declare no conflicts of interest.

References

1. Buono FJ, Feldman ML (1979) Kirk-Othmer, encyclopedia of chemical technology. In: Mark HF, Othmer DF, Overberger CG, Seaborg GT (eds) 3rd edn, vol 8. Wiley, New York, p 34
2. Bossert RG (1950) Ultrasonic study of molecular interactions and compressibility behaviour of strontium soaps in chloroform-propylene glycol mixture. J Chem Educ 27:10
3. Huang TC, Toraya H, Blanton TN, Wu Y (1993) X-ray powder diffraction analysis of silver behenate, a possible low-angle diffraction standard. J Appl Crystallogr 26:180
4. Matthews FW, Warren GG, Michell JH (1950) Derivatives of fatty acids identification by X-ray diffraction powder patterns. Anal Chem 22:514
5. Keiderling U, Gilles R, Wiedenmann A (1999) Application of silver behenate powder for wavelength calibration of a SANS instrument—a comprehensive study of experimental setup variations and data processing techniques. J Appl Crystallogr 32:456
6. Engels T, von Rybinski W (1998) Liquid crystalline surfactant phases in chemical applications. J Mater Chem 8:1313–1320
7. Bonekamp K, Hegemann B, Jonas J (2007) Polarization microscopy of short chain sodium carboxylate mesophases. Mol Cryst Liq Cryst 87:13–28
8. Mehrotra KN, Kachhwaha R, Singh M (1983) Studies on beryllium soaps. Infrared absorption spectra and thermogravimetric analysis. Thermochim Acta 62:179–185
9. Burrows HD (1990) The structure, dynamics and equilibrium properties of colloidal systems. Kluwer Academic Publ., Netherlands, pp 415–426
10. Donnio B, Bruce DW (1999) Liquid crystal II: structure and bonding. Springer, New York, pp 193–247
11. Ohta K, Shibuya T, Ando M (2006) Flying-seed-like liquid crystals. J Mater Chem 16:3635–3639
12. Lee SJ, Han SW, Choi HJ, Kim K (2002) Structure and thermal behavior of a layered silver carboxylate. Phys Chem B 106:2892–2900
13. Klose G, Volke F, Hentschel M, Mops A (1983) Molecular motions and mesophases in anhydrous thallium (I) soaps studied by ^{1}H- and ^{205}TL-NMR. The ^{205}TL chemical shift tensor. Mol Cryst Liq Cryst 90:245–254
14. Wilkes JS (2002) A short history of ionic liquids-from molten salts to neoteric solvents. Green Chem 4(2):73
15. Seddon KR (1998) Molten salt chemistry forum. In: Wendt H (ed) Proceedings of the 5th international conference on molten salt chemistry and technology, vol 53, pp 5–6
16. Welton T (1999) Room-temperature ionic liquids. Solvents for synthesis and catalysis. Chem Rev 99:2071
17. Kambe H, Ozawa T, Onoue M, Igarashi S (1962) Physico-chemical studies on cobalt salts of higher fatty acids. VI. Some observations on thermal transitions of cobalt soaps, by differential thermal analysis, thermogravimetry, and magnetic measurement. Bull Chem Soc Jpn 35:81
18. Akanni MS, Burrows HD, Ellis HA, Asongwed DN, Babalola HB, Ojo PO (1984) Solution behavior of some divalent metal carboxylates in organic solvents. J Chem Technol Biotechnol 34A:127–135
19. Carrell CJ, Carrell HL, Erlebacher J, Glusker JP (1988) Structural aspects of metal ion carboxylate interactions. J Am Chem Soc 110:8651–8656
20. Akanni MS, Okoh EK, Burrows HD, Ellis HA (1992) The thermal behaviour of divalent and higher valent metal soaps: review. Thermochim Acta 208:1–41
21. Adeosun SO, Sime S (1976) Properties of molten carboxylates. Part 4: a quantitative differential thermal analysis study of

melting and mesophase formation in some lead(II) carboxylates. J Thermochim Acta 17:351
22. Adeosun SO, Sime SJ (1978) The properties of molten carboxylates: part 7. Odd–even variations in melting and mesophase formation in the lead(III) carboxylates. Thermochim Acta 27:319
23. Ellis HA (1986) Thermotropic phase transitions in some lead(II) carboxylates. Mol Cryst Liq Cryst 139:281–290
24. Amorim da Costa AM, Burrows HD, Geraldes CFGC, Teixeira-Dias JJC, Bazium CG, Gullion D, Skoulios A, Blackmore E, Tiddy GJT (1986) Mesophase formation in lead(II) decanoate. Liq Cryst 1:215
25. Bazuin CG, Guillon D, Skoulios A, Amorim da Costa AM, Burrows HD, Geraldes CFGC, Teixeira-Dias JJC, Blackmore E, Tiddy GJT (1988) Thermotropic polymorphism in liquid-crystalline lead(II) alkanoates. Liq Cryst 3:1655
26. Ellis HA (1997) An X-ray and microscopic investigation of the thermal phases of lead(II) decanoate. Mol Cryst Liq Cryst 308:111–120
27. Martinez Casado FJ, Ramos Riesco M, Sanchez Arenas A, Garcia Perez MV, Redondo MI, Lopez-Andres S, Garrido L, Cheda JAR (2008) A novel rotator glass in lead(II) pentanoate: calorimetric and spectroscopic study. J Phys Chem B 112:16601–16609
28. Attard GS, Cullum PR (1990) Thermotropic mesomorphism in a branched chain copper(II) carboxylate and its pyrazine complex. Liq Cryst 8:299
29. Akopova OB, Shabyshev LS, Bobrov VI (1995) Synthesis, structure, and discotic mesomorphism of the new series of copper carboxylates. Russ Chem Bull 44(7):1210–1214
30. Taylor RA, Ellis HA (2009) Polymorphism, crystal–crystal and liquid crystalline thermotropic transition behaviour of even chain length zinc(II) *n*-alkanoates. Liq Cryst 36(3):257–268
31. Taylor RA, Ellis HA (2013) Zinc carboxylate liquid crystals: solid state structures and thermotropic phase transition behaviour of a homologous series of zinc(II) n-alkanoates. LAP Lambert Academic Publishing
32. Roberts K, Friberg S (1969) Phase transitions of adsorbed carboxylic acids on zinc oxide and of zinc soaps. Infrared and X-ray diffraction investigations. Kolloid Z Z Polym 230:357
33. Zou CF, Sahyun MRV, Levy B, Serpone N (1996) Mechanisms of latent image formation in photothermographic silver imaging media. J Imaging Sci Technol 40:94
34. Sahyun MRV (1998) Thermally developable photographic materials (TDPM): a review of the state-of-the-art in mechanistic understanding. J Imaging Sci Technol 42:23
35. Geuens I, Vanwelkenhuysen I (1999) Physical characterization of silver behenate as a tool for the development of thermographic and photothermographic materials. J Imaging Sci Technol 43:521
36. Zavlin PM, Batrakov AN, Velinzon PZ, Gaft SI, Kuznetsov LL (1999) Thermally developed photographic materials based on silver organic salts. J Imaging Sci Technol 43:540
37. Ellis HA (1981) Kinetics and reaction mechanism for the thermal decomposition of some even chain lead(II) carboxylates. Thermochim Acta 47:261–270
38. Ellis HA, Okoh EK (1982) Thermal decomposition kinetics of lead(II) decanoate. J Chem Soc Perkin Trans 2:1497–1501
39. Galwey AK, Jamiesson DM, Brown ME (1974) Thermal decomposition of three crystalline modifications of anhydrous copper(II) formate. J Phys Chem 78:2664
40. Reichert C, Fung DKC, Lin DCK, Westmore JB (1968) Thermal decomposition of copper(II) carboxylates: mass spectra of binuclear copper(I) carboxylates. Chem Commun 1094–1095. doi:
41. Piper SH (1929) An X-ray examination of some salts of the fatty acids. J Chem Soc 234–239. doi:
42. Kovalenko VI, Gayfutdinova DN, Furer VL (2002) Ionic layers of potassium and sodium stearates at liquid-crystal transitions according to the data from infra-red spectroscopy. Physicochem Investig 2(6):37
43. Glazer J, McRoberts TS, Schulman JH (1950) The preparation of a stable aluminium dodecanoate (laurate) with no gelling properties in hydrocarbons. J Chem Soc 426:2082–2083
44. Smith GH, Pomeroy HH, McGee CG, Mysels KJ (1053) Preparation of aluminium di-soaps. J Am Chem Soc 1948:70
45. Scott FA, Goldenson J, Wiberley SE, Bauer WH (1954) Spectra of aluminium soaps and soap–hydrocarbon gels. J Phys Chem 58:61–64
46. Mehrotra RC, Pande KC (1956) Studies in aluminium soaps—I aluminium tri-soaps. J Inorg Nucl Chem 2:60–65
47. Mehrotra RC, Rai AK (1962) Studies in heavy metal soaps. Synthesis of chloride soaps of aluminium. J Indian Chem Soc 39:1–4
48. Wood JA, Seddon AB (1981) Identification of the chromium salt of stearic acid. Thermochim Acta 45:365–368
49. Eshel M, Bino A, Felner I, Johnston DC, Luban M, Miller LL (2000) Polynuclear chromium(III) carboxylates. 1. Synthesis, structure, and magnetic properties of an octanuclear complex with a ring structure. Inorg Chem 39:1376–1380
50. Eshel M, Bino A (2002) Polynuclear chromium(III) carboxylates. 3. Cyclic and cubane type hexachromium acetates. Inorg Chim Acta 329:45–50
51. Atkinson IM, Benelli C, Murrie M, Parsons S, Winpenny REP (1999) Turning up the heat: synthesis of octanuclear chromium(III) carboxylates. Chem Commun 285–286. doi:
52. Rai AK, Parashar GK (1979) Thermogravimetric analysis of some higher carboxylate derivatives of chromium(III). Thermochim Acta 29:175–179
53. Misra SN, Misra TN, Mehrotra RC (1963) Organic salts of lanthanide elements—I. Preparation of tri-acylates of lanthanum and cerium(III) from aqueous solutions. J Inorg Nucl Chem 25:195–199
54. Misra SN, Misra TN, Mehrotra RC (1963) Organic salts of lanthanide elements—II: preparation of lanthanum and cerium(III) acylates from a non-aqueous medium. J Inorg Nucl Chem 25:201–203
55. Demus D, Richter L (1978) Textures of liquid crystals. Verlag Chemie, New York
56. De Vries A, Ekachai A, Spielberg N (1979) Why the molecules are tilted in all smectic a phases, and how the layer thickness can be used to measure orientational disorder. Mol Cryst Liq Cryst 49:143–452
57. Binnemans K, Görller-Walrand C (2002) Lanthanide-containing liquid crystals and surfactants. Chem Rev 102:2303–2345
58. Binnemans K (2005) Ionic liquid crystals. Chem Rev 105:4148–4204
59. Binnemans K, Jongen L, Bromant C, Hinz D, Meyer G (2000) Structure and mesomorphism of neodymium(III) alkanoates. Inorg Chem 39:5938–5945
60. Marques EF, Burrows HD, da Graca Miguel M (1998) The structure and thermal behaviour of some long chain cerium(III) carboxylates. J Chem Soc Faraday Trans 94(12):1729–1736
61. Taylor RA, Ellis HA, Maragh PT, White NAS (2006) The room temperature structures of anhydrous zinc(II) hexanoate and pentadecanoate. J Mol Struct 787:113–120
62. Taylor RA, Ellis HA (2007) Room temperature molecular and lattice structures of a homologous series of anhydrous zinc(II) *n*-alkanoate. Spectrochim Acta 68:99–107

63. Ellis HA, White NA, Hassan I, Ahmad R (2002) A room temperature structure for anhydrous lead(II) decanoate. J Mol Struct 642:71–76
64. Ellis HA, White NAS, Taylor RA, Maragh PT (2005) Infrared, X-ray and microscopic studies on the room temperature structure of anhydrous lead(II) *n*-alkanoates. J Mol Struct 738:205–210
65. Corkery RW (2004) A variation on Luzzati's soap phases. Room temperature thermotropic liquid crystals. Phys Chem Chem Phys 6:1534–1546
66. Nakamoto K, Fujita J, Tanaka S, Kobayashi M (1957) Infrared spectra of metallic complexes. IV. Comparison of the infrared spectra of unidentate and bidentate metallic complexes. J Am Chem Soc 79:4904–4908
67. Ouchi A, Suzuki Y, Ohki Y, Koizumi Y (1988) Structure of rare earth carboxylates in dimeric and polymeric forms. Coord Chem Rev 92:29–43
68. Edwards DA, Hayward RN (1968) Transition metal acetates. Can J Chem 46:3443–3446
69. Cotton FA, Wilkinson G (1980) Advanced inorganic chemistry: a comprehensive text, 4th edn, vol 170. Wiley-Interscience, New York, pp 58–602
70. Nelson PN, Ellis HA (2012) Structural, odd–even chain alternation and thermal investigations of a homologous series of anhydrous silver(I) *n*-alkanoates. Dalton Trans 41:2632
71. Shoeb ZE, Hammad SM, Yousef AA (1999) Oleochemicals I: studies on the preparation and the structure of lithium soaps. Grasas Aceites 50:426–434
72. Ferloni P, Westrum EF Jr (1992) Thermodynamic properties of lithium *n*-alkanoates. Pure Appl Chem 64(1):73–78
73. Bui LH (2013) Alkali metal C1–C12 *n*-alkanoates. Masters of Science in Engineering Thesis, University of Alberta
74. Lewis ELV, Lomer TR (1969) The refinement of the crystal structure of potassium caprate (form A). Acta Cryst B25:702
75. Lomer TR (1952) The unit-cell dimensions of potassium soaps. Acta Cryst 5:11
76. White NAS, Ellis HA, Nelson PN, Maragh PT (2011) Thermal and odd–even behaviour in a homologous series of lithium *n*-alkanoates. J Chem Thermodyn 43:584–590
77. Casado FJM, Riesco MR, Perez MVG, Redondo MI, Lopez-Andres S, Cheda JAR (2009) Structural and thermodynamic study on short metal alkanoates: lithium propanoate and pentanoate. J Phys Chem B 113:2896–12902
78. Bremmer L, Burgard MD, Rodger CS, Swearingen JK, Rath NP (2001) Silver(I) carboxylates: versatile inorganic analogs of carboxylic acids for supramolecular network formation. Chem Commun 2468–2469. doi:
79. Binnemans K, Deun RV, Thijs B, Vanwelkenhuysen I, Geuens I (2004) Structure and mesomorphism of silver alkanoates. Chem Mater 16(10):2021–2027
80. Tolochko BP, Chernov SV, Nikitenko SG, Whitcomb DR (1998) Nucl Instrum Methods Phys Res 405:428
81. Olson L, Whitcomb DR, Rajeswaran M, Blanton TN, Stwertka BJ (2006) The simple yet elusive crystal structure of silver acetate and the role of the Ag–Ag bond in the formation of silver nanoparticles during the thermally induced reduction of silver carboxylates. Chem Mater 18:1667–1674
82. Vold MJ, Funakoshi H, Vold RD (1976) The polymorphism of lithium palmitate. J Phys Chem 80:1753–1761
83. Lomer TR, Perera K (1974) Anhydrous copper(II) octanoate. Acta Cryst B30:2913–2915
84. Bird MJ, Lomer TR (1972) The crystal and molecular structure of anhydrous copper butyrate. Acta Cryst B28:242–246
85. Burrows H, Ellis HA (1982) The thermal behaviour and spectral properties of some long chain copper(II) carboxylates. Thermochim Acta 52:121–129
86. Berkesi O, Katona T, Dreveni I, Andor JA, Mink J (1995) Temperature-dependent Fourier transform infrared and differential scanning calorimetry studies of zinc carboxylates. Vib Spectrosc 8:167–174
87. Dreveni I, Berkesi O, Andor JA, Mink J (1996) Influence of the spatial structure of the alkyl chain on the composition of the product of the direct neutralization reaction between aliphatic carboxylic acids and zinc hydroxide. Inorg Chim Acta 249:17–23
88. Šegedin P, Lah N, Žefran M, Leban I, Golič L (1999) Synthesis and characterization of bis(carboxylato)zinc(II) (C_6–C_{18})—crystal structure of bis(haxanoato)zinc(II) $Zn(O_2CC_5H_{11})_2$—form A. Acta Chim Slov 46:173–184
89. Berkesi O, Dreveni I, Andor JA, Mink J (1992) FT-IR studies on the formation of tetrazinc long straight-chain (even-numbered C_6–C_{18}) μ4-oxo-hexa-μ-carboxylates from the corresponding bis(carboxylato)zinc compounds. Inorg Chim Acta 195:169–173
90. Peultier J, François M, Steinmetz J (1999) Anhydrous polymeric zinc(II) heptanoate. Acta Cryst C55:2064–2065
91. Lacouture F, Peultier J, François M, Steinmetz J (2000) Anhydrous polymeric zinc(II) octanoate. Acta Cryst C56:556–557
92. Clegg W, Little IR, Straughan BP (1987) Orthorhombic anhydrous zinc(II) propionate. Acta Cryst C43:456–457
93. Blair J, Howie RA, Wardell JL (1993) Structure of monoclinic zinc(II) *n*-butanoate. Acta Cryst C49:219–221
94. Clegg W, Little IR, Straughan BP (1986) Monoclinic anhydrous zinc(II) acetate. Acta Cryst C42:1701–1703
95. Ishioka T, Shibata Y, Takahashi M, Kanesaka I, Kitagawa Y, Nakamura KT (1998) Vibrational spectra and structures of zinc carboxylates I. Zinc acetate dehydrate. Spectrochim Acta Part A 54:1827–1836
96. Ishioka T, Maeda K, Watanabe I, Kawauchi S, Harada M (2000) Infrared and XAFS study on structure and transition behavior of zinc stearate. Spectrochim Acta Part A 56:1731–1737
97. Mesbah A, Juers C, Francois M, Ricca E, Steinmetz J (2007) Structures of magnesium and zinc long aliphatic chains carboxylates. Z Kristallogr Suppl 26:593–598
98. Nelson PN, Taylor RA, Ellis HA (2013) The effects of molecular and lattice structures on thermotropic phase behaviour of zinc(II) undecanoate and isomeric zinc(II) undecynoates. J Mol Struct 1034:75–83
99. Harrison PG, Steel AT (1982) Lead(II) carboxylate structures. J Organomet Chem 239:105
100. Mesubi MA (1961) J Mol Struct 11:61
101. Baxter DV, Cayton RH, Chisholm MH, Huffman JC, Putilina EF, Tagg SL, Wesemann JL, Zwanziger JW, Darrington FD (1994) Multiple bonds between metal atoms in ordered assemblies. 2. Quadrupole bonds in the mesomorphic state. J Am Chem Soc 116:4551–4566
102. Jongen L, Binnemans K, Hinz D, Meyer G (2001) Mesomorphic behaviour of praseodymium(III) alkanoates. Liq Cryst 28(6): 819–825
103. Thalladi VR, Boese R (2000) Why is the melting point of propane the lowest among *n*-alkanes. New J Chem 24:579–581
104. Boese R, Weiss H-C, Bläser D (1999) The melting point alternation in the short-chain *n*-alkanes: single-crystal X-ray analyses of propane at 30 K and of *n*-butane to *n*-nonane at 90 K. Angew Chem Int Ed 38:988–992
105. Thalladi VR, Boese R, Weiss H-C (2000) The melting point alternation in α, ω-alkanedithiols. J Am Chem Soc 122: 1186–1190
106. Polishchuk AP, Timofeeva TV (1993) Metal-containing liquid-crystal phases. Russ Chem Rev 62:291–321
107. Lawrence ASC (1938) The metal soaps and the gelation of their paraffin solutions. Trans Faraday Soc 34:660
108. Hattiangdi GS, Vold MJ, Vold RD (1949) Differential thermal analysis of metal soaps. Ind Eng Chem 41:2320

109. Koga Y, Matsuura R (1961) Mem Fac Sci Kyushu Univ Ser C 4:1
110. Koga Y, Matsuura R (1965) Chem Abstr 62:13391c
111. Skoda W (1969) Dilatometric investigation of phase transitions and structures of anhydrous soaps. Kolloid-Z Polym 234:1128
112. Gallot B, Skoulios A (1966) Structure des savons de rubidium. Mol Cryst 1:263
113. White NAS, Ellis HA (2009) Thermal behaviour of even chain length lithium n-alkanoates. Mol Cryst Liq Cryst 501:28–42
114. Gallot B, Skoulios A (1966) Structure des savons alcalins. Kolloid-Z Polym 209:164
115. Seurin P, Guillon D (1981) Skoulios, Smectogènes Dissymétriques III. Synthèse et Propriétés Mésomorphes des p, *n*-Alkoxybenzydèneanilines Parasubstituées, A. Mol Cryst Liq Cryst 65:85
116. Luzzati V (1968) Biological membranes. In: Chapman D (ed) Academic Press, New York, p 71
117. Skoulios A, Luzzati V (1961) La structure des colloïdes d'association. III. Description des phases mésomorphes des savons de sodium purs, rencontrées au-dessus de 100 °C. Acta Cryst 14:278
118. Benton DP, Howe PG, Farnard R, Puddington IE (1955) The mesomorphic behaviour of anhydrous soaps: part II. Densities of alkali metal stearates. Can J Chem 33:1798
119. Vold RD (1941) Anhydrous sodium soaps. Heats of transition and classification of the phases. J Am Chem Soc 63:2915
120. Vold RD, Vold MJ (1939) Successive phases in the transformation of anhydrous sodium palmitate from crystal to liquid. J Am Chem Soc 61:808
121. Sokolov NM (1954) Zh Obshch Khim 24:1150
122. Baum E, Demus D, Sackmann H (1970) Wiss Z Univ Halle 19:37
123. Ubbelohde AR, Michels HJ, Duruz JJ (1970) Liquid crystals in molten salt systems. Nature 228:50
124. Ferlani P, Sanesi M, Franzosini PZ (1975) Naturforsch Teil A 30:1447
125. Storonkin AV, Vasil'kova IV, Vestn SS (1974) Leningr Univ Fiz Khim 10:84
126. Cingolani A, Spinolo G, Sanesi MZ (1979) Naturforsch Teil A 34:575
127. Roth J, Meisel T, Seybold K, Halmos Z (1976) Investigation of the thermal behaviour of fatty acid sodium salts. J Therm Anal 10:223
128. Busico V, Ferraro A, Vacatello M (1985) Thirmotropic smectic liquid crystals of ionic amphiphilic compounds: a general discussion. Mol Cryst Liq Cryst 128:243–261
129. Duruz JJ, Michels HJ, Ubbelohde AR (1971) Molten fatty acid salts as model ionic liquids. I. Thermodynamic and transport parameters of some organic sodium salts. Proc R Soc Lond A 322:1550
130. Franzosini P, Westrum EF Jr (1984) Thermophysics of metal alkanoates IV. Heat capacities and thermodynamic properties of potassium 2-methylpropanoate. J Chem Thermodyn 16:127–130
131. Tse WS, Chiang PY, Lin SJ (1986) A Raman spectral study of the phase transitions in crystalline potassium acetate. Chin J Phys 24(1):63–68
132. Bolduan OEA, McBain JW, Ross S (1943) Diffraction of X-rays by sodium laurate and sodium palmitate at higher temperatures. J Phys Chem 47:528
133. Aleixo AI, Oliveira PH, Diogo HP, da Piedade MEM (2005) Enthalpies of formation and lattice enthalpies of alkaline metal acetates. Thermochim Acta 428:131–136
134. Johnson DA (1982) Some thermodynamic aspects of inorganic chemistry, 2nd edn. Cambridge University Press, Cambridge, pp 29–81
135. Dasent WE (1982) Inorganic energetics, 2nd edn. Cambridge University Press, Cambridge (chapter 3)
136. Jenkins HDB, Roobottom HK (2002). In: Lide DR (ed) CRC handbook of chemistry and physics, 83rd edn. CRC Press, Boca Raton, pp 22–36 (section 12)
137. Boerio-Goates J, de Lopez La Fuente FL, Cheda JAR, Westrum EF Jr (1985) Thermodynamics of thallium alkanoates I. Heat capacity and thermodynamic functions of thallium(I) *n*-hexanoate. J Chem Thermodyn 17:401–408
138. Fernandez-Martin F, Lopez de La Fuente FL, Cheda JAR (1984) DSC thermophysics on some lower thallium(I) *N*-alkanoates. Thermochim Acta 73:109
139. Ngeyi SP, de Lopez La Fuente FL, Cheda JAR, Fernandez-Martin F, Westrum EF Jr (1985) Thermodynamics of thallium alkanoates II. Heat capacity and thermodynamic functions of thallium(I) *n*-heptanoate. J Chem Thermodyn 17:409
140. Ngeyi SP, de Lopez La Fuente FL, Cheda JAR, Fernandez-Martin F, Westrum EF Jr (1987) Thermodynamics of thallium alkanoates III. Heat capacity and thermodynamic functions of thallium(I) *n*-tetradecanoate from 7 to 450 K. J Chem Thermodyn 19:327
141. de Lopez La Fuente FL, Westrum EF Jr, Cheda JAR, Fernandez-Martin F (1987) Thermophysics of thallium alkanoates IV. Heat capacity and thermodynamic functions of thallium(I) *n*-dodecanoate from 7 to 470 K. J Chem Thermodyn 19:1261
142. Meisel T, Seybold K, Halmos Z, Roth J, Melykuti C (1976) Thermal behaviour or thallium(I) fatty acid salts, I. J Therm Anal 10:419
143. Lindau J, Diele S, Kruger H, Dorfler HD (1981) Phys Chem 262:775
144. de Lopez La Fuente FL, Cheda JAR, Fernandez-Martin F, Westrum EF (1988) Thermodynamics of thallium alkanoates V. Heat capacity and thermodynamic functions of thallium(I) *n*-decanoate from 6 to 480 K. J Chem Thermodyn 20:1137–1148
145. Peltzl G, Sackmann H (1971) Birefringene of smectic modifications of the homologous thallium soaps. Mol Liq Liq Cryst 15:75
146. Fernández-García M, García MV, Redondo MI, Cheda JAR, Fernandez-García MA, Westrum EF Jr, Fernandez-Martin F (1997) Molecular association of normal alkanoic acids with their thallium(I) salts: a new homologous series of fatty acid metal soaps. J Lipid Res 38:361
147. Fernández-García M, Cheda JAR, Westrum EF Jr, Fernández-Martin F (1997) Interactions in the thallium(I) heptanoate and heptanoic acid system: association, aggregation, and phase behavior. J Colloid Interface Sci 185:371
148. Cheda JAR, Fernandez-García M, Ugarelli P, Ferloni P, Fernández-Martin F (2000) Phase behavior of the *n*-decanoic acid + thallium(I) *n*-decanoate system. Langmuir 16:5825
149. Andreev VM, Burleva LP, Boldyrev VV (1984) Izv Sib Otd Akad Nauk SSSR Ser Khim Nauk 5:3
150. Andreev VM, Burleva LP, Boldyrev VV, Mikhailov YuI (1983) Izv Sib Otd Akad Nauk SSSR Ser Khim Nauk 4:58
151. Uvarov NF, Burleva NF, Mizen MB, Whitcomb DR, Zou C (1998) Solid State Ion 107:31
152. Bokhonov BB, Burleva LP, Whitcomb DR, Usanov YE (2001) Formation of nano-sized silver particles during thermal and photochemical decomposition of silver carboxylates. J Imaging Sci Technol 45:259
153. Iwig F, Hecht O (1886) Ber 19:238
154. Judd MD, Plunkett BA, Pope MI (1974) The thermal decomposition of calcium, sodium, silver and copper(II) acetates. J Therm Anal 6:555
155. Burrows HD, Geraldes CFGC, Pinheiro TJT, Harris RK, Sebald A (1988) 207Pb NMR of lead(II) soaps in solid, liquid-crystalline and liquid phases. Liq Cryst 3:853
156. Adeosun SO, Kehinde AO, Odesola GA (1979) Electrical conductance, density, viscosity and thermal behaviour of lead(II)

and zinc(II) salts of some unsaturated fatty acids. Thermochim Acta 28:133
157. Attard GS, West YD (1990) Lead(II) alkadiynoates. A novel class of liquid-crystalline precursors to poly(diacetylenes). Liq Cryst 7:487
158. Baraldi P (1981) Thermal behavior of metal carboxylates—II. Lead formate. Spectrochim Acta 37A:99
159. Freeman ES, Caroll B (1958) The application of thermoanalytical techniques to reaction kinetics: the thermogravimetric evaluation of the kinetics of the decomposition of calcium oxalate monohydrate. J Phys Chem 62:394
160. Rasheed A, Bhobe RA (1976) J Indian Chem Soc LIII:442
161. Anderson DA, Freeman ES (1961) The kinetics of the thermal degradation of polystyrene and polyethylene. J Polym Sci 54:253
162. Abied H, Guillon D, Skoulios A, Weber P, Giroud-Godquinn AM, Marchon JC (1987) Investigation of the structures of the crystalline and columnar phases of linear chain copper(II) alkanoates. Liq Cryst 2:269
163. Marchon JC, Maldivi P, Giroud-Godquin AM, Guillon D, Skoulios A, Strommen DP (1990) Columnar liquid crystals derived from long-chain carboxylates of transition metal ions [and discussion]. Philos Trans R Soc Lond Ser A 330:109
164. Giroud-Godquin AM, Marchon JC, Guillon A, Skoulios A (1984) Discotic mesophases of dirhodium tetracarboxylates. J Phys Lett 45:L681
165. Takekoshi M, Watanabe N, Tamamaushi B (1975) Copper stearate as a thermotropic liquid crystal. Colloid Polym Sci 256:588
166. Abied H, Guillon D, Skoulios A, Giroud-Godquin AM, Maldivi P, Marchon JC (1988) Dilatometric investigation of thermotropic linear chain copper(II) alkanoates. Colloid Polym Sci 266:579
167. Strommen DP, Giroud-Godquin AM, Maldivi P, Marchon JC, Marchon B (1987) Vibrational studies of some dicopper tetracarboxylates which exhibit a thermotropic columnar mesophase. Liq Cryst 2:689
168. Giroud-Godquin AM, Latour JM, Marchon JC (1985) Magnetic susceptibility as a probe of the solid-discotic phase transition in binuclear copper(II) *n*-alkanoates. Inorg Chem 24:4452
169. Abied H, Guillon D, Skoulios A, Dexpert H, Giroud-Godquin AM, Marchon JC (1988) J Phys Fr 49:345
170. Grant RF (1964) Proton magnetic resonance absorption in copper(II) stearate. Can J Chem 42:951
171. Ramos Riesco M, Martinez Casado FJ, Lopez-Andres S, Garcia Perez MV, Redondo Yelamos MI, Torres MR, Garrido L, Cheda JAR (2008) Monotropic polymorphism in copper(II) decanoate. Cryst Growth Des 8(7):2547–2554
172. Busico V, Cernichiaro P, Corradini P, Valcatello M (1983) Polymorphism in anhydrous amphiphilic systems. Long-chain primary *n*-alkylammonium chlorides. J Phys Chem 87:1631–1635
173. Busico V, Ferraro A, Valcatello M (1984) J Phys Chem 88:4055–4058
174. Van My L, Perinet G (1965) Bull Soc Chim Fr 1379
175. Akanni MS, Ajayi OB, Lambi JN (1986) Pyrolytic decomposition of some even chain length copper (II) carboxylates. J Therm Anal 31:131
176. Konkoly-Thege I, Ruff I, Adeosun SO, Sime SJ (1978) Properties of molten carboxylates part 6. A quantitative differential thermal analysis study of phase transitions in some zinc and cadmium carboxylates. Thermochim Acta 24:89
177. Mesubi MA (1982) An infrared study of zinc, cadmium, and lead salts of some fatty acids. J Mol Struct 81:61
178. Wood JA, Seddon AB (1985) The solubility of zinc dodecanoate in xylene and tetrachloromethane. Colloid Polym Sci 263:600
179. Dollimore D, Tonge KH (1967) The thermal decomposition of zinc and manganous formats. J Inorg Nucl Chem 29:621
180. Rasheed A, Bhobe RA (1976) J Indian Chem Soc 53:442
181. Odilora CA (1989) On the thermal degradation of poly(vinyl chloride). Synergistic effect of mixtures of lead and zinc carboxylates. Acta Polym 40:541
182. Vold RD, Hattiangdi GS (1949) Characterization of Heavy Metal Soaps by X-Ray Diffraction. Ind Eng Chem 41:2311
183. Ellis HA (1986) Thermal Behaviour of Mercury(II) Carboxylates. Mol Cryst Liq Cryst 138:321
184. Adeosun SO (1978) A differential thermal analysis study of phase transitions in some mercury(II) carboxylates. J Therm Anal 14:235
185. Akanni MS, Borrows HD, Begun PB (1984) Product analysis, reaction mechanism and kinetics of the thermal decomposition of some even chain-length mercury(II) carboxylates. Thermochim Acta 81:45
186. Spegt PA, Skoulios AE (1963) La structure des colloïdes d'association. VIII. Description de la structure des savons de cadmium à température élevée. Acta Crystallogr 16:301
187. Prasad S, Ranganathchar T, Krishna V (1985) J Indian Chem Soc 35:267
188. Mirnaya TA, Sudovtsova LS, Yaremchuk GG, Tolochko AS, Lisetskii LN (2004) Russ J Inorg Chem 49:1440
189. Yakerson VI (1967) Izv Akad Nauk SSSR Otd Khim Nauk 1003
190. Wilcox DE, Bromley LA (1963) Computer estimation of heat and free energy of formation for simple inorganic compounds. Ind Eng Chem 55:32
191. Methrotra KN, Sharma N (1996) Studies on molar volume and rheology of terbium soaps in a benzene-DMF mixture. Monatsh Chem 127:257
192. Upadhyaya SK, Sharma PSJ (1993) Indian Chem Soc 70:735
193. Binnemans K, Jongen L, Görller-Walrand C, D'Olieslanger W, Hinz D, Meyer G (2000) Lanthanide(III) dodecanoates: structure, thermal behaviour, and ion-size effects on the mesomorphism. Eur J Inorg Chem 2000(7):1429–1436
194. Li H, Bu W, Qi W, Wu L (2005) Self-assembled multibilayers of europium alkanoates: structure, photophysics, and mesomorphic behavior. J Phys Chem B 109:21669–21676
195. Nagle JF (1980) Theory of the main lipid bilayer phase transition. Annu Rev Phys Chem 2:157
196. Binnemans K, Heinrich B, Guillon D, Bruce DW (1999) On the mesomorphism of lanthanum(III) alkanoates. Liq Cryst 26(11):1717–1721
197. Giroud-Godquin AM (1998) My 20 years of research in the chemistry of metal containing liquid crystals. Coord Chem Rev 178–180:1485–1499
198. Cukiernik FD, Ibn-Elhaj M, Chaia ZD, Marchon J-D, Giroud-Godquin A-M, Guillon D, Skoulios A, Maldavi P (1998) Mixed-valent diruthenium (II, III) long-chain carboxylates. 1. Molecular design of columnar liquid-crystalline order. Chem Mater 10:83–91
199. Chisholm MH, Wilson PJ, Woodward PM (2002) Intermolecular recognition and crystal packing in molybdenum and tungsten coordination polymers as deduced from powder X-ray diffraction data. Chem Commun 566–567. doi:
200. Montmitonnet P, Monasse B, Haudin JM, Delamore F (1985) Calorimetric and microscopic study of barium stearate. Mater Lett 3:98
201. Spegt P, Skoulios A (1970) Acad C R Sci Paris 251:2199
202. Spegt P, Skoulios A (1962) Acad C R Sci Paris 254:4316
203. Spegt PA, Skoulios AE (1964) La structure des colloïdes d'association. X. Description de la structure des savons de calcium à température ordinaire et à température élevée. Acta Crystallogr 17:198

204. Spegt P, Skoulios A (1965) J Chim Phys 62:377
205. Spegt P, Skoulios A (1965) J Chim Phys 62:418
206. Spegt PA, Skoulios AE (1966) Structure des savons de strontium en fonction de la température. Acta Crystallogr 21:892
207. Winsor PA (1975) Liquid crystals and plastic crystals, vol 1, chapter 5. Ellis Harwood, Chichester
208. Reichmuth G, Dubler E (1985) Synthetic and biogenic calcium fatty acid salts: stoichiometry and phase transition behavior. Thermochim Acta 89:485
209. Montmitonnet P, Delamore F (1982) Determination of the plastic and viscoelastic properties of a metallic soap by a hot hardness measurement. J Mater Sci 17:121
210. Shiba S (1961) Differential thermal analysis of aluminium soaps. Bull Chem Soc Jpn 34:804
211. Adeosun SO (1979) Electrical conductance and thermal behaviour of some manganese(II) carboxylates. Can J Chem 57:151
212. Ghosh SK, Pathak GK, Chaudhuri AK (1978) Indian J Chem 16A:670
213. Giroud-Godquin AM, Marchon JC, Guillon D, Skoulios A (1986) Discotic mesophases of dirhodium tetracarboxylates. J Phys Chem 90:5502
214. Poizat O, Strommen DP, Maldivi P, Giroud-Godquin AM, Marchon JC (1990) Raman spectroscopy: a convenient probe of solid-liquid crystal-phase transitions in binuclear rhodium(II) *n*-alkanoates. Inorg Chem 29:4851
215. Maldivi P, Giroud-Godquin AM, Marchon JC, Guillon D, Skoulios A (1989) Diruthenium(II, II) tetra-μ-alkylcarboxylates: magnetic susceptibility studies of their electronic configuration and thermotropic liquid crystalline mesophase. Chem Phys Lett 157:552
216. Hagino K (1964) Kogyo Kagaku Zasshi 67:597
217. Adeosun SO, Ellis HA (1979) Binary phase diagrams of some bivalent metal carboxylate systems. Thermochim Acta 28:313–321
218. Jongen L, Hinz D, Meyer G, Binnemans K (2001) Induced mesophases in binary mixtures of lanthanide(III) dodecanoates. Chem Mater 13:2243–2246
219. Gray GW (1962) Molecular structure and the properties of liquid crystals. Academic Press, New York, p 314
220. Martınez Casado FJ, Ramos Riesco M, da Silva I, Labrador A, Redondo MI, Garcıa Pérez MV, López-Andrés S, Rodrıguez Cheda JA (2010) Thermal and structural study of the crystal phases and mesophases in the lithium and thallium(I) propanoates and pentanoates binary systems: formation of mixed salts and stabilization of the ionic liquid crystal phase. J Phys Chem B 114:10075–10085
221. Mirnaya TA, Polishchuk AP, Bereznitski YV, Ferloni P (1996) Phase diagram of the binary system of barium and sodium *n*-butanoates. J Chem Eng Data 41(6):1337
222. Ellis HA (1988) Thermal behaviour of the system: lead(II) decanoate/lead(II) octadecanoate. Thermochim Acta 130:281
223. Burrows HD, Geraldes CFGC, Miguel MGM, Pinheiro TJT, Cruz Pinto JJC (1992) The phase behaviour of some mixtures of lead(II) carboxylates with phosphatidylcholines. Thermochim Acta 206:203
224. Adeosun SO, Sime WJ, Sime SJ (1977) Properties of molten carboxylates part 5. A quantitative differential thermal analysis study of mesophase formation in the systems lead(II)dodecanoate/lead(II) oxide and lead(II)dodecanoate/hendecane. Thermochim Acta 19:275
225. Adeosun SO (1978) Thermal behaviour of the system lead(II) dodecanoate/lead acetate. Thermochim Acta 25:333
226. Adeosun SO, Akanni MS (1978) Differential thermal analysis study of the system lead(II) dodecanoate/dodecanoic acid. Thermochim Acta 27:133
227. O'Regan B, Grfitzeli M (1991) A low-cost, high-efficiency solar cell based on dye-sensitized colloidal TiO_2 films. Nature 353:737
228. Wang P, Zakeeruddin SM, Moser J-E, Grfitzeli M (2003) A new ionic liquid electrolyte enhances the conversion efficiency of dye-sensitized solar cells. J Phys Chem B 107:13280

3

Acyclic diene metathesis polymerization and precision polymers

Nicolas F. Sauty · Lucas Caire da Silva ·
Michael D. Schulz · Chip S. Few · Kenneth B. Wagener

Abstract The history of and major advances in the acyclic diene metathesis (ADMET) reaction are described. Because precise branch identity and frequency can be achieved by ADMET polymerizations of symmetrical α,ω-dienes, polyethylenes with precisely spaced alkyl branches of specified length have been prepared. Investigations of their morphologies and thermal properties have provided valuable insight into the behavior of polyethylene. ADMET preparation of ethylene copolymers and telechelic oligomers, as well as the properties of these materials, is also discussed.

Keywords ADMET polymerization · Olefin metathesis · Precision polymers · Polyethylene

Introduction

The development of olefin metathesis has led to remarkable achievements in organic and polymer chemistry. From allowing facile synthesis of complex natural products to producing intricate polymer architectures, olefin metathesis has had an undeniable impact on organic and polymer synthesis. There are two distinct modes of metathesis polymerization: ring-opening metathesis polymerization (ROMP) and acyclic diene metathesis (ADMET) polymerization [1, 2] (Fig. 1). The development of the ADMET reaction is the subject of this review.

N. F. Sauty · L. C. da Silva · M. D. Schulz ·
C. S. Few · K. B. Wagener
The George and Josephine Butler Polymer Research Laboratory, Department of Chemistry and Center for Macromolecular Science and Engineering, University of Florida, Gainesville, FL 32611-7200, USA

Historical background

The development of well-defined, efficient olefin metathesis catalysts (Fig. 2) was paramount to the development of the ADMET reaction. Although classical olefin metathesis catalysts were effective in some cases, the exact nature of the catalytically active species was not known. Early work by Grubbs [3–7] and Schrock [8–12] produced some of the first spectroscopically identifiable olefin metathesis catalysts.

The introduction of Schrock's catalysts (e.g., **[Mo]1**) enabled significant progress in olefin metathesis and its use in organic and polymer synthesis. However, these catalysts are sensitive to air, water, and some polar functional groups, and this sensitivity has limited their application in some cases. To address this issue, late transition metal catalysts were developed both by Schrock [13] and Grubbs [14]. Of these, the Grubbs ruthenium-based catalysts have been particularly successful. The first generation of these ruthenium complexes was reported in 1992 and 1993 [15–18]. Although first-generation ruthenium catalysts such as **[Ru]1** have less activity than the Mo-based Schrock catalysts, the Grubbs' complexes are more tolerant to moisture, oxygen, and polar functional groups. A second-generation of ruthenium catalysts, principally **[Ru]2**, resulted from the replacement of a phosphine with an *N*-heterocyclic carbene (NHC) ligand. These catalysts are very reactive in olefin metathesis reactions and possess greater functional group tolerance than the first-generation ruthenium catalysts [19–21]. The activity of some second-generation ruthenium catalysts is comparable to the activity of **[Mo]1** [22]. In 1999, a new class of olefin metathesis catalysts with an isopropoxy group on the benzylidene moiety, **[Ru]3**, was reported by Hoveyda and coworkers [23]. Soon thereafter, the second-generation Hoveyda–Grubbs catalysts were

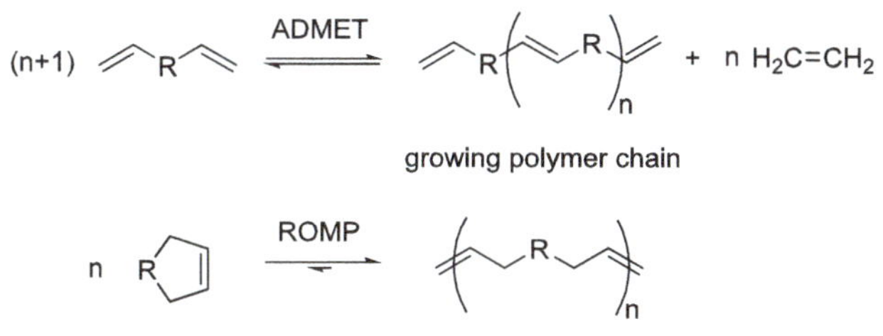

Fig. 1 The ADMET reaction

reported, all incorporating an NHC ligand (**[Ru]4** is one example). These catalysts are very effective for many different olefin metathesis reactions [22, 24].

These and other catalyst developments [25] allowed ADMET to be a useful and feasible reaction. The Wagener group began investigating unconjugated α,ω-diene polycondensation, later known as ADMET polymerization, using the $WCl_6/EtAlCl_2$ catalyst system in the late 1980s [26]. These reactions resulted in a combination of soluble oils and intractable solids. It was determined that the oil was a low molecular weight polymer—the result of olefin metathesis. It was believed that the insoluble product resulted from a vinyl addition reaction, which, in competition with metathesis, would result in a crosslinked polymer, but the solid product was not rigorously characterized. In order to test this hypothesis, Wagener and coworkers reacted $WCl_6/EtAlCl_2$ with styrene. The predicted product of olefin metathesis for this reaction would be stilbene; however, the reaction also yielded polystyrene. This result demonstrated that vinyl addition and olefin metathesis are competing reactions when these catalyst systems are used. To achieve high polymer, the competing vinyl addition reaction must be eliminated. Serendipitously, this work coincided with the report of catalyst **[Mo]1**. The concurrence of these advances enabled ADMET to be applied as a means of synthesizing high molecule weight polymer.

The ADMET reaction

Acyclic diene metathesis polymerization is an equilibrium step-growth condensation of α,ω-dienes promoted by a suitable metathesis catalyst. The reaction produces a growing unsaturated polymer chain that retains the symmetry and functionality of the monomer.

Over the years, monomer design and the development of suitable catalysts have opened the way to a variety of well-defined polymer structures which are not readily achievable by other common techniques. The ADMET polymerization proceeds via a metal-carbene mechanism, in which the metal methylidene is the true catalyst formed during the polymerization cycle (Fig. 3). This is the same metal-carbene mechanism operating in cross-metathesis [27]. Coordination of the metallic center of the catalyst with a terminal olefin from either the monomer or polymer is followed by a reversible 2 + 2 cycloaddition to produce a metallocyclobutane intermediate. This undergoes further reversion leading to either a productive or a non-productive pathway. Chain growth is promoted by the productive metathesis pathway resulting from cycloreversion, which will not restore the coordinating olefin and free catalyst. This pathway is characterized by production of ethylene, which should be removed from the reaction mixture in order to drive the equilibrium towards chain growth. ADMET polymerization affords PDIs around 2, which are typical for step-growth polymerization. Also, molecular weights up to 50,000 g/mol can be achieved, depending primarily on catalyst life time, monomer purity, and reaction conditions. It is fair to say that catalyst compatibility and monomer design/purification are the main challenges to be overcome in order to achieve high polymer with well-defined primary structure.

ADMET polymerization usually requires relatively long reaction times in order to guarantee high conversion and consequently high molecular weight polymers. Because some reactions require days to be completed, catalyst lifetime can be a limiting parameter. Catalyst lifetime depends primarily on the rate of deactivation by catalyst poisoning and can be a major problem if a compatible metathesis catalyst is not available. Grubbs' ruthenium-based and Schrock's molybdenum-based catalysts [28] are among the most used catalysts for ADMET. Grubbs' first-generation catalyst is very common due to its high tolerance towards moisture, oxygen and coordinating functional groups, even though its reactivity is not as high as Schrock's catalyst [29]. The highly reactive Schrock's catalyst

Fig. 2 Five examples of well-defined olefin metathesis catalysts

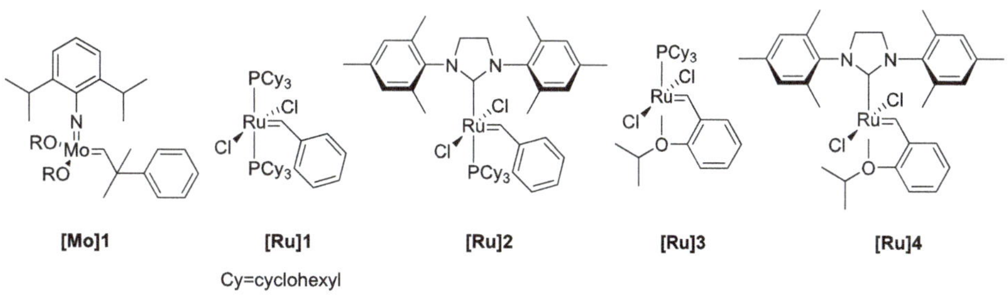

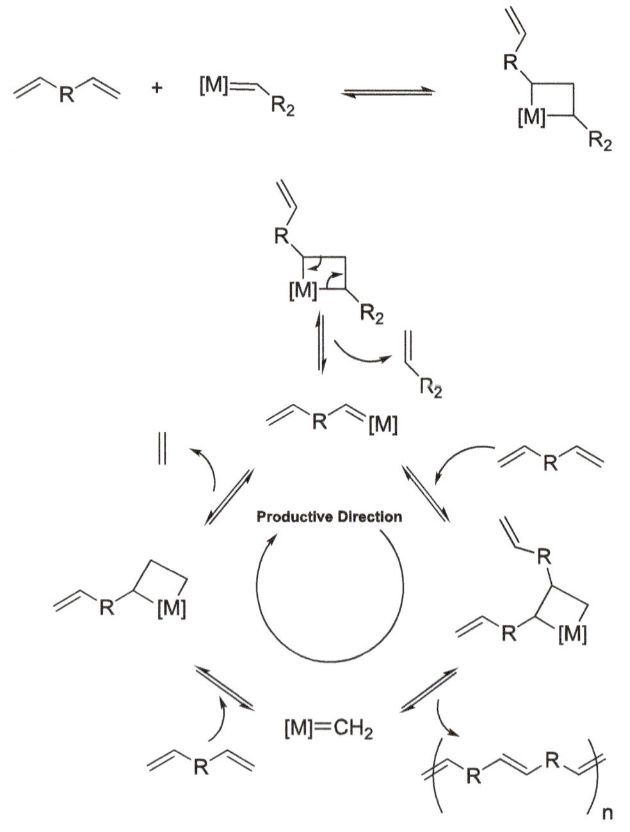

Fig. 3 The ADMET mechanism

is much more sensitive and requires careful handling and extremely dry and pure monomers. The high oxidation state of the molybdenum metal center makes the catalyst highly vulnerable to polar functional groups, water and oxygen. Traces of impurities can deactivate the catalyst rendering very poor conversion/molecular weights.

Even though there are a large number of metathesis catalysts available, many are known to promote isomerization, which ultimately results in loss of control over the polymer primary structure. Grubbs' first generation and Schrock's catalysts are known to promote minimal isomerization under regular ADMET reaction conditions, while studies on Grubbs' second generation, as well as other ruthenium-based metathesis catalysts, shows significant isomerization under similar conditions [30, 31]. Isomerization is a drawback, especially when the choice of a catalyst is limited by monomer functionality. Fortunately, several methods are available to minimize isomerization, usually by use of special additives such as benzoquinones [32, 33] or boron-containing Lewis acids [34].

Virtually any α,ω-diene can be polymerized via ADMET if a suitable catalyst is available that tolerates the desired monomer functionality. The mechanism governing metathesis ensures control over primary structure based on retention of monomer symmetry/asymmetry in the final polymer. Saturation of the internal olefins of the unsaturated ADMET polymer eliminates the *cis/trans* distribution, allowing for polymers with well-defined primary structures. Tacticity in ADMET polymers is not yet fully controlled and is being currently addressed by the Wagener group.

ADMET polymerization is ideally carried out in the bulk with monomer-to-catalyst ratios typically in the range of 100–500:1 under high vacuum. The increased viscosity during polymerization is overcome by performing the reaction at temperatures ranging from 40 to 75 °C under constant stirring. Where solubility is not an issue, solvents can be used to maintain low viscosity levels, as long as the catalyst is not deactivated in solution. Plenio and coworkers showed that high-boiling solvents such as 1,2-dichlorobenzene can be used in ADMET polymerization when a slight vacuum is desirable [35]. More recently, Simocko et al. [36] demonstrated that ionic liquids are excellent media to perform ADMET polymerizations. Grubbs' catalyst is usually employed when solvents need to be used, given its higher tolerance towards contaminants and polar functional groups. During the polymerization, removal of ethylene condensate can be achieved by applying high vacuum, or by purging the reaction vessel with an inert gas, in order to drive the reaction to completion. This can be achieved by applying high vacuum or by purging the reaction vessel with an inert gas in order to push the reaction to completion. Reaction progress can be easily followed by comparing the relative ^{1}H NMR signal intensities of the terminal olefin on the monomer and internal olefin of the resulting unsaturated polymer.

Modeling polyethylene

Polyethylene (**PE**) is the most produced polymer and still dominates the polymer market because of its low cost, versatility, and ease of manufacture [37]. Conventional industrial methods inherently afford polyethylenes with random branches of various lengths along the polymer chain, and this branching greatly affects **PE** properties [38]. Over the years, considerable effort has been devoted to controlling branching and ideally to producing linear polyethylene. In addition, copolymerization of ethylene with polar vinyl monomers often leads to a higher branching content in conjunction with the low incorporation of the polar entity in the polymer backbone. This eventually produces a gradient polymer rather than a statistical polymer.

ADMET polymerization of symmetrical diene monomers is a remarkable tool to afford flawless and impeccably regular polyolefins. Subsequent hydrogenation produces polyethylene copolymers possessing a functionality (either within the polymer backbone or side branching) at

Fig. 4 Monomer synthesis

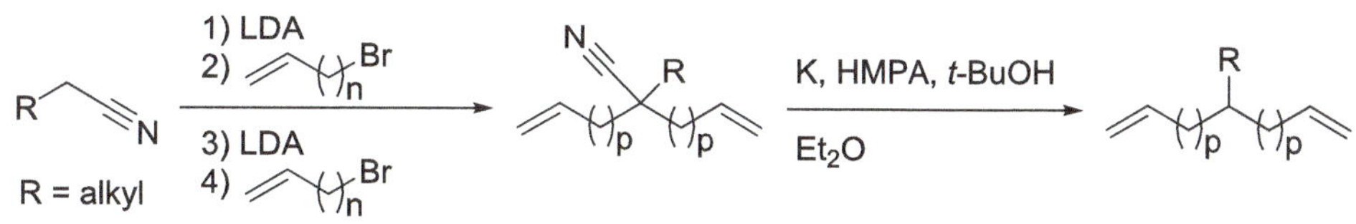

Fig. 5 Synthesis of precision alkyl-branched polyethylene

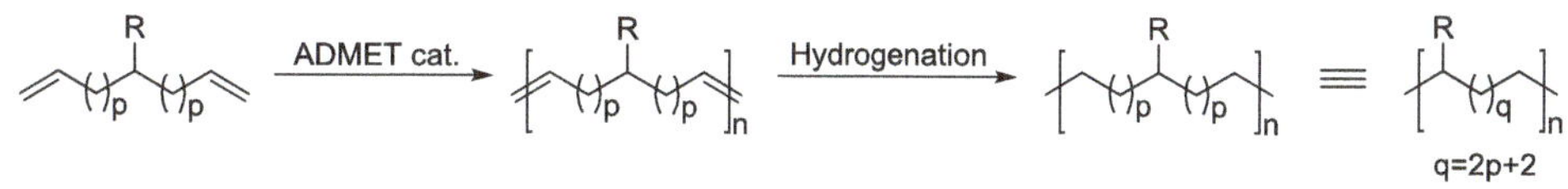

perfectly unvarying intervals (i.e., separated by a constant number of methylene spacers). Depending on the catalyst, molecular weights (M_n) of ADMET polyethylene copolymers usually range from 5,000 to 50,000 g/mol. These M_ns are sufficient to be accepted as polyethylene models, given that the entanglement molecular weight of linear polyethylene is about 1,000 g/mol [39].

ADMET polymerization offers a unique way to model branched polyethylene, as well as any copolymers of ethylene and α-alkenes (C_nH_{2n}). A two-step route involving a dialkylation of a nitrile intermediate followed a reductive decyanation [40] has been used to prepare a collection of symmetrical alkyl-branched α,ω-dienes (Fig. 4).

Subjecting these monomers to ADMET polymerization and subsequent exhaustive hydrogenation afforded polyethylenes with precise placement (every *n*th carbon) of alkyl branches on their backbones (Fig. 5).

Conversely, statistical placement of alkyl branches can be obtained by copolymerization of the aforementioned symmetrical monomer with a linear α,ω-diene (typically 1,9-decadiene) [41]. Precision alkyl-branched polyethylenes were found to possess enhanced thermal properties due to higher crystallinity, leading to sharper melting transitions at lower temperatures and greater heats of fusion. In precision polymers, two factors determine the thermal properties and morphologic parameters of these materials: the branch identity and the branch frequency. The crystalline structures have been studied using wide and small angle X-ray scattering (WAXS and SAXS), transmission electron microscopy (TEM) and infrared (IR) spectroscopy. Morphology is greatly influenced by the alkyl branch size and spacing.

Regularity in primary structure predictably translates to uniformity in the form of narrow lamellar thickness distributions in polyethylene containing precisely spaced alkyl branches [42]. The lamellar thickness corresponds to the distance between branch sites in these materials, which suggests that the chains are folding at the branch sites, with the branches residing at the interface of the lamellae [43]. As a consequence of the narrow lamellar thickness distribution, ADMET polymers have a narrower melt temperature range compared to other polyethylenes.

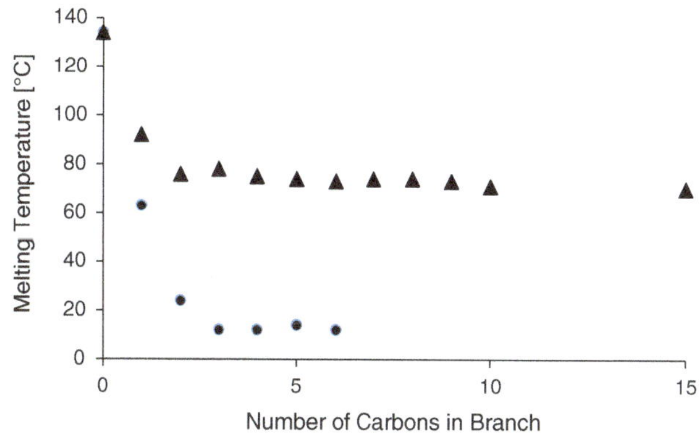

Fig. 6 The effect of alkyl branch size on T_m can be seen for ADMET polyethylenes with branches on every 21st (*filled circle*) and 39th (*filled triangle*) carbon along the polymer backbone

Multifarious branch sizes, ranging from methyl to pentadecyl, have been incorporated in a precise fashion on the polyethylene backbone. It was discovered that the resulting fusion temperatures were highly dependent on the nature of the branching (Fig. 6). As a general rule, the melting temperature dwindles with longer alkyl branches. The effect of alkyl branch size on the melting temperature can be seen for ADMET polyethylenes with branches on every 21st [44] and 39th [45] carbon. Polymers with branches excluded from the lamellae exhibit similar fusion temperatures. This expulsion of alkyl chains from the lamellae is affected by defect size, as ADMET polymers bearing smaller alkyl groups (e.g., methyl) crystallize with a greater degree of branch inclusion. This is true of precision polymers of varying branch spacing, and the maximum included branch size decreases with increasing branch interval. Indeed, when the branch is placed on every 21st carbon, methyl and ethyl branches are included in the unit cell and afford polymers displaying melting transitions at 63 and 24 °C, respectively. Conversely, polymers with branches larger than the maximum included branch size exhibit similar melting temperatures. In other words, the transition becomes independent of the branch identity with T_ms of approximately 12 °C. Similarly, when an alkyl defect is placed every 39th carbon, only a methyl branch can be included in the unit cell, and as a consequence,

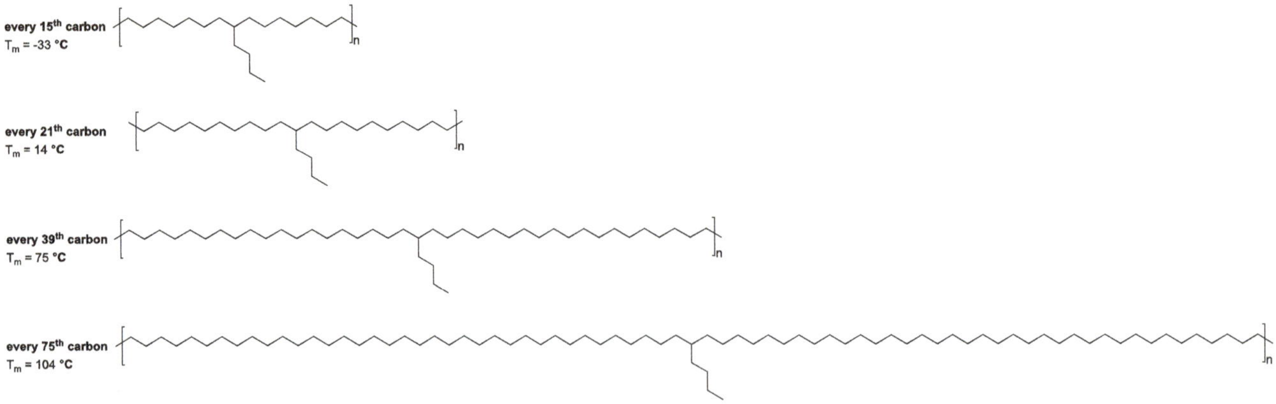

Fig. 7 Branch frequency: alkyl branch every *n*th carbon

similar melting temperatures are found for the polymers bearing longer branches.

Various branch frequencies have been investigated with the goal of diminishing the frequency (increasing the spacing) as much as possible. The general observation is that for a given branch identity, thermal transitions occur at higher temperatures with diminished frequencies. The Wagener group has synthesized butyl-branched polyethylenes in which the branch appears every 9th, 21st, 39th, and 75th [46] carbon (Fig. 7).

This has led to the discovery of a linear relationship between the branch frequency and fusion temperature of the aforementioned polymers. It also suggested a melting temperature of 136 °C for an infinitely low frequency which is quite close to that of linear ADMET polyethylene (134 °C).

The distance between alkyl branches also has an effect on the crystalline lattice, as less frequent branching leads to more ordered structure. The extreme case, with no branching, leads to orthorhombic structure. The introduction of branches leads to less-ordered crystalline structures, of which there are several examples in the literature [47, 48]. When precisely spaced alkyl groups are placed every 7th carbon or closer, the resulting polymers are amorphous.

The morphological effects described in the preceding paragraphs can be further understood by considering the crystallization behavior of representative polymers. As is the case in polyethylenes synthesized by other means, ADMET polyethylenes crystallize through the formation of a hexagonal mesophase, which expands as the chains slide past one another in the all *trans* conformation until the equilibrium lamellar thickness is reached [49]. While crystallization of longer methylene sequences occurs before shorter sequences in metallocene PE, forcing shorter methylene sequences into the amorphous phase, all methylene sequences are of the same length in ADMET polyethylene, allowing crystallites of equilibrium lamellar thickness to form from the transient hexagonal mesophase [50].

By varying the size and interval of alkyl branches on polyethylene, the resulting morphological effects are delineated. Uniform ethylene sequences allow a metastable mesophase to form, at which point the chains can reposition themselves into an energetically stable morphology, resulting in a material with narrow lamellar thickness distribution. Depending on defect size, the branches may be contained within the lamellar stem in the case of smaller branches, or excluded to the surface of the crystallite in the case of longer branches. The lamellar thickness and crystalline lattice type depend on the distance between defects, with longer intervals leading to more ordered polymorphisms.

Beyond the thermal and morphologic elucidation of precision alkyl-branched polyethylene, molecular dynamics were probed utilizing deuterium quadrupolar echo NMR in the solid state. The ^{2}H-NMR spectra of precision CD_3-branched polyethylene were acquired for a wide range of temperatures [51, 52]. A T_1-fitting methodology was applied to deconvolute the spectral contributions of the amorphous and crystalline phases, allowing isolation of two distinctive dynamic behaviors. This is the consequence of a bi-exponential behavior of the spin–lattice relaxation in the CD_3 branches. The amorphous phase exhibited a similar behavior to that of linear polyethylene [53]. Conversely, the crystalline domains displayed a harmonic librational axial rotation reorientation model with a mean jump angle up to 40° when approaching the melting transition. Various other deuterated precision polymers are being analyzed by deuterium quadrupolar echo methods with the intention of elucidating the dynamics of precision polymers.

Ethylene copolymers

The advent of ADMET polymerization not only promoted modeling of polyethylene, but also stimulated the understanding of numerous classes of polymers. Various

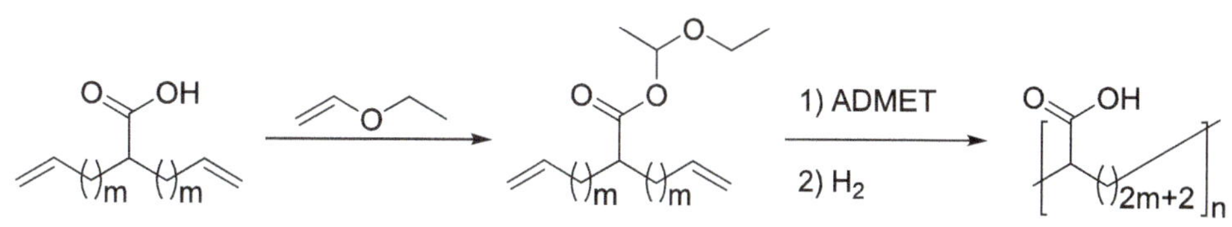

Fig. 8 Synthesis of precision poly(ethylene-*co*-acrylic acid)

copolymers of ethylene and polar vinyl intermediates were scrutinized and compared to commercial samples. All syntheses relied on the ADMET polymerization of symmetrical α,ω-diene monomers followed by exhaustive hydrogenation to afford the desired polyethylene precision copolymers. Relevant examples are briefly described in this section.

Ethylene–vinyl acetate copolymers (EVA) are widespread materials, with hot-melt adhesives and plastic wraps being common applications [54]. Although free-radical copolymerizations usually produce gradient or block polymers, perfectly random copolymers are obtained from the polymerization of ethylene and vinyl acetate due to similar reactivities [55]. However, undesired and random branching is inherently promoted via a back-biting process, significantly altering the characteristics and performance of the copolymers. However, perfectly linear ethylene–vinyl acetate copolymers can be prepared by means of ADMET polymerization [56]. A number of monomers with various methylene run lengths ranging from 8 to 12 were subjected to polymerization and subsequent hydrogenation, furnishing copolymers with different vinyl acetate contents (acetate branches separated by 18–26 carbons). Commercial samples with similar vinyl acetate ratios were compared with the ADMET copolymers. Similar to the commercial copolymers, the precision polymers displayed a linear relationship between the melting temperature and vinyl acetate content, but with dissimilar slopes. A larger slope, underlining the precision factor, evidenced the greater order in precision polymers. With a similar approach, copolymers of ethylene and vinyl alcohol were made without difficulty.

While metathesis catalysts such as **[Ru]1** can tolerate a variety of functional groups, functionalities such as primary or secondary amines, sulfides, or carboxylic acids deactivate the catalyst, thereby precluding preparation of high molecular weight materials (oligomers to about 3,000 g/mol were produced). In the case of carboxylic acids, this problem can be circumvented by protecting the acid functionality. Various agents can be employed, but ethyl vinyl ether was found to be extremely convenient [57]. The resulting 1-ethoxyethyl ester, inert to ruthenium catalysts, is quantitatively cleaved during the hydrogenation of the resulting polyolefin (Fig. 8).

The ethylene–acrylic acid copolymers were found to be amorphous when only 8 or 14 methylene units separated two consecutive branching sites. Conversely, increasing the spacing to 20 methylenes resulted in a semicrystalline material displaying dimerized acid moieties in an orthorhombic lattice. Simple treatment of the copolymer with zinc reagents, produced a zinc-neutralized material containing 22 % free acid and 66 % zinc dicarboxylates. The zinc dicarboxylate components aggregated into a self-assembled structures adopting a cubic crystalline lattice, an unprecedented observation for ionomers having a purely hydrocarbon backbone [58].

Precision ethylene–vinyl halide copolymers have also been examined [59]. Contrary to many other systems, it was discovered that regardless of the halide identity (except iodide) and appearance frequency on the polymer backbone, the halogen atoms were included in the unit cell. This yielded an orthorhombic system in the case of fluorine, and triclinic systems in the case of the larger chlorine and bromine substituents.

Motivated by the recent advances in fuel cell membranes, precision polymers incorporating sulfonic [60] and phosphonic [61, 62] acids have been investigated for potential uses as proton carriers. As with carboxylic acids, the aforementioned functionalities must be protected prior to polymerization to prevent poisoning of the metathesis catalyst. While ethyl phosphonates could straightforwardly be deprotected post-hydrogenation, the sulfonate analogues have not led to comparable results. The significant insolubility of the poly(ethylene-*co*-vinylsulfonic acid)s promotes premature precipitation of the material and prevents complete deprotection. This problem is currently being addressed by the Wagener group. With sufficient spacing (20 methylene spacers or greater), the ethylene–vinylphosphonic acid copolymers show a semicrystalline behavior. Amorphous materials are promoted by shorter methylene segments.

Finally, among others, ADMET polymerization has been utilized to produce precision materials containing ether [63], silicon [64], or amine [65] moieties.

Telechelic oligomers via ADMET

Many telechelic oligomers (oligomers with functionalized end groups) have been synthesized by ADMET. These materials are often macromonomers used to synthesize larger polymers, and they have many other uses as well. Telechelic oligomers of polybutadiene (Fig. 9) have been prepared via the reaction of 1,5-hexadiene and a mono- or

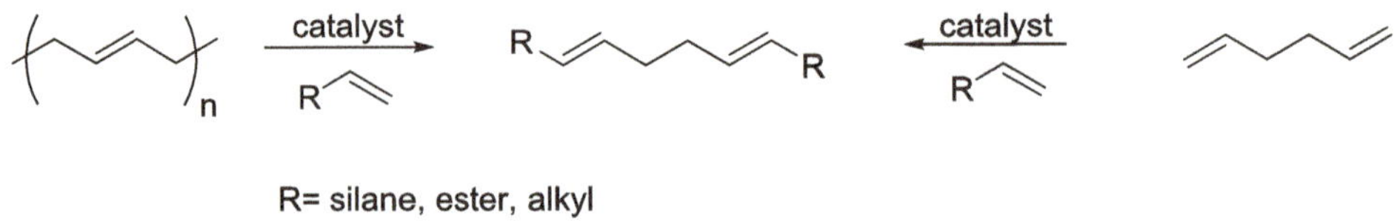

Fig. 9 Synthesis of telechelics via ADMET

di-functional olefin [66, 67]. ADMET polybutadiene may alternatively be achieved by the polymerization of 1,5-hexadiene. This polymer can then be used to prepare telechelic oligomers by reacting with a suitable olefin. Sometimes the reaction is incomplete, producing telechelic oligomers. Additionally, polybutadiene produced by cationic polymerization can undergo ADMET depolymerization with acrylates to yield ester-terminated telechelics [68]. This approach was also used to produce a polyamide by depolymerizing polybutadiene with acryloyl chloride and subsequently reacting with 1,6-diaminohexane. Ester-end-capped polyethylene oligomers have been made by ADMET by reacting 1,9-decadiene and 9-decenyl acetate and subsequently hydrogenating the unsaturated units. Telechelic polyacetylenes have been synthesized by ADMET polymerization of 2,4-hexadiene with 1,2-disubstituted olefins [69]. Polyacetylene oligomers terminated with a variety of end groups, including silanes, alkyl groups, and esters were prepared with this method.

Hydrophobic, amorphous telechelic diols have also been synthesized by ADMET [70]. These materials could potentially be used in UV- and hydrolysis-resistant polyurethanes. The hydrocarbon backbone was produced by ADMET polymerization followed by hydrogenation of a gem-dimethyl substituted α,ω-diene, and this material was then endcapped with alcohol-containing monoenes with varied numbers of methylenes between the alcohol group and olefin. Some molecular weight control was achieved in the resulting telechelics, which had 2.0 functionality.

In another report, 1,9-decadiene was polymerized using **[Ru]1** with epoxy-containing monoenes [71].The resulting telechelic polymer was then reacted with toluene diisocyanate resulting in a copolymer containing oxazolidone linkages. ADMET has been used to synthesize ester-terminated telechelics with a similar approach [72].

Conclusion

Over the last two decades, ADMET polymerization has been extensively studied as a unique and valuable means to access and synthesize precision polymers. These materials have provided better understanding of structure–property relationships in various systems, as they serve as models for their ill-defined commercial counterparts. While significant advances have resulted from precision polymers, many more systems have yet to be explored [73].

References

1. Wagener KB, Boncella JM, Nel JG (1991) Acyclic diene metathesis (ADMET) polymerization. Macromolecules 24:2649–2657
2. Mutlu H, de Espinosa LM, Meier MAR (2011) Acyclic diene metathesis: a versatile tool for the construction of defined polymer architectures. Chem Soc Rev 40:1404–1445
3. Howard TR, Lee JB, Grubbs RH (1980) Titanium metallacarbene-metallacyclobutane reactions: stepwise metathesis. J Am Chem Soc 102:6876–6878
4. Lee JB, Gajda GJ, Schaefer WP, Howard TR, Ikariya T, Straus DA, Grubbs RH (1981) Structures of titanacyclobutanes. J Am Chem Soc 103:7358–7361
5. Lee JB, Ott KC, Grubbs RH (1982) Kinetics and stereochemistry of the titanacyclobutane–titanamethylene interconversion. Investigation of a degenerate olefin metathesis reaction. J Am Chem Soc 104:7491–7496
6. Gilliom LR, Grubbs RH (1986) A titanacyclobutane precursor to alkyl-substituted titanium-carbene complexes. Organometallics 5:721–724
7. Gilliom LR, Grubbs RH (1986) Titanacyclobutanes derived from strained cyclic olefins: the living polymerization of norbornene. J Am Chem Soc 108:733–742
8. Schaverien CJ, Dewan JC, Schrock RR (1986) A well-characterized, highly active, lewis acid free olefin metathesis catalyst. J Am Chem Soc 108:2771–2773
9. Schrock RR, DePue RT, Feldman J, Schaverien CJ, Dewan JC, Liu AH (1988) Preparation and reactivity of several alkylidene complexes of the type $W(CHR')(N\text{-}2,6\text{-}C_6H_3\text{-}i\text{-}Pr_2)(OR)_2$ and related tungstacyclobutane complexes. Controlling metathesis activity through the choice of alkoxide ligand. J Am Chem Soc 110:1423–1435
10. Schrock RR, DePue RT, Feldman J, Yap KB, Yang DC, Davis WM, Park L, DiMare M, Schofield M, Anhaus J, Walborsky E, Evitt E, Krüger C, Betz P (1990) Further studies of imido alkylidene complexes of tungsten, well-characterized olefin metathesis catalysts with controllable activity. Organometallics 9:2262–2275
11. Murdzek JS, Schrock RR (1987) Well-characterized olefin metathesis catalysts that contain molybdenum. Organometallics 6:1373–1374
12. Schrock RR, Murdzek JS, Bazan GC, Robbins J, DiMare M, O'Regan M (1990) Synthesis of molybdenum imido alkylidene complexes and some reactions involving acyclic olefins. J Am Chem Soc 112:3875–3886
13. Toreki R, Schrock RR (1990) A well-defined rhenium(VII) olefin metathesis catalyst. J Am Chem Soc 112:2448–2449

14. Trnka TM, Grubbs RH (2001) The development of $L_2X_2 = CHR$ olefin metathesis catalysts: an organometallic success story. Acc Chem Res 34:8–29
15. Nguyen ST, Johnson LK, Grubbs RH, Ziller JW (1992) Ring-opening metathesis polymerization (ROMP) of norbornene by a group VIII carbene complex in protic media. J Am Chem Soc 114:3974–3975
16. Nguyen ST, Grubbs RH, Ziller JW (1993) Synthesis and activities of new single-component ruthenium-based olefin metathesis catalysts. J Am Chem Soc 115:9858–9859
17. Schwab P, Grubbs RH, Ziller JW (1993) Synthesis and applications of $RuCl_2(=CHR')(PR_3)_2$. J Am Chem Soc 118:100–110
18. Wu Z, Nguyen ST, Grubbs RH, Ziller JW (1995) Reactions of ruthenium carbenes of the type $(PPh_3)_2(X)_2Ru{=}CH{-}CH{=}CPh_2$ ($X = Cl$ and CF_3COO) with strained acyclic olefins and functionalized olefins. J Am Chem Soc 117:5503–5511
19. Huang J, Stevens ED, Nolan SP, Petersen JL (1999) Olefin metathesis-active ruthenium complexes bearing a nucleophilic carbene ligand. J Am Chem Soc 121:2674–2678
20. Weskamp T, Schattenmann WC, Spiegler M, Herrmann WA (1998) A novel class of ruthenium catalysts for olefin metathesis. Angew Chem Int Ed 37:2490–2493
21. Scholl M, Trnka TM, Morgan JP, Grubbs RH (1999) Increased ring closing metathesis activity of ruthenium-based olefin metathesis catalysts coordinated with imidazolin-2-ylidene ligands. Tetrahedron Lett 40:2247–2250
22. Garber SB, Kingsbury JS, Gray BL, Hoveyda AH (2000) Efficient and recyclable monomeric and dendritic Ru-based metathesis catalysts. J Am Chem Soc 122:8168–8179
23. Kingsbury JS, Harrity JPA, Bonitatebus PJ Jr, Hoveyda AH (1999) A recyclable Ru-based metathesis catalyst. J Am Chem Soc 121:791–799
24. Gessler S, Randl S, Blechert S (2000) Synthesis and metathesis reactions of a phosphine-free dihydroimidazole carbene ruthenium complex. Tetrahedron Lett 41:9973–9976
25. Vougioukalakis GC, Grubbs RH (2010) Ruthenium-based heterocyclic carbene-coordinated olefin metathesis catalysts. Chem Rev 110:1746–1787
26. Lindmark-Hamburg M, Wagener KB (1987) Acyclic metathesis polymerization. The olefin metathesis reaction of 1,5-hexadiene and 1,9-decadiene. Macromolecules 20:2949–2951
27. Opper K, Wagener KB (2011) ADMET: metathesis polycondensation. J Polym Sci Part A Polym Chem 49:821–831
28. Oskam JH, Fox HH, Yap KB, McConville DH, O'Dell R, Lichtenstein BJ, Schrock RR (1993) Ligand variation in alkylidene complexes of the type $Mo(CHR)(NR')(OR'')_2$. J Organomet Chem 459:185–198
29. Wagener KB, Brzezinska K, Anderson JD, Younkin TR, Steppe K, DeBoer W (1997) Kinetics of acyclic diene metathesis (ADMET) polymerization. Influence of the negative neighboring group effect. Macromolecules 30:7363–7369
30. Lehman SE Jr, Schwendeman JE, O'Donnell PM, Wagener KB (2003) Olefin isomerization promoted by olefin metathesis catalysts. Inorg Chim Acta 345:190–198
31. Courchay FC, Sworen JC, Wagener KB (2003) Metathesis activity and stability of new generation ruthenium polymerization catalysts. Macromolecules 36:8231–8239
32. Fokou PA, Meier MAR (2010) Studying and suppressing olefin isomerization side reactions during ADMET polymerizations. Macromol Rapid Commun 31:368–373
33. Hong SH, Sanders DP, Lee CW, Grubbs RH (2005) Prevention of undesirable isomerization during olefin metathesis. J Am Chem Soc 127:17160–17161
34. Simocko C, Wagener KB (2013) Effects of boron-containing lewis acids on olefin metathesis. Organometallics 32:2513–2516
35. Weychardt H, Plenio H (2008) Acyclic diene metathesis polymerization of divinylarenes and divinylferrocenes with Grubbs-type olefin metathesis catalysts. Organometallics 27:1479–1485
36. Simocko C, Yang Y, Swager TM, Wagener KB (2013) Metathesis step-growth polymerization in ionic liquids. ACS Macro Lett 2:1061–1064
37. Knuuttila H, Lehtinen A, Nummila-Pakarinen A (2004) Advanced polyethylene technologies—controlled material properties, long-term properties of polyolefins. Adv Polym Sci 169:13–27
38. Fawcet EW, Gibson RQ, Perrin MH, Patton JG, Williams EG (1937) Imperial Chemical Industries Ltd. Pat 2, pp 816–883
39. Baughman TW, Wagener KB (2005) Recent advances in ADMET polymerization. In: Buchmeiser M (ed). Springer, Berlin, pp 1–42
40. Rojas G, Wagener KB (2008) Avoiding olefin isomerization during decyanation of alkylcyano α, ω-dienes: a deuterium labeling and structural study of mechanism. J Org Chem 73:4962–4970
41. Rojas G, Wagener KB (2009) Precisely and irregularly sequenced ethylene/1-hexene copolymers: a synthesis and thermal study. Macromolecules 42:1934–1947
42. Hosoda S, Nozue Y, Kawashima Y, Suita K, Seno S, Nagamatsu T, Wagener KB, Inci B, Zuluaga F, Rojas G, Leonard JK (2011) Effect of the sequence length distribution on the lamellar crystal thickness and thickness distribution of polyethylene: perfectly equisequential ADMET polyethylene vs ethylene/α-olefin copolymer. Macromolecules 44:313–319
43. Hosoda S, Nozue Y, Kawashima Y, Utsumi S, Nagamatsu T, Wagener K, Berda E, Rojas G, Baughman T, Leonard J (2009) Perfectly controlled lamella thickness and thickness distribution: a morphological study on ADMET polyolefins. Macromol Symp 282:50–64
44. Rojas G, Inci B, Wei Y, Wagener KB (2009) Precision polyethylene: changes in morphology as a function of alkyl branch size. J Am Chem Soc 131:17376–17386
45. Inci B, Lieberwirth I, Steffen W, Mezger M, Graf R, Landfester K, Wagener KB (2012) Decreasing the alkyl branch frequency in precision polyethylene: effect of alkyl branch size on nanoscale morphology. Macromolecules 45:3367–3376
46. Inci B, Wagener KB (2011) Decreasing the alkyl branch frequency in precision polyethylene: pushing the limits toward longer run lengths. J Am Chem Soc 133:11872–11875
47. Qiu W, Sworen J, Pyda M, Nowak-Pyda E, Habenschuss A, Wagener KB, Wunderlich B (2006) Effect of the precise branching of polyethylene at each 21st CH_2 group on its phase transitions, crystal structure, and morphology. Macromolecules 39:204–217
48. Lieser G, Wegner G, Smith JA, Wagener KB (2004) Morphology and packing behavior of model ethylene/propylene copolymers with precise methyl branch placement. Colloid Polym Sci 282:773–781
49. Matsui K, Seno S, Nozue Y, Shinohara Y, Amemiya Y, Berda EB, Rojas G, Wagener KB (2013) Influence of branch incorporation into the lamella crystal on the crystallization behavior of polyethylene with precisely spaced branches. Macromolecules 46:4438–4446
50. Nozue Y, Kawashima Y, Seno S, Nagamatsu T, Hosoda S, Berda EB, Rojas G, Baughman TW, Wagener KB (2011) Unusual crystallization behavior of polyethylene having precisely spaced branches. Macromolecules 44:4030–4034
51. Bowers CR, Wei Y, Aitken BS, Reeg CR, Akel CD (2012) Molecular dynamics in precision deuteriomethyl branched polyethylene from solid-state deuterium NMR. Polymer 53:2633–2642
52. Wei Y, Graf R, Sworen JC, Cheng C-Y, Bowers CR, Wagener KB, Spiess HW (2009) Local and collective motions in precise

polyolefins with alkyl branches: a combination of ^{2}H and ^{13}C solid-state NMR spectroscopy. Angew Chem Int Ed 48:4617–4620
53. Spiess WH (1983) Molecular dynamics of solid polymers as revealed by deuteron NMR. Colloid Polym Sci 261:193–209
54. Zutty NL, Faucher JA, Bonotto S, Mark HF (eds) (1967) Encyclopedia of polymers science and technology. Wiley, New York
55. Odian G (1991) Principles of polymerization, 3rd edn. Wiley, New York, p 460
56. Watson MD, Wagener KB (2000) Ethylene/vinyl acetate copolymers via acyclic diene metathesis polymerization. Examining the effect of "long" precise ethylene run lengths. Macromolecules 33:5411–5417
57. Baughman TW, Chan CD, Winey KI, Wagener KB (2007) Synthesis and morphology of well-defined poly(ethylene-co-acrylic acid) copolymers. Macromolecules 40:6564–6571
58. Seitz ME, Chan CD, Opper KL et al (2010) Nanoscale morphology in precisely sequenced poly(ethylene-co-acrylic acid) zinc ionomers. J Am Chem Soc 132:8165–8174
59. Boz E, Ghiviriga I, Nemeth AJ, Jeon K, Alamo RG, Wagener KB (2008) Random, defect-free ethylene/vinyl halide model copolymers via condensation polymerization. Macromolecules 41:25–30
60. Opper KL, Wagener KB (2009) Precision sulfonic acid ester copolymers. Macromol Rapid Commun 30:915–919
61. Opper KL, Markova D, Klapper M, Müllen K, Wagener KB (2010) Precision phosphonic acid functionalized polyolefin architectures. Macromolecules 43:3690–3698
62. Markova D, Opper KL, Wagner M, Klapper M, Wagener KB, Müllen K (2013) Synthesis of proton conducting phosphonic acid-functionalized polyolefins by the combination of ATRP and ADMET. Polym Chem 4:1351–1363
63. Sydlik SA, Delgado PA, Inomata S et al (2013) Triptycene-containing polyetherolefins via acyclic diene metathesis polymerization. J Polym Sci Part A Polym Chem 51:1695–1706
64. Matloka PP, Wagener KB (2006) The acyclic diene metathesis (ADMET) polymerization approach to silicon containing materials. J Mol Catal A Chem 257:89–98
65. Leonard JK, Wei Y, Wagener KB (2012) Synthesis and thermal characterization of precision poly(ethylene-co-vinyl amine) copolymers. Macromolecules 45:671–680
66. Marmo JC, Wagener KB (1993) Acyclic diene metathesis (ADMET) depolymerization. Synthesis of mass-exact telechelic polybutadiene oligomers. Macromolecules 26:2137–2138
67. Marmo JC, Wagener KB (1995) ADMET depolymerization. Synthesis of perfectly difunctional ($f = 2.0$) telechelic polybutadiene oligomers. Macromolecules 28:2602–2606
68. Schulz MD, Ford RR, Wagener KB (2013) Insertion metathesis depolymerization. Polym Chem 4:3656–3658
69. Tao D, Wagener KB (1993) Acyclic diene metathesis (ADMET) polymerization. Synthesis of telechelic polyacetylene. Polym Prepr 34:469–470
70. Schwendeman JE, Wagener KB (2009) Synthesis of amorphous hydrophobic telechelic hydrocarbon diols via ADMET polymerization. Macromol Chem Phys 210:1818–1833
71. Tamura H, Nakayama A (2002) Functional telechelic polymer synthesis via ADMET polymerization. J Macromol Sci Pure Appl Chem A 39:745–758
72. Tamura H, Maeda N, Matsumoto R, Nakayama A, Hayashi H, Ikushima K, Kuraya M (1999) Synthesis of ester terminated telechelic polymer via ADMET polymerization. J Macromol Sci Pure Appl Chem A 36:1153–1170
73. Atallah P, Wagener KB, Schulz MD (2013) ADMET: the future revealed. Macromolecules 46:4735–4741

4

Characterization of single-chain polymer folding using size exclusion chromatography with multiple modes of detection

Peter Frank · Alka Prasher · Bryan Tuten · Danming Chao · Erik Berda

Abstract We highlight here recent work from our laboratory on the subject of fabricating nanostructures from single polymer chains. These so-called single-chain nanoparticles are synthesized by inducing intra-molecular cross-linking on discrete macromolecules in dilute solution. Among the biggest challenges in this rapidly expanding area of research is reliable and accurate means to characterize this process. In this paper, we review our preferred method of characterization: size exclusion chromatography featuring multiple modes of detection. Multi-angle light scattering in conjunction with a concentration detector can provide absolute molecular weight data; viscometric detection can provide information about solution size and conformation. Correlation of these data provides a simple and robust way to quantify the process by which we fold single polymer coils into architecturally defined unimolecular nanostructures.

Keywords Single-chain nanoparticles · Polymer folding · TEM · SEC

P. Frank · A. Prasher · E. Berda
Department of Chemistry, University of New Hampshire, Durham, USA

B. Tuten · E. Berda
Materials Science Program, University of New Hampshire, Durham, USA

D. Chao
Alan G. MacDiarmid Institute, College of Chemistry, Jilin University, Changchun, People's Republic of China

Introduction

A fast growing research area, single-chain nanoparticles (SCNP) have demonstrated promise for use in catalysis [1], nanoreactors [2], controlled release [3], chemical sensors [4], enzyme mimetics [5–7], diagnostics [8] and optics [9]. This breadth of application is attributed to the precise control over the composition, functionality, geometry and topology this technology permits for accessing architectures in the sub-20 nm size regime. These features are programmed into SCNP a priori by means of robust living polymerization techniques including ring-opening metathesis polymerization [10] (ROMP), reversible addition-fragmentation chain-transfer [11] (RAFT), and atom transfer radical polymerization [12] (ATRP). Such techniques afford parent polymers with narrow molecular weight distributions ($Đ$), targeted molecular weights, tunable functionality and diverse architectures [13–17]. These parent chains are then subjected to intra-molecular cross-linking under ultra-dilute conditions, thus folding the chains into architecturally defined nanoparticles with dimensions smaller than the original dimensions of the solvated coil (Fig. 1) [18].

Although young, this field is developing rapidly; a number of important contributions have appeared in the recent literature. [1, 12, 19–21] The Meijer group showed that the molecular weight and backbone rigidity of the parent polymer does not significantly influence the SCNP fabrication induced by supramolecular interactions. Instead, the solvent plays a key role in mediating intra-molecular folding [21]. Meijer and coworkers also expanded the scope of possible architectures by successfully synthesizing SCNP from cylindrical brush block polymers as well as an orthogonally collapsible ABA triblock copolymer [6, 22]. Additionally, this group studied

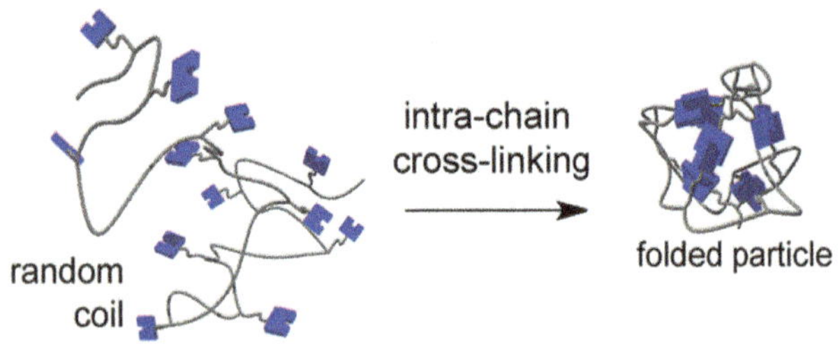

Fig. 1 Schematic representation of SCNP synthesis

temperature and co-solvent induced changes in the secondary structures of benzene-1,3,5-tricarboxamides (BTA)-based SCNPs, and developed a compartmentalized metal ion SCNP sensor that shows impressive binding affinities [4, 23]. The Barner-Kowollik group demonstrated a photo-induced Diels–Alder ligation to create SCNP in addition to using metal–ligand complexation to induce controlled folding by linking chain ends [1, 24]. Other work has demonstrated that SCNP can be formed from a host of reactions including benzocylcobutene coupling [25], olefin cross-metathesis [26], azide–alkyne click chemistry [19, 27–30], thiol-ene click chemistry [31, 32], Bergman cyclization [33–35], Curtius rearrangement [36], Glasser–Hay coupling [37], nitrene-mediated C-alkylation (sulfonyl azide) [38], oxidative polymerization through pendant moieties [12], benzosulfone chemistry [39], carbodiimide coupling [40], benzoxazine chemistry [11], amide bond formation [10, 31], cinnamoyl [40] and coumarin [2] photodimerization, guest–host interactions [41], hydrogen bonding [20, 42–44], and vulcanization [45].

These methods draw synthetic parallel to nature's elegant biomacromolecules, albeit in a crude and simplistic fashion [3, 5]. Given that synthetic polymers are always subject to distributions in molecular weights and heterogeneities in microstructure [46, 47], it is important to develop convenient and consistent methods for characterizing SCNP fabrication. We discuss here the analysis of SCNPs using size exclusion chromatography (SEC), specifically the application of multi-detection SEC. We have recently shown that when applied in conjunction with microscopy and dynamic light scattering (DLS), multi-detection SEC is a powerful tool for studying the folding of individual polymer chains into architectural defined nanostructures.

Polymer design and SCNP synthesis

There are a variety of synthetics strategies that can be used to fabricate SCNP [48]. The first methodology successfully employed in our labs involves the addition of an external, bifunctional cross-linker to a chain decorated with complementary reactive handles (Scheme 1) [10]. Polymer chains decorated with cyclic anhydride units were folded by the introduction of *p*-aminophenyl disulfide as the bifunctional cross-linker. We chose to work with poly(norbornene-exo-anhydride), as the requisite reactive monomer is commercially available and polymerization (or copolymerization with cyclooctene) via ROMP proceeds rapidly with excellent control. In series **1a**, the number of intra-chain linkages can be controlled by modulating the amount of bifunctional diamine cross-linker added to a solution of

Scheme 1 Synthetic approach to series **1** via addition of an external bifunctional cross-linker

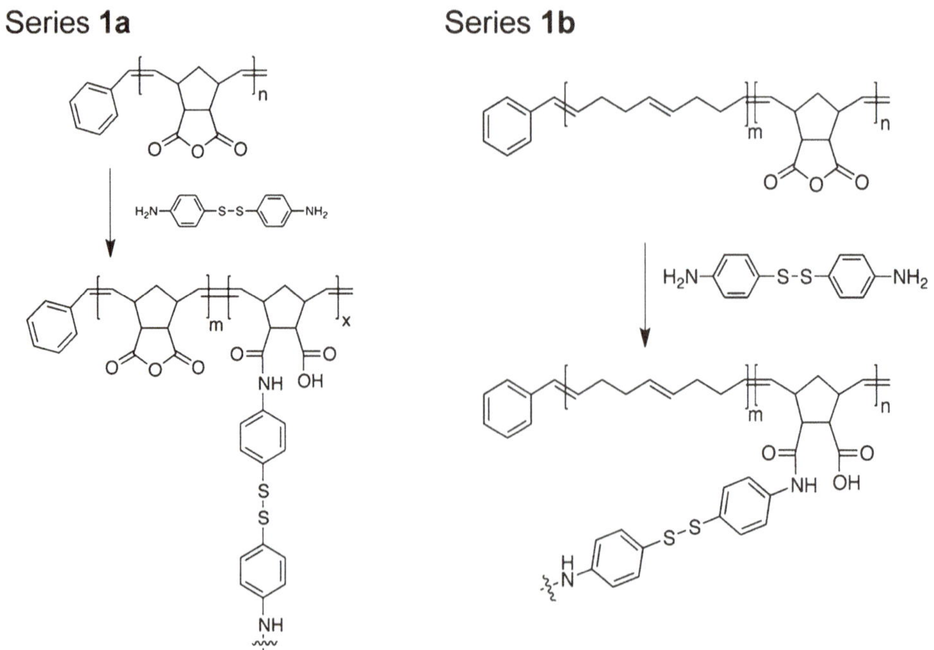

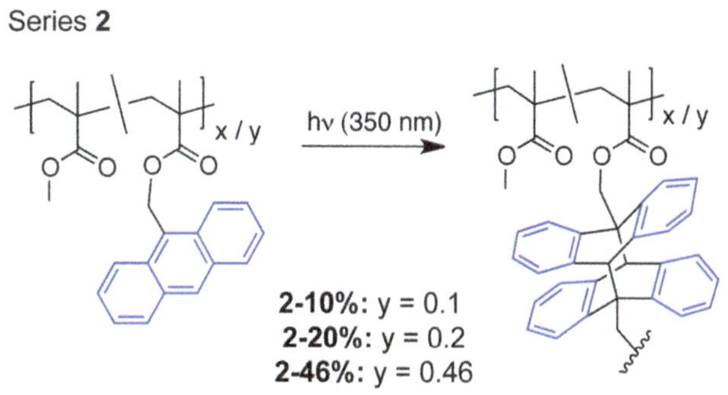

Scheme 2 Synthetic approach to series **2** using photo cross-linkable pendant groups

the polyanhydride. In series **1b**, the extent of intra-chain cross-links is controlled by changing the feed ratio of anhydride co-monomer relative to cyclooctene, followed by the addition of stoichiometric amounts of diamine. Using a cross-linker with a reversible covalent bond permits the unfolding and refolding of the chain and allows us to better characterize this coil to globule transition [10].

The second route we are using to fabricate SCNPs involves the homocoupling of a built in functional unit [49]. For series **2**, foldable methacrylic chains bearing pendant anthracene units were synthesized by RAFT polymerization (Scheme 2). Anthracene is well known for its ability to undergo a photo-

Scheme 3 Synthetic approach to series **3** using sequentially activated orthogonal intra-chain interactions

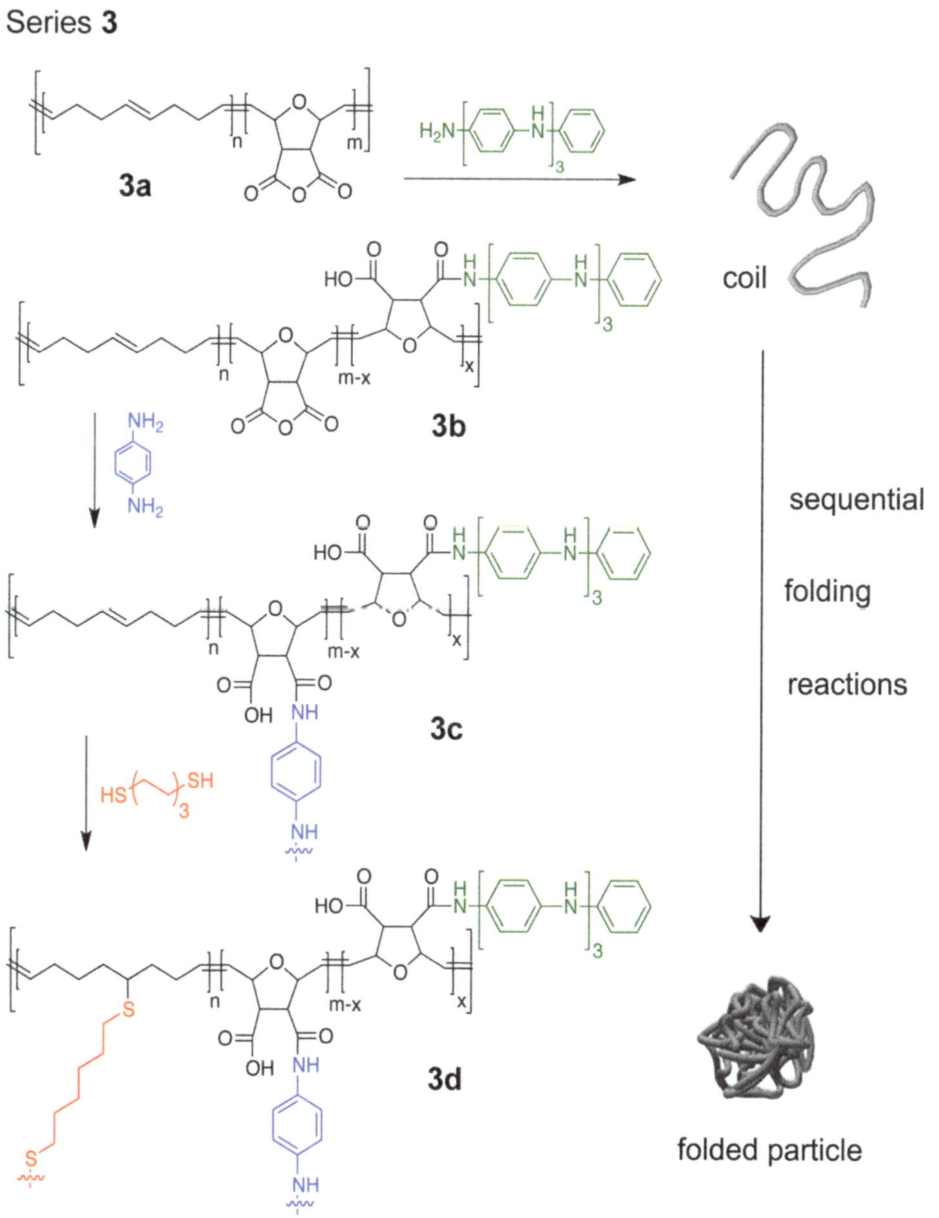

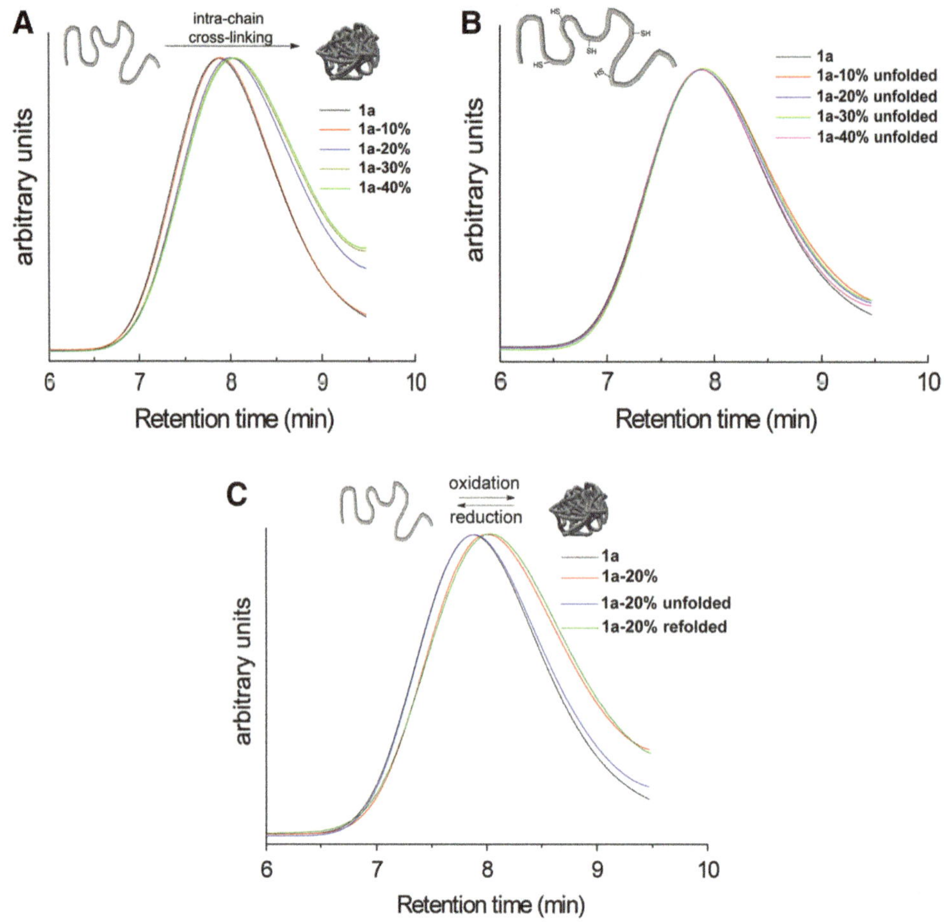

Fig. 2 SEC traces for series **1a**. **a** Addition of varying amounts of cross-linker, **b** after unfolding via disulfide reduction, **c** reversible folding and unfolding via redox chemistry (Tuten et al. [10]—adapted with permission of The Royal Society of Chemistry)

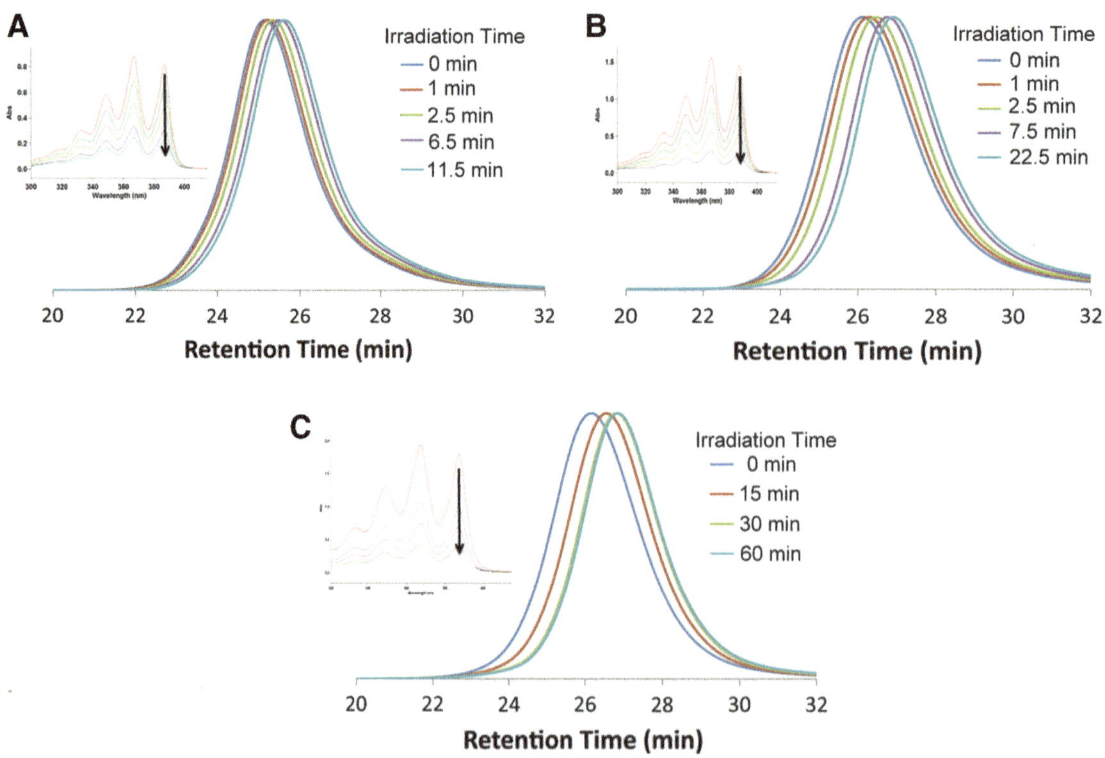

Fig. 3 SEC traces and corresponding UV–Vis spectra for series **2** photodimerization studies (Frank et al. [49]—adapted with permission of John Wiley And Sons Inc)

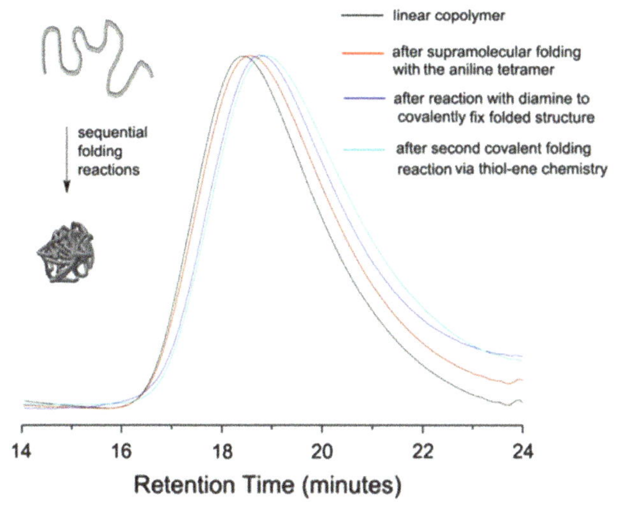

Fig. 4 SEC traces displaying the decrease in hydrodynamic volume with each sequential folding reaction for series 3 (Chao et al. [31]—adapted with permission of The Royal Society of Chemistry)

Table 1 SEC data for series **1b**

	Peak retention time (min)[a]	M_w (kg/mol)[b]	$Đ$	$[\eta]$ (mL/g)	R_h (nm)
Parent polymer (**1b**)	17.8	50.1	1.22	13.9	4.5
Folded nanoparticle	18.8	59.3	1.32	8.7	3.9

[a] Retention times are from MALS detector trace

[b] Absolute molecule weight calcultated by MALS

Table 2 SEC data for series **2** before and after complete photo-induced folding

Polymer	Irradiation time (min)	Peak retention time (min)	M_w (kg/mol)	$Đ$	R_h (nm)	$[\eta]$
2a-10	0	24.9	30.8	1.17	4	13.3
	11.5	25.4	33.5	1.16	3.4	7.5
2b-20	0	25.6	30.6	1.14	3.6	9.5
	22.5	26.5	29.9	1.14	2.9	5.1
2c-46	0	25.7	42.9	1.16	3.5	6.6
	60	26.4	36.7	1.14	2.3	2.4

Table 3 Comparison of predicted vs. actual absolute M_w for series **3**

Polymer	Predicted molecular weights M_w (kg/mol)	Actual molecular weights M_w (kg/mol)	R_h (nm)
3a	16.1	15.0	6.5
3b	19.7	19.9	5.7
3c	20.5	22.6	4.9
3d	23.9	23.3	4.3

R_h values from DLS confirm the efficacy of each sequential folding reaction

induced 4 + 4 cycloaddition when irradiated with 350 nm-centered UV light, resulting in photodimers that can be reversed via irradiation with wavelengths <300 nm or thermally at temperatures >180 °C [50], providing an attractive route to synthesizing thermally responsive and photosensitive nanomaterials. In addition, using light as stimulus for SCNP fabrication offers additional functional group tolerance that other cross-linking strategies lack [49].

A third method for SCNP fabrication takes advantage of multiple orthogonal chemistries triggered sequentially to fold synthetic chains into particles [31]. Scheme 3 shows a series of poly(oxanorbornene anhydride-*co*-cyclooctadiene) materials synthesized using ROMP. Analogous to series **1**, the polymer structures contained in series **3** feature anhydride groups as reactive handles that can be easily modified with a variety of nucleophiles, simultaneously inducing folding and installing functionality. In this system, aniline tetramer (AT) was used to fold the chains via supramolecular interactions (π–π and hydrogen bonding) between these pendant groups. The chains were then reacted with *p*-aminoaniline, a bifunctional covalent cross-linker that pulls in the remaining anhydride units and further collapses the chain. Finally, internal olefin units installed within the backbones during ROMP were connected using photoinitiated thiol-ene click chemistry with 1,6-hexanedithiol as a bifunctional cross-linker. This sequential folding approach demonstrates that the application of orthogonal chemistries in a synergistic fashion results in more tightly collapsed structures with the goal of more closely mimicking natural folded polymeric constructs. [31]

SCNP characterization

Conventional SEC

SEC with conventional calibration is the standard method used to characterize SCNPs formation. Comparing the retention time of the parent polymer chains to the retention time of the folded nanoparticles after intramolecular cross-linking is induced typically reveals shifts to longer retention times. This indicates a decrease in hydrodynamic volume based on the inherent principles of size exclusion chromatography [51–53]. This is clearly demonstrated in all of the polymer series discussed here [10, 31, 49].

SEC traces for series **1a** below show the effect of adding increasing amounts of bifunctional cross-linker: as more intra-chain linkages are formed the SEC traces shift to longer retention times as expected (Fig. 2a). To ensure this is truly an effect of conformational changes and not due to other factors, such as interactions between the SEC columns and the analyte, we take advantage of the cleavable disulfide linkage that was installed during single-chain nanoparticles formation. Cleaving this linkage after the initial folding reaction takes place allows us to determine that the shifts in the retention time is actually due to the folding of the chain and not other behavior that might affect retention time. This is indeed the case, as shown in Fig. 2b. The reduction of the disulfide bridges via dithiothreitol (DTT) results in the unfolding of the SCNPs, confirmed by SEC traces showing that retention times (and therefore hydrodynamic volume) revert to those of the parent polymer chain, regardless of the amount of cross-linker added. Furthermore, refolding of the unfolded chains via thiol oxidation in the presence of catalytic $FeCl_3$ revealed shifts to longer retention times, indicating the reduction in the solvent dynamic volume (refolding), confirming the utility of this characterization method as well as demonstrating this chemistry as a route to responsive nanostructures (Fig. 2c).

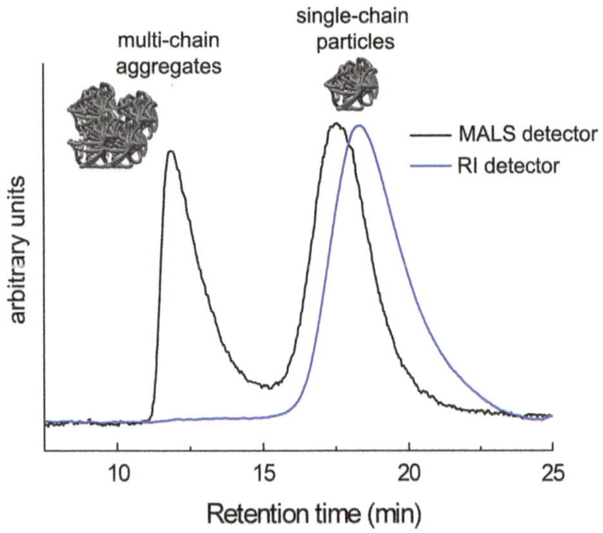

Fig. 5 MALS and RI detection traces for SEC of **1b** highlighting the ability of MALS to detect large, multi-chain aggregates (Tuten et al. [10]—Reproduced by permission of The Royal Society of Chemistry)

Analyzing the SEC traces from series **2** shows that the extent of intra-molecular folding or collapse can be controlled very easily using built-in reactive co-monomer. Both reaction time and the amount of co-monomer incorporation can be used to control the extent of intra-chain cross-links when employing this method. Figure 3a–c highlights this behavior. All samples show steady shifts to longer retention times with increasing irradiation times, an

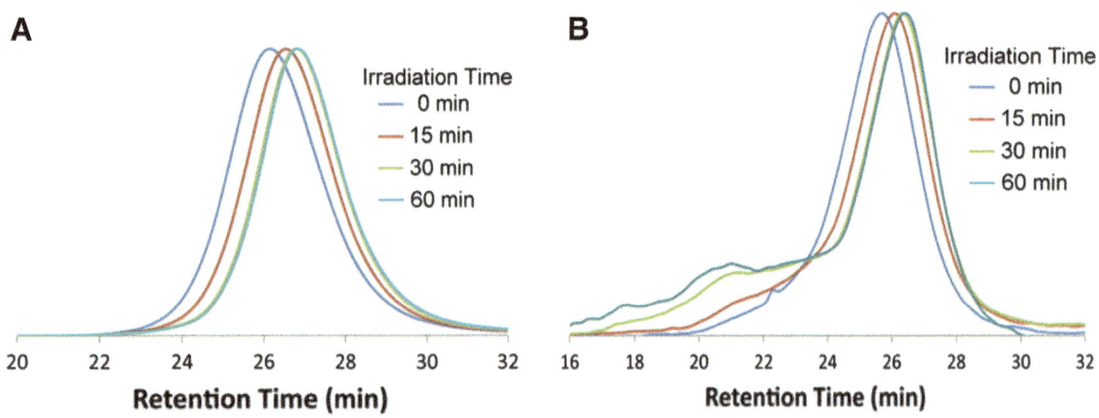

Fig. 6 SEC traces for **2c-46** photodimerization studies showing the appearance of multi-chain aggregates. **a** UV detector traces, **b** MALS detector traces (Frank et al. [49]—Reproduced by permission of John Wiley And Sons Inc)

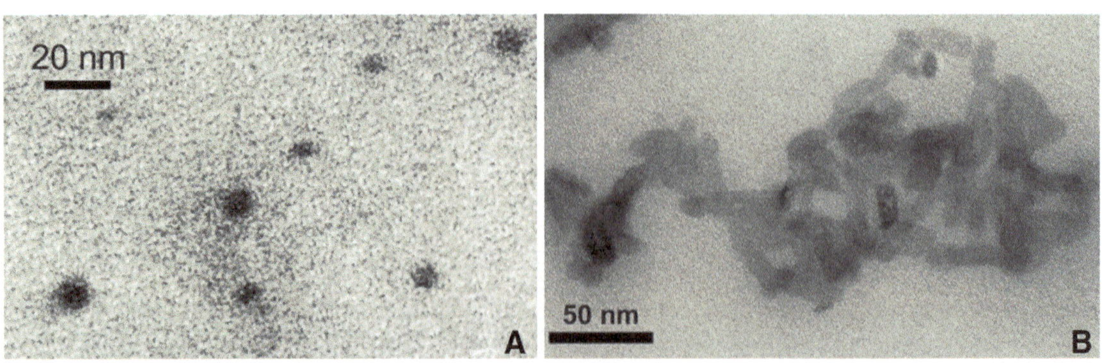

Fig. 7 TEM images of SCNP obtained from **1b** (**a**) and **2a-10** (**b**)

indication that folding is occurring with the formation of anthracene photodimers. Anthracene dimerization was confirmed by the disappearance of the characteristic anthracene absorption peak in the UV–Vis spectra (insets in Fig. 3). The effect of reactive co-monomer incorporation is clearly observed here: samples with 10 % (**2a-10**) anthracene co-monomer showed the least pronounced changes in retention time when compared to the samples with 20 % (**2a-20**) and 46 % (**2a-46**) anthracene co-monomer (Fig. 3).

Conventional SEC reveals the same behavior even when multiple orthogonal intra-chain cross-linking chemistries are used. Figure 4 shows that for polymers in series **3**, the expected shifts in retention time confirm the efficacy of each folding reaction when engaged sequentially. As expected, the SEC trace of the final nanoparticle shifted to the longest retention time indicating a more tightly wrapped structure consistent with theory and experiment [6].

Triple detection SEC

The application of standard SEC as described above provides useful qualitative information about changes in the solution conformation of polymer chains during SCNP fabrication. Still this method is not without ambiguities, especially with regards to actual changes in molecular weight (or lack thereof) when triggering intra-chain cross-linking. Since relative molecular weight and retention time are linked in traditional SEC, it is impossible to formally characterize SCNP molecular weight with this technique. Furthermore, while shifts in retention time certainly signify changes in hydrodynamic volume it is impossible to quantify these changes with standard SEC measurements alone. Dynamic light scattering (DLS) in conjunction with SEC can be a very useful tool for SCNP characterization, but lacks the chromatographic separation provided by SEC.

To alleviate some of these issues our group implemented SEC with multiple modes of detection for SCNP characterization. The use of multi-angle light scattering (MALS) detection with SEC allows the determination of polymer and SCNP molecular weight independent of retention time [54]. The addition of an inline differential viscometer permits characterization of molecular conformation, including hydrodynamic radius, for each thin chromatographic slice as the sample elutes from the SEC column. With this setup, the absolute molecular weight of the parent polymers can be easily measured and used to calculate the expected SCNP molecular weight. Comparing this value with the experimental absolute SCNP molecular weight can provide useful quantitative information about the chemistry involved in this process. Changes in retention time, which are assumed to correlate with changes in solvated volume, can likewise be readily quantified by this method. Recent work from our lab demonstrating the utility of this technique is highlighted below.

Table 1 summarizes the triple detection data for series **1b.** As expected, the MALS detector trace shows a shift to longer retention time after addition of cross-linker similar to what we observed using standard SEC. Examining the absolute molecular weight reveals that although the SCNP curve shifted to a longer retention time the molecular weight actually increased by the amount predicted with the addition of stoichiometric cross-linker. In addition, the viscometric data confirm and quantify the decrease in hydrodynamic radius. It further reveals a decrease in intrinsic viscosity after SCNP formation consistent with theory and experiment [55].

The SEC data for series **2** showed similar behavior where shifts to longer retention times occur after folding as expected. The viscometric data reveals a decrease in both intrinsic viscosity and hydrodynamic radius. The absolute molecular weight, predicted to remain constant, changes only slightly and is consistent with a primarily unimolecular phenomenon (Table 2). This series also displays the trend, although unsurprising, that the extent of reactive co-monomer incorporation directly dictates how much folding will occur.

SEC MALS data from series **3**, shown in Table 3, again highlight how useful this technique can be as a quantitative tool. For these polymers, the comparison between the predicted and the measured absolute molecular weights demonstrates that the folding chemistry proceeded as designed. In addition, DLS measurements confirmed that retention time shifts witnessed in SEC measurements (Table 3) are indeed due to folding.

Intra-chain versus inter-chain reactions

A crucial bit of qualitative data provided by SEC MALS is the distinction between intra-chain folding and inter-chain coupling. In our experience, MALS is capable of detecting larger nanoaggregates in very minute concentrations while traditional SEC detectors such as UV and RI do not show the characteristic aggregate peak [10, 49]. For series **1**, multi-chain aggregates were detectable via MALS while virtually unseen in RI traces (Fig. 5) [10].

Likewise for series **2**, the UV detector fails to detect inter-chain aggregates that are readily observed by MALS, evident in the emergence of peak shoulders in MALS traces. These data also suggest that at a certain level of reactive co-monomer incorporation inter-chain coupling becomes unavoidable (Fig. 6) [49]. Thus, triple detection SEC provides a robust method for analyzing polymer folding chemistry and confirming single-chain behavior.

Correlating solution measurements with other characterization tools

We highlighted in the previous sections how triple detection SEC can be applied to effectively characterize single-chain folding. Additional secondary characterization techniques can offer invaluable information about SCNP size and shape, including small angle neutron scattering (SANS) [3, 23], thorough DLS measurements [4, 23], atomic force microscopy (AFM) [11, 19, 20, 26, 29, 30, 33, 39, 42, 43, 56, 57], and transmission electron microscopy (TEM) [2, 7, 10, 11, 27, 38, 49]. We found TEM particularly useful as a tool to provide visual evidence for SCNP formation. Figure 7 shows representative images from series **1** (**1b**) and series **2** (**2a-10**). In each case, the images reveal nanostructures with dimensions on the same order of magnitude as numbers obtained from solution measurements. Interestingly, series **2** SCNP adopt an oblong ellipsoid geometry, rather than a spherical conformation. We are still investigating the cause of this behavior, which is consistent with form factor measurements made by others [3, 23].

Summary

In summary, conventionally calibrated SEC is a satisfactory method to characterize SCNP via interpretation of comparative shifts in retention time. Evolving this tool by marrying SEC with MALS and viscometric detection affords a precise and reliable method for the qualitative and quantitative study of SCNP fabrication. This methodology facilitates the detailed investigation of structure property relationships using absolute molecular weight, intrinsic viscosity, and hydrodynamic radius of SCNP systems. During SCNP formation, the decrease in both the intrinsic viscosity and hydrodynamic radius is observed with, or in the absence of, concurrent changes in the absolute molecular weight depending on the folding technique employed. Triple detection SEC in conjunction with additional characterization tools including TEM and DLS provides a variety of useful information for studying the fascinating and rapidly expanding research area of single-chain folding processes and single-chain nanoparticles.

References

1. Willenbacher J, Altintas O, Roesky PW, Barner-Kowollik C (2013) Macromol Rapid Commun 1521–3927
2. He J, Tremblay L, Lacelle S, Zhao Y (2011) Preparation of polymer single chain nanoparticles using intramolecular photodimerization of coumarin. Soft Matter 7:2380–2386
3. Sanchez-Sanchez A, Akbari S, Etxeberria A, Arbe A, Gasser U, Moreno AJ, Colmenero J, Pomposo JA (2013) "Michael" nanocarriers mimicking transient-binding disordered proteins. ACS Macro Lett 2:491–495
4. Gillissen MAJ, Voets IK, Meijer EW, Palmans ARA (2012) Single chain polymeric nanoparticles as compartmentalised sensors for metal ions. Polym Chem 3:3166–3174
5. Perez-Baena I, Barroso-Bujans F, Gasser U, Arbe A, Moreno AJ, Colmenero J, Pomposo JA (2013) Endowing single-chain polymer nanoparticles with enzyme-mimetic activity. ACS Macro Lett 2:775–779
6. Hosono N, Gillissen MAJ, Li Y, Sheiko SS, Palmans ARA, Meijer EW (2013) Orthogonal self-assembly in folding block copolymers. J Am Chem Soc 135:501–510
7. Terashima T, Mes T, De Greef TFA, Gillissen MAJ, Besenius P, Palmans ARA, Meijer EW (2011) Single-chain folding of polymers for catalytic systems in water. J Am Chem Soc 133:4742–4745
8. Murray BS, Fulton DA (2011) Dynamic covalent single-chain polymer nanoparticles. Macromolecules 44:7242–7252
9. Wang Y, Dong L, Xiong R, Hu AJ (2013) Practical access to bandgap-like N-doped carbon dots with dual emission unzipped from PAN@PMMA core–shell nanoparticles. Mater Chem C 1:7731–7735
10. Tuten BT, Chao D, Lyon CK, Berda EB (2012) Single-chain polymer nanoparticles via reversible disulfide bridges. Polym Chem 3:3068–3071. doi:
11. Wang P, Pu H, Jin MJ (2011) Single-chain nanoparticles with well-defined structure via intramolecular crosslinking of linear polymers with pendant benzoxazine groups. Polym Sci Part A Polym Chem 49:5133–5141
12. Dirlam PT, Kim HJ, Arrington KJ, Chung WJ, Sahoo R, Hill LJ, Costanzo PJ, Theato P, Char K, Pyun J, Kim J, Chung J (2013) Single chain polymer nanoparticles via sequential ATRP and oxidative polymerization. Polym Chem 4: 3765–3773
13. Moad G, Chong YK, Postma A, Rizzardo E, Thang SH (2005) Advances in RAFT polymerization: the synthesis of polymers with defined end-groups. Polymer 46:8458–8468
14. Moad G, Rizzardo E, Thang SH (2005) Living radical polymerization by the RAFT process. Aust J Chem 58:379–410
15. Alfred SF, Lienkamp K, Madkour AE, Tew GN (2008) Water-soluble ROMP polymers from amine-functionalized norbornenes. Polymer 6672–6676
16. Bielawski CW, Grubbs RH (2007) Living ring-opening metathesis polymerization. Prog Polym Sci 32:1–29
17. Matyjaszewski K, Tsarevsky NV (2009) Nanostructured functional materials prepared by atom transfer radical polymerization. Nat Chem 1:276–288
18. Altintas O, Barner-Kowollik C (2012) Single chain folding of synthetic polymers by covalent and non-covalent interactions: current status and future perspectives. Macromol Rapid Commun 33:958–971
19. Ormategui N, García I, Padro D, Cabañero G, Grande HJ, Loinaz I (2012) Synthesis of single chain thermoresponsive polymer nanoparticles. Soft Matter 8:734–740
20. Foster EJ, Berda EB, Meijer EW (2009) Metastable supramolecular polymer nanoparticles via intramolecular collapse of single polymer chains. J Am Chem Soc 131:6964–6966
21. Stals PJM, Gillissen MAJ, Nicolaÿ R, Palmans ARA, Meijer EW (2013) The balance between intramolecular hydrogen bonding, polymer solubility and rigidity in single-chain polymeric nanoparticles. Polym Chem 4:2584–2597

22. Stals PJM, Li Y, Burdyńska J, Nicolaÿ R, Nese A, Palmans ARA, Meijer EW, Matyjaszewski K, Sheiko SS (2013) How far can we push polymer architectures? J Am Chem Soc 135:11421–11424
23. Gillissen MAJ, Terashima T, Meijer EW, Palmans ARA, Voets IK (2013) Sticky supramolecular grafts stretch single polymer chains. Macromolecules 46:4120–4125
24. Altintas O, Willenbacher J, Wuest KNR, Oehlenschlaeger KK, Krolla-Sidenstein P, Gliemann H, Barner-Kowollik C (2013) A mild and efficient approach to functional single-chain polymeric nanoparticles via photoinduced Diels–Alder ligation. Macromolecules 46:8092–8101
25. Harth E, Van Horn B, Lee VY, Germack DS, Gonzales CP, Miller RD, Hawker CJ (2002) A facile approach to architecturally defined nanoparticles via intramolecular chain collapse. J Am Chem Soc 124:8653–8660
26. Cherian AE, Sun FC, Sheiko SS, Coates GW (2007) Formation of nanoparticles by intramolecular cross-linking: following the reaction progress of single polymer chains by atomic force microscopy. J Am Chem Soc 129:11350–11351
27. De Luzuriaga AR, Ormategui N, Grande HJ, Odriozola I, Pomposo JA, Loinaz I (2008) Intramolecular click cycloaddition: an efficient room-temperature route towards bioconjugable polymeric nanoparticles. Macromol Rapid Commun 29:1156–1160
28. Schmidt BVKJ, Fechler N, Falkenhagen J, Lutz J-F (2011) Controlled folding of synthetic polymer chains through the formation of positionable covalent bridges. Nat Chem 3:234–238
29. Perez-Baena I, Loinaz I, Padro D, García I, Grande HJ, Odriozola I (2010) Single-chain polyacrylic nanoparticles with multiple Gd(iii) centres as potential MRI contrast agents. J Mater Chem 20:6916–6922
30. Oria L, Aguado R, Pomposo JA, Colmenero J (2010) A versatile "click" chemistry precursor of functional polystyrene nanoparticles. Adv Mater 22:3038–3041
31. Chao D, Jia X, Tuten B, Wang C, Berda EB (2013) Controlled folding of a novel electroactive polyolefin via multiple sequential orthogonal intra-chain interactions. Chem Commun 49:4178–4180. doi:
32. Ding L, Yang G, Xie M, Gao D, Yu J, Zhang Y (2012) More insight into tandem ROMP and ADMET polymerization for yielding reactive long-chain highly branched polymers and their transformation to functional polymer nanoparticles. Polymer 53:333–341
33. Zhu B, Ma J, Li Z, Hou J, Cheng X, Qian G, Liu P, Hu A (2011) Formation of polymeric nanoparticles via Bergman cyclization mediated intramolecular chain collapse. J Mater Chem 21: 2679–2683
34. Zhu B, Sun S, Wang Y, Deng S, Qian G, Wang M, Hu AJ (2013) Preparation of carbon nanodots from single chain polymeric nanoparticles and theoretical investigation of the photoluminescence mechanism. Mater Chem C 1:580–586
35. Zhu B, Qian G, Xiao Y, Deng S, Wang M, Hu AJ (2011) A convergence of photo-bergman cyclization and intramolecular chain collapse towards polymeric nanoparticles. Polym Sci Part A Polym Chem 49:5330–5338
36. Zheng H, Ye X, Wang H, Yan L, Bai R, Hu W (2011) A facile one-pot strategy for preparation of small polymer nanoparticles by self-crosslinking of amphiphilic block copolymers containing acyl azide groups in aqueous media. Soft Matter 7:3956–3962
37. Sanchez-Sanchez A, Asenjo-Sanz I, Buruaga L, Pomposo JA (2012) Naked and self-clickable propargylic-decorated single-chain nanoparticle precursors via redox-initiated RAFT polymerization. Macromol Rapid Commun 33:1262–1267
38. Jiang X, Pu H, Wang P (2011) Polymer nanoparticles via intramolecular crosslinking of sulfonyl azide functionalized polymers. Polymer 52:3597–3602
39. Adkins CT, Muchalski H, Harth E (2009) Nanoparticles with individual site-isolated semiconducting polymers from intramolecular chain collapse processes. Macromolecules 42:5786–5792
40. Njikang G, Liu G, Curda SA (2008) Tadpoles from the intramolecular photo-cross-linking of Diblock copolymers. Macromolecules 41:5697–5702
41. Appel EA, Dyson J, del Barrio J, Walsh Z, Scherman OA (2012) Formation of single-chain polymer nanoparticles in water through host-guest interactions. Angew Chem Int Ed Engl 51:4185–4189
42. Foster EJ, Berda EB, Meijer EWJ (2011) Tuning the size of supramolecular single-chain polymer nanoparticles. Polym Sci Part A Polym Chem 49:118–126
43. Berda EB, Foster EJ, Meijer EW (2010) Toward controlling folding in synthetic polymers: fabricating and characterizing supramolecular single-chain nanoparticles. Macromolecules 43:1430–1437
44. Seo M, Beck BJ, Paulusse JMJ, Hawker CJ, Kim SY (2008) Polymeric nanoparticles via noncovalent cross-linking of linear chains. Macromolecules 41:6413–6418
45. Cheng C, Qi K, Germack DS, Khoshdel E, Wooley KL (2007) Synthesis of core-crosslinked nanoparticles with controlled cylindrical shape and narrowly-dispersed size via core-shell brush block copolymer templates. Adv Mater 19:2830–2835
46. Berda EB, Baughman TW, Wagener KBJ (2006) Precision branching in ethylene copolymers: Synthesis and thermal behavior. Polym Sci Part A Polym Chem 44:4981–4989
47. Odian G (2004) Principles of polymerization, 4th edn. Wiley, New Jersey, p 832
48. Aiertza MK, Odriozola I, Cabañero G, Grande H-J, Loinaz I (2012) Single-chain polymer nanoparticles. Cell Mol Life Sci 69:337–346
49. Frank PG, Tuten BT, Prasher A, Chao D, Berda EB (2014) Intra-chain photodimerization of pendant anthracene units as an efficient route to single-chain nanoparticle fabrication. Macromol Rapid Commun 35:249–253. doi:
50. Becker HD (1993) Unimolecular photochemistry of anthracenes. Chem Rev 93:145–172
51. Sun T, Chance RR, Graessley WW, Lohse DJ (2004) A study of the separation principle in size exclusion chromatography. Macromolecules 37:4304–4312
52. Grubisic Z, Rempp P, Benoit HJ (1967) A universal calibration for gel permeation chromatography. Polym Sci Part B Polym Lett 5:753–759
53. Moore JCJ (1964) Gel permeation chromatography. I. A new method for molecular weight distribution of high polymers. Polym Sci Part A Gen Pap 2:835–843
54. Wyatt PJ (1993) Light scattering and the absolute characterization of macromolecules. Anal Chim Acta 272:1–40
55. Beck JB, Killops KL, Kang T, Sivanandan K, Bayles A, Mackay ME, Wooley KL, Hawker CJ (2009) Facile preparation of nanoparticles by intramolecular crosslinking of isocyanate functionalized copolymers. Macromolecules 42:5629–5635
56. Kim Y, Pyun J, Fréchet JMJ, Hawker CJ, Frank CW (2005) The dramatic effect of architecture on the self-assembly of block copolymers at interfaces. Langmuir 21:10444–10458
57. Croce TA, Hamilton SK, Chen ML, Muchalski H, Harth E (2007) Alternative o—quinodimethane cross-linking precursors for intramolecular chain collapse nanoparticles. Macromolecules 40:6028–6031

5

A kinetic model for ethylene oligomerization using zirconium/aluminum- and nickel/zinc-based catalyst systems in a batch reactor

Adil A. Mohammed · Seif-Eddeen K. Fateen · Tamer S. Ahmed · Tarek M. Moustafa

Abstract The aim of this work is to develop a kinetic model of the oligomerization of ethylene to linear alpha olefins (LAOs) for zirconium/aluminum and nickel/zinc catalyst systems. The development of such model helps in the study of the behavior of industrial LAOs reactors as well as in the optimization of their operation. The kinetic model was developed based on a four-step mechanism: site activation, initiation and propagation, chain transfer and site deactivation. A novel stochastic optimization algorithm, Intelligent Firefly Algorithm, was used to obtain the kinetic model parameters that best fit the available experimental data that were obtained from published sources. The values of the kinetic parameters were obtained for the developed kinetic models for two catalyst systems. The performance of the model with the estimated parameters was tested against the experimental data. The proposed kinetic model predicts the product distribution for the zirconium/aluminum catalyst system with suitable accuracy. The model can also predict the product distribution for the nickel/zinc catalyst system with good accuracy for all products. As expected, the accuracy of the model to predict the concentration of the higher carbon products decreases with the carbon number.

Keywords Ethylene · Oligomerization · Modeling · Zirconium/aluminum catalyst · Nickel/zinc catalyst

A. A. Mohammed · S.-E. K. Fateen · T. S. Ahmed
T. M. Moustafa
Chemical Engineering Department, Faculty of Engineering, Cairo University, Giza 12613, Egypt

Abbreviation

Notation

A	Pre-exponential factor
C_{CAT}	Catalyst concentration, mol/l
C^k_{CAT}	Active catalyst concentration, mol/l
$C_{CAT^k \cdot M}$	Complex active catalyst/ethylene concentration, mol/l
C_{decy}	Moles of deactivated catalysts, mol
C_M	Concentration of ethylene monomer, mol/l
C^k_M	Concentration of active ethylene monomer, mol/l
$C_{M^k \cdot TEA}$	Concentration of complex active ethylene monomer/co-catalyst, mol/l
C_{P_0}	Concentration of active site, mol/l
C_{P_i}	Concentration of living polymers, mol/l
$C_{P^k_i}$	Concentration of active living polymers, mol/l
$C_{P^k_i \cdot TEA}$	Concentration of complex active living polymers/co-catalyst, mol/l
C_{TEA}	Co-catalyst concentration, mol/l
C^k_{TEA}	Active co-catalyst concentration, mol/l
$C_{TEA^k \cdot CAT}$	Complex active co-catalyst/catalyst concentration, mol/l
$C_{TEA^k_1 \cdot CAT}$	Complex active co-catalyst/catalyst concentration-catalyst, mol/l
D	Moles of dead oligomer, mol
E	Activation energy, cal/mol
k_1	Rate constant of active site
k_2	Rate constant of chain initiation
k_3	Rate constant of chain propagation
k_4	Rate constant of chain transfer
k_5	Rate constant of deactivation
k_{+1}	Rate constant of attachment of the catalyst in the site activation
k_{-1}	Rate constant of detachment of the catalyst in the site activation

k_{+2}	Rate constant of attachment of the monomer in the site activation
k_{-2}	Rate constant of detachment of the monomer in the site activation
k_{+4}	Rate constant of attachment in the chain propagation reaction
k_{-4}	Rate constant of detachment in the chain propagation reaction
k_{+5}	Rate constant of attachment in the chain transfer reaction
k_{-5}	Rate constant of detachment in the chain transfer reaction
$k_c^{'}$	Rate constant of the first reaction of co-catalyst in the site activation
$k_c^{''}$	Rate constant of the second reaction of co-catalyst in the site activation
K_A	Equilibrium constant of the catalyst in the site activation
K_B	Equilibrium constant of the monomer in the site activation
K_C	Equilibrium constant of the co-catalyst in the site activation
K_D	Equilibrium constant in the chain propagation reaction
K_E	Equilibrium constant in the chain transfer reaction
M	Ethylene monomer
M^k	Active ethylene monomer
P_i	Living oligomers
P_i^k	Active living oligomers
P_0	Active site
R	Gas constant, mol/cal K
T	Temperature, K
T_r	Reference temperature, K
v	Volume, l

Greek letters

α Alpha position

Subscripts

c	Calculated mole fractions of product
e	Experimental mole fractions of product
CAT	Catalyst
CAT^k	Active catalyst
TEA	Co-catalyst
TEA^k	Active co-catalyst

Introduction

Linear alpha olefins (LAOs) are linear hydrocarbons having a double bond between the first and second carbon atoms. Linear α-olefins have found wide application in various areas of petrochemical synthesis. They are used for copolymerization with ethylene for the purpose of obtaining low- and high-density linear polyethylene and for the preparation of detergents and synthetic lubricants [12, 14, 41, 48, 54, 61, 62]. They can also be used in the production of surfactants, agents for enhanced oil recovery, corrosion inhibitors, high-performance lubricating oils, linear alkylbenzenes, oxo-synthesis alcohols, α-olefin sulfonates, oil additives, and as drilling fluids [1, 3, 6, 28, 35, 51, 59]. The worldwide growth rate of LAOs production between 1980 and 1990 amounted to 10–15 wt% per year. From 1990 to 2000, the annual increase in LAO production came to 4.5–5 % wt%. Their average annual demand over the next 10–15 years is expected to increase by 3.5 wt %. The total amounts of LAOs manufactured worldwide of years 1999, 2003, and 2005 were 2.6, 3.5, and 4.2 million tons, respectively [7].

Ethylene oligomerization is one of the vital industrial processes in which linear alpha olefins can be produced. The products of Ethylene oligomerization can be further separated into different LAOs based the carbon number of the product [9, 38]. Different industrial processes are used to produce LAOs and extensive research studies are under way on the oligomerization of ethylene in different catalysts and processes conditions. The companies that produce alpha olefins via ethylene oligomerization include Ethyl, Chevron, Shell, Idemitsu, Mitsubishi, IFP-Axens, UOP, and Sabic-Linde [7, 15, 38, 45]. These processes are catalyzed by different Ziegler–Natta catalysts. Investigations of catalytic oligomerization of ethylene have been mainly focused on the complexes of titanium, zirconium, chromium, and nickel for decades. Complexes of titanium and nickel are the most often used catalysts for ethylene oligomerization [47, 49, 58]. Zirconium complexes have also been found to be very active catalyst in this reaction [5, 33, 34, 42]. The use of other transition metal complexes in ethylene oligomerization is uncommon. [3, 4, 17, 23, 30, 33, 43, 44, 53, 63]

Several authors proposed models to describe the kinetics of the polymerization processes on different catalysts. Zacca and Ray [57] discussed a new model for the polymerization of olefins in a loop reactor. Their reactor was modeled as a two-tubular interconnected reactor. They studied the effects of recycle rate, axial dispersion and heat transfer. Sharma and coworkers [8] developed a mathematical model for the isothermal, slurry polymerization of ethylene using solid Ziegler–Natta catalysts. They discussed the effect of gas liquid mass transfer limitations on the overall rate and polymer properties. Kiparissides [36] produced synthetic polymers via a multitude of reaction mechanisms and processes, including addition and step growth polymerizations. He provided an overview of the different polymerization processes and the mathematical modeling approaches and addressed the problems related with the computer-aided design, monitoring, optimization and control of polymerization reactors.

Dube et al. [19] developed a practical methodology for the computer modeling of multi-component chain growth polymerizations, which is applicable to many multi-component systems. Various co-monomer systems were used to illustrate the development of practical semi-batch and continuous reactor operational policies for the manufacture of copolymers with high quality and productivity. Hatzantonis et al. [24] reviewed recent developments in modeling gas phase olefin polymerization fluidized-bed (FBR) reactors in the presence of a multi-site Ziegler–Natta catalyst. In addition, they developed an FBR model to account for the effect of varying of bubble size with the bed height on the reactor dynamics and the molecular properties of the polymer product. They studied the effect of important reactor parameters on the dynamic and steady-state behavior of the FBR. Alhumaizi [2] developed a dynamic mathematical model for the ethylene dimerization reactor to simulate the distribution of butene-1 and other oligomers in the products under different operating conditions. He studied the effects of the recycle flow rate and the cooler heat transfer coefficient for optimum conditions for production of butene-1. Soares [46] reviewed the principal of mathematical modeling techniques for describing the microstructure of polyolefins produced by coordination polymerization. Fernandes and Lona [21, 22] developed a heterogeneous model describing the behavior of the three-phase fluidized-bed reactor in polymer production. The model incorporated the interactions between phases and also predicted the physicochemical properties of the polymer.

Yoon et al. [56] reviewed recent developments in modeling techniques for the calculation of polymer properties and the application of process models to the design of model-based reactor optimizations and controls. Jin et al. [29] developed a mathematical model to simulate the ethylene polymerization in continuous stirred tank reactor. The molecular properties of various polymers were calculated based on the method of moment. Valencia and Soares [52] developed a simulation model for the polymerization of ethylene in a process with (*n*) reactors working in series, which could predict raw material conversions and product properties. Their model parameters were obtained from laboratory data. Zhang et al. [60] developed a mathematical model to describe the effect of different polymerization conditions on the preparation of ethylene-1-hexene copolymerization with a tandem catalysis system. The general feature of the model simulation agreed well with both their experimentation data and other results reported in the literature. Ghasem and coworkers [26, 27] developed a modified dynamic three-phase structure model that took into account the presence of particles participating in the reaction with emulsion and catalyst phases. The control system was discussed using of neural network controller. In addition, they modified the model to include heat and mass transfer between the bubble and the cloud as well as between the cloud and the emulsion phases. Touloupides et al. [50] developed a comprehensive mathematical model, which was capable of simulating the dynamic operation of an industrial slurry phase cascade loop reactor series under different plant operating conditions. They used multi-site ZN kinetic and method of moments for calculation of polymer molecular properties.

However, to the best of our knowledge, there has been no published study that proposed kinetic models for the oligomerization of ethylene with zirconium-based catalyst with aluminum-based co-catalyst or with DPA(Na)/ $NiCl_2{\cdot}6H_2O/Zn$ catalyst system. The objective of this work is to develop two kinetic models for the said catalyst systems. The remainder of this paper is divided as follows. "Model formulation" introduces the model formulation. The experimental data are introduced briefly in "Experimental data". "Parameters estimation" describes the methodology of estimating the parameters of the kinetic model using the Intelligent Firefly Algorithm. "Result and discussion" presents the results and discusses its significance. The conclusions of this study are summarized in "Conclusion".

Model formulation

Many studies focused on the kinetics of olefin polymerization using Ziegler–Natta catalysts. The polymerization reactions occur at several reactive sites on the catalyst. Due to the complexity of the process of polymerization and oligomerization of olefins with Ziegler–Natta catalysts, the mechanism of the reaction is usually divided into a number of reaction steps [10, 11, 13, 16, 25, 31, 32, 37, 39, 57]. These steps are:

Site activation

The active site is formed by the reaction between the catalyst with co-catalyst. The co-catalyst acts as an alkylating and reducing agent.

Initiation and propagation

The active site having a coordination vacancy attracts the electrons in the olefin π-bond. Coordination is followed by the insertion into the polymer chain (R) and the re-establishment of the coordination vacancy for further monomer insertion.

Chain transfer

Several chain transfer mechanisms are applied in Z–N catalyst such as transfer by β-hydride elimination, transfer to monomer, and transfer to co-catalyst.

Site deactivation

For some polymerizations, the rate of reaction decreases with time due to the catalyst deactivation reaction. The two active sites form the deactivated catalyst, which is inactive for monomer oligomerization. Another deactivation reaction forms a dead polymer chain and a dead catalyst site.

Model mechanism

The proposed model is limited to the production of α-olefins up to C_{20}. The oligomerization reaction can be broken into five parts; catalyst activation, chain initiation, chain propagation, chain transfer and catalyst deactivation.

Model features

The following points summarize the main features of the model.

1. The concentrations of the catalyst (CAT), ethylene (M), co-catalyst (TEA), and oligomers (P_i^k) are divided into active and inactive stages.
2. The catalyst, ethylene, and co-catalyst are involved in the reactions as follows:

$$\text{active catalyst}\left(\mathrm{CAT}^k\right) + \text{ethylene}(\mathrm{M}) \overset{K_\mathrm{A}}{\leftrightarrow} \mathrm{CAT}^k \cdot \mathrm{M}$$

$$\text{at equilibrium}\, k_{+1} * C_{\mathrm{CAT}}^k * C_\mathrm{M} - k_{-1} C_{\mathrm{CAT}^k \cdot \mathrm{M}} = 0$$

$$\therefore C_{\mathrm{CAT}^k \cdot \mathrm{M}} = K_\mathrm{A} C_{\mathrm{CAT}}^k * C_\mathrm{M}, K_\mathrm{A} = \frac{k_{+1}}{k_{-1}} \quad (1)$$

$$\text{balance of } C_\mathrm{CAT} : C_\mathrm{CAT} = C_\mathrm{CAT}^k + C_{\mathrm{CAT}^k \cdot \mathrm{M}}$$

$$C_\mathrm{CAT}^k = \frac{C_\mathrm{CAT}}{K_\mathrm{A} C_\mathrm{M} + 1}$$

$$\text{active ethylene}\left(\mathrm{M}^k\right) + \text{cocatalyst (TEA)} \overset{K_\mathrm{B}}{\leftrightarrow} \mathrm{M}^k \cdot \mathrm{TEA}$$

$$\text{at equilibrium}\, k_{+2} * C_\mathrm{M}^k * C_\mathrm{TEA} - k_{-2} C_{\mathrm{M}^k \cdot \mathrm{TEA}} = 0$$

$$\therefore C_{\mathrm{M}^k \cdot \mathrm{TEA}} = K_\mathrm{B} C_\mathrm{M}^k * C_\mathrm{TEA}, K_\mathrm{B} = \frac{k_{+2}}{k_{-2}}$$

$$\text{balance of } C_\mathrm{M} : C_\mathrm{M} = C_\mathrm{M}^k + C_{\mathrm{M}^k \cdot \mathrm{TEA}}$$

$$C_\mathrm{M}^k = \frac{C_\mathrm{M}}{K_\mathrm{B} C_\mathrm{TEA} + 1} \quad (2)$$

$$\text{active cocatalyst}\left(\mathrm{TEA}^k\right) + \text{catalyst(CAT)} \overset{K_\mathrm{c}'}{\rightarrow} \mathrm{TEA}^k \cdot \mathrm{CAT}$$

$$\text{active complex } \left(\mathrm{TEA}^k \cdot \mathrm{CAT}\right) + \text{catalyst(CAT)} \overset{K_\mathrm{c}''}{\rightarrow} \mathrm{TEA}_1^k \cdot \mathrm{CAT}$$

$$C_{\mathrm{TEA}^k \cdot \mathrm{CAT}} = k_\mathrm{c}' C_\mathrm{TEA}^k C_\mathrm{CAT}$$

$$C_{\mathrm{TEA}_1^k \cdot \mathrm{CAT}} = k_\mathrm{c}'' C_{\mathrm{TEA}^k \cdot \mathrm{CAT}} C_\mathrm{CAT} = k_\mathrm{c}'' k_\mathrm{c}' C_\mathrm{TEA}^k C_\mathrm{CAT}^2$$

$$\begin{aligned} \text{balance of } C_\mathrm{TEA} : C_\mathrm{TEA} &= C_\mathrm{TEA}^k + C_{\mathrm{TEA}^k \cdot \mathrm{CAT}} + C_{\mathrm{TEA}_1^k \cdot \mathrm{CAT}} \\ &= C_\mathrm{TEA}^k + k_\mathrm{c}' C_\mathrm{TEA}^k C_\mathrm{CAT} + k_\mathrm{c}'' k_\mathrm{c}' C_\mathrm{TEA}^k C_\mathrm{CAT}^2 \\ &= C_\mathrm{TEA}^k \left(1 + k_\mathrm{c}' C_\mathrm{CAT} + k_\mathrm{c}'' k_\mathrm{c}' C_\mathrm{CAT}^2\right) \end{aligned} \quad (3)$$

$$\text{Assuming that}\, k_\mathrm{c}' = 2K_\mathrm{c}, k_\mathrm{c}'' = \frac{k_\mathrm{c}'}{4}$$

$$C_\mathrm{TEA}^k = \frac{C_\mathrm{TEA}}{(K_\mathrm{C} C_\mathrm{CAT} + 1)^2}$$

3. The oligomers of propagation proceed as in the following reaction

$$\mathrm{TEA} + \text{active oligomers}\left(P_i^k\right) \overset{K_\mathrm{D}}{\leftrightarrow} P_i^k \cdot \mathrm{TEA}$$

$$\text{at equilibrium}\, k_{+4} * C_{P_i}^k * C_\mathrm{TEA} - k_{-4} C_{P_i^k \cdot \mathrm{TEA}} = 0$$

$$\therefore C_{P_i^k \cdot \mathrm{TEA}} = K_\mathrm{D} C_{P_i}^k * C_\mathrm{TEA}, K_\mathrm{D} = \frac{k_{+4}}{k_{-4}}$$

$$\text{balance of } C_{P_i} : C_{P_i} = C_{P_i}^k + C_{P_i^k \cdot \mathrm{TEA}} C_{P_i^k} - \frac{C_{P_i}}{K_\mathrm{D} C_\mathrm{TEA} + 1} \quad (4)$$

4. The oligomers of chain transfer proceed as in the following reaction:

$$\mathrm{TEA} + \text{active oligomers } \left(P_i^k\right) \overset{K_\mathrm{E}}{\leftrightarrow} P_i^k \cdot \mathrm{TEA}$$

$$\text{at equilibrium}\, k_{+5} * C_{P_i}^k * C_\mathrm{TEA} - k_{-5} C_{P_i^k \cdot \mathrm{TEA}} = 0$$

$$\therefore C_{P_i^k \cdot \mathrm{TEA}} = K_\mathrm{E} C_{P_i}^k * C_\mathrm{TEA}, K_\mathrm{E} = \frac{k_{+5}}{k_{-5}} \quad (5)$$

$$\text{balance of } C_{P_i} : C_{P_i} = C_{P_i}^k + C_{P_i^k \cdot \mathrm{TEA}}$$

$$C_{P_i^k} = \frac{C_{P_i}}{K_\mathrm{E} C_\mathrm{TEA} + 1}$$

5. The activator affects the rate equations of the chain propagation and chain transfer by rate law of C_TEA^a and C_TEA^b, respectively.
6. Only one single site is involved in the catalytic process.
7. The co-catalyst does not deactivate.
8. The only transfer mechanism included is the transfer to monomer. The transfer reactions are assumed to produce monomer (M) and catalyst (C_CAT).

The mechanism of the reaction is as follows:

Site activation

The active catalyst (CAT^k), active ethylene (M^k), and active co-catalyst $\left(TEA^k\right)$ react to form active site (P_0). The active center for olefins in Ziegler–Natta catalyst systems is a complex acting as an alkylating and reducing agent.

$$C^k_{\mathrm{CAT}} + C^k_{\mathrm{M}} + \frac{1}{2} C^k_{\mathrm{TEA}} \xrightarrow{k_1} P_0 + C^k_{\mathrm{M}} + \frac{1}{2} C^k_{\mathrm{TEA}} \\ r_1 = k_1 C^k_{\mathrm{CAT}} C^k_{\mathrm{M}} \left(C^k_{\mathrm{TEA}}\right)^{\frac{1}{2}} \tag{6}$$

Chain initiation

After the catalyst is activated, it is ready to polymerize with monomer. The active site that has a coordination vacancy reacts with an ethylene monomer to form a live oligomer chain of length one (P^k_1).

$$P_0 + \mathrm{M} \xrightarrow{k_2} P^k_1 \quad r_2 = k_2 C_{P_0} C_{\mathrm{M}} \tag{7}$$

Chain propagation

Coordination is followed by the insertion of a monomer into the chain, which increases the chain length by one monomer unit until chain transfer takes place.

$$P^k_i + \mathrm{M} \xrightarrow{k_3} P_{i+1} \quad r_3 = k_3 C_{P^k_i} C_{\mathrm{M}} C^a_{\mathrm{TEA}} \tag{8}$$

Chain transfer

Chain transfer to monomer may take place if the growing chain abstracts an atom from an unreacted monomer existing in the reaction medium. This leads to the death of the active chain (P^k_i) to form a dead chain (De_i) and begins a new one.

$$P^k_i + \mathrm{M} \xrightarrow{k_4} \mathrm{De}_i + \mathrm{M} + C_{\mathrm{CAT}} \quad r_4 = k_4 C_{P^k_i} C_{\mathrm{M}} C^b_{\mathrm{TEA}} \tag{9}$$

Site deactivation

Catalyst deactivation happens when active sites form a stable complex that is inactive for monomer oligomerization. Most Ziegler–Natta catalysts deactivate to form deactivation site (C_d) and/or dead chain (De_i).

$$P_0 \xrightarrow{k_5} C_{\mathrm{d}} \quad r_{5,0} = k_5 C_{P_0} \tag{10}$$

$$P_i \xrightarrow{k_5} C_{\mathrm{d}} + \mathrm{De}_i \quad r_{5,i} = k_5 P_i \tag{11}$$

Kinetic rate constants are known to be temperature dependent; this dependence is usually described using Arrhenius relationships. The comparative form of the Arrhenius equation was used to estimate kinetic rate constants to improve the convergence in parameter estimation; the rate constants take the form:

$$k_i = \mathrm{A}_i \exp\left(\frac{\mathrm{E}_i}{\mathrm{R}}\left(\frac{1}{\mathrm{T}} - \frac{1}{\mathrm{T}_r}\right)\right) \tag{12}$$

Population balance equations in batch reactors

Population balances are defined for all active species in a batch reactor. Population balance equations describe how the concentrations of living and dead chains of different lengths vary in time during the polymerization in a batch reactor. The material balance for driving population balances in batch reactor for the volume reactor (v) is:

rate of accumulation = rate of generation by chemical reaction

Population balance for active site:

$$\frac{1}{v}\frac{dP_0}{dt} = \left(r_1 - r_2 - r_{5,0}\right) \tag{13}$$

For living oligomer chains with $i \geq 2$, the following population balance can be derived:

$$\frac{1}{v}\frac{dP_i}{dt} = \left(r_{3,i-1} - r_{3,i} - r_{5,i}\right) \tag{14}$$

For living oligomer chains with $i = 1$, a slightly different equation is applied:

$$\frac{1}{v}\frac{dP_1}{dt} = \left(r_2 - r_{3,1} - r_{5,1}\right) \tag{15}$$

Similarly, the equations can be derived for dead oligomer formation for $i = 2$–10

$$\frac{1}{v}\frac{dD_i}{dt} = \left(r_{4,i} + r_{5,i}\right) \tag{16}$$

The concentration of the catalyst deactivation can be derived as follows:

$$\frac{1}{v}\frac{dC_{\mathrm{decy}}}{dt} = \sum_{i=1}^{10} r_{5i} \tag{17}$$

Experimental data

Kinetic modeling in the current study was performed using experimental data obtained from the literature. All experimental data were done in slurry batch reactors. Two sets of experimental data were used in this study to estimate the model parameters for two different catalyst systems. Experimental data (1) are for zirconium-based catalyst and aluminum-based co-catalyst system with molar ratio of 17–45 between catalyst and co-catalyst, range of temperature 65–95 °C and pressure 2–4 MPa [40]. Experimental data (2) are for DPA(Na)/$NiCl_2 \cdot 6H_2O$/Zn catalyst system with ratio of 1.5 between DPA

(Na) and $NiCl_2 \cdot 6H_2O$ and a molar ratio of 2–20 between zinc and $NiCl_2 \cdot 6H_2O$ at 95 °C and 5.5–6.0 MPa [18].

Parameters estimation

Regression was performed to obtain the parameters that best fit the model to the experimental data through minimization of the difference between experimental points and model predictions. The objective function for estimation of the model parameters is based on the difference in mole fractions of the products, and it takes the form

$$\text{Min} \quad \sum_{i=1}^{n}\sum_{j=1}^{m}\left(y_{e,i,j} - y_{c,i,j}\right)^2\Bigg) \tag{18}$$

The minimization routine used was a stochastic global optimization routine named the Intelligent Firefly Algorithm (IFA) [20]. Stochastic global optimization algorithms have significant advantages over deterministic optimization methods and thereby deemed more appropriate for our application. IFA is a metaheuristic algorithm, inspired by the flashing behavior of fireflies. The primary purpose for a firefly's flash is to act as a signal system to attract other fireflies, a metaheuristic is a procedure designed to find a good solution to a difficult optimization problem. The Firefly algorithm was developed by Xin-She Yang [55] and it is based on idealized behavior of the flashing characteristics of fireflies. All fireflies are unisex, so that one firefly is attracted to other fireflies regardless of their sex. Attractiveness is proportional to their brightness, thus for any two flashing fireflies, the less bright one will move towards the brighter one. The attractiveness is proportional to the brightness and they both decrease as their distance increases. If no one is brighter than a particular firefly, it will move randomly. The brightness or light intensity of a firefly is determined by the value of the objective function of a given problem.

Results and discussion

The values of the kinetic parameters for both experimental data used are tabulated and discussed as follows.

Results of experimental data for zirconium/aluminum catalyst system

Values of the proposed model kinetic parameters of experimental data for zirconium/aluminum catalyst system are shown in Table (1).

Table 1 Parameters estimated value for experimental data (1)

Parameter	Estimated value	Unit	Expression
A1	4.525 × 105	l/(mol s)	A factor of activation
E1	3.985 × 103	cal/mol	Activation energy of activation and deactivation
A2	5.101 × 108	l/(mol s)	A factor of initiation
E2	5.267 × 104	cal/mol	Activation energy of initiation
A3	4.949 × 108	l/(mol s)	A factor of propagation
E3	1.514 × 104	cal/mol	Activation energy of propagation
A4	4.650 × 105	l/(mol s)	A factor of Chain transfer
E4	1.433 × 104	cal/mol	Activation energy of Chain transfer
A5	1.255	1/s	A factor of deactivation
K_A	4.812 × 102	l/mol	Equilibrium constant of the catalyst in the site activation
K_B	2.771 × 102	l/mol	Equilibrium constant of the monomer in the site activation
K_C	6.906 × 103	l/mol	Equilibrium constant of the co-catalyst in the site activation
K_D	4.083 × 104	l/mol	Equilibrium constant in the chain propagation reaction
K_E	5.955 × 101	l/mol	Equilibrium constant in the chain transfer reaction
a	1.116		
b	0.954		

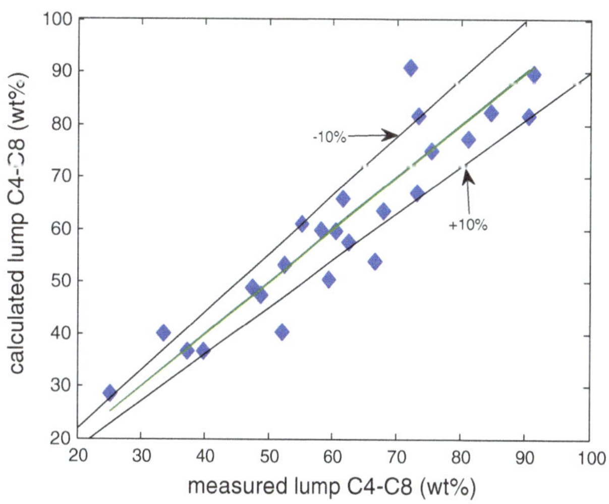

Fig. 1 Comparison calculated and measured data of lumped C_4, C_6 and C_8 product

Figure (1) shows the parity plot for lumped butane, hexane and octane using the estimated parameters. Each point represents one experimental run. Almost 80 % of the points are predicted with <10 % error lines. Figure (2) shows the experimental and calculated points for C_{10+}. It is clear that many runs are predicted outside the 20 % error boundary lines.

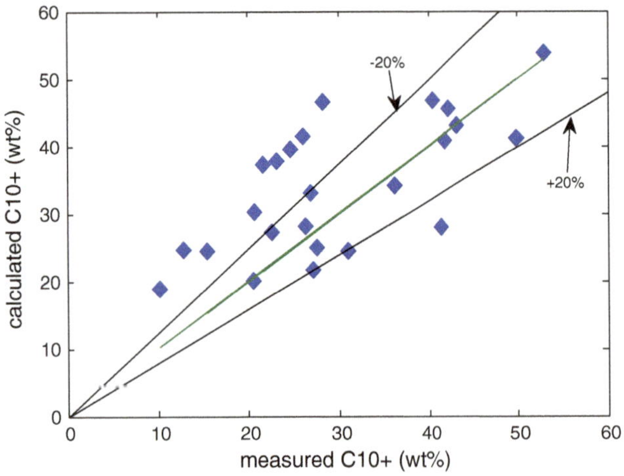

Fig. 2 Comparison calculated and measured data of C_{10+} product

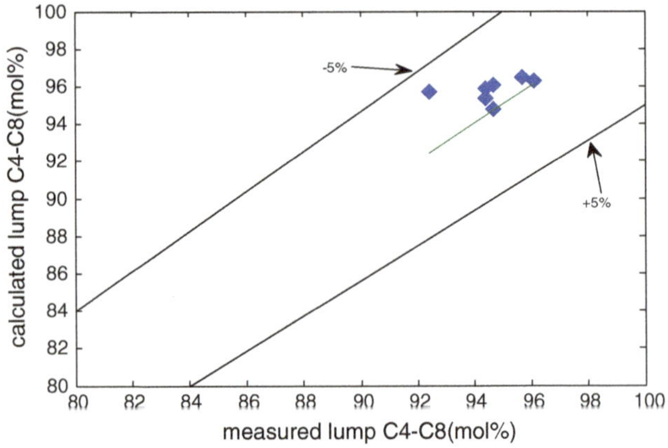

Fig. 3 Comparison calculated and measured data of lumped C_4, C_6 and C_8 product

Table 2 Kinetic parameters of experimental data (2)

Parameter	Estimated value	Unit	Expression
A1	4.447 × 106	l/(mol s)	A factor of activation
E1	1.455 × 103	cal/mol	Activation energy of activation and deactivation
A2	6.842 × 108	l/(mol s)	A factor of initiation
E2	4.757 × 104	cal/mol	Activation energy of initiation
A3	5.928 × 105	l/(mol s)	A factor of propagation
E3	1.237 × 104	cal/mol	Activation energy of propagation
A4	2.634 × 105	l/(mol s)	A factor of Chain transfer
E4	1.137 × 104	cal/mol	Activation energy of Chain transfer
A5	0.973	1/s	A factor of deactivation
K_A	8.546 × 104	l/mol	Equilibrium constant of the catalyst in the site activation
K_B	1.464 × 104	l/mol	Equilibrium constant of the monomer in the site activation
K_C	9.750 × 103	l/mol	Equilibrium constant of the co-catalyst in the site activation
K_D	7.946 × 104	l/mol	Equilibrium constant in the chain propagation reaction
K_E	9.604 × 103	l/mol	Equilibrium constant in the chain transfer reaction
a	0.845		
b	0.919		

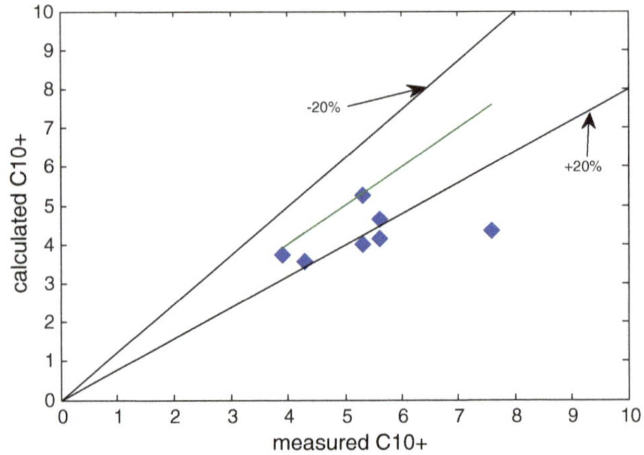

Fig. 4 Comparison calculated and measured data of C_{10+} product

Results for experimental data for nickel/zinc catalyst system

Values of the proposed models' kinetic parameters for experimental data for nickel/zinc catalyst system are shown in Table (2).

Figure (3) shows the parity plot of lumped butane, hexane and octane using the estimated parameters. All the points lay between the 5 % error lines for all three products. Figure (4) shows the experimental and calculated points for C_{10+} production. Some of the runs are predicted outside the 20 % error boundary lines.

The results show that the proposed model predicts both experimental data for zirconium/aluminum and nickel/zinc catalyst systems in a different way. The prediction of the model is better for lump C_4–C_8 compared to C_{10+}. For lump C_4–C_8, the model predictions for the zirconium/aluminum catalyst system lie mostly between the 10 % confidence limits. On the other hand, for the nickel/zinc catalyst system, the model predictions for lump C_4–C_8 lie between the 5 % confidence limits. As for the C_{10+}, the prediction of the model is lower for both systems. In both catalyst systems, most of the points lie between the 20 % confidence limits.

The results show that the proposed model fits the experimental data for nickel/zinc catalyst system with higher accuracy as compared to the experimental data for zirconium/aluminum catalyst system. This may be attributed to that for the zirconium/aluminum catalyst system, the experimental data for the reaction are available over a wide temperature range, while for the nickel/zinc catalyst system, the experimental data are only available at a constant temperature.

Finally, the model prediction is better for low carbon atoms. As the number of carbon atoms increases, the concentration of the polymer decreases and the model fails to predict the product concentration. The formation of higher carbon number polymer is a complex process that may not be well represented by this simple model. Relaxing some of the model assumptions may positively affect the results of its prediction.

Conclusions

A kinetic model of the oligomerization of ethylene was proposed in this study. Two sets of experimental data were used for the estimation of the model parameters. The performance of the model with estimated parameters has been tested against the experimental data. The proposed model predicts the product distribution for experimental data for zirconium/aluminum catalyst system with a suitable accuracy. The model predictions of the concentration for the components of C_4 up to C_8 fit within the 10 % confidence limits, while the absolute errors for C_{10+} were higher between the 20 % confidence limits. Similarly, the model could predict the experimental data for nickel/zinc catalyst system well. The prediction accuracy was better in comparison to the prediction of the zirconium/aluminum catalyst kinetic model. This may be attributed to that for the zirconium/aluminum catalyst system, the experimental data for the reaction are available over a wide temperature range, while for the nickel/zinc catalyst system, the experimental data are only available at a constant temperature. Finally, the model prediction is better for low carbon atoms. Relaxing some of the model assumptions may positively affect the results of its prediction. In all cases, the importance of the model lies in its ability to predict the concentration of the low carbon number oligomers.

References

1. Agapie T (2011) Selective ethylene oligomerization: recent advances in chromium catalysis and mechanistic investigations. Coord Chem Rev 255:861–880
2. Alhumaizi KI (2000) Stability Analysis of the Ethylene Dimerization Reactor for the Selective Production of Butene-1. Chem Eng Res Des 78:492–497
3. Aliyev V, Mosa F, Al-Hazmi M (2010) Catalyst composition for oligomerization of ethylene oligomerization process and method for its preparation. US 2010/0292423 A1 1–5
4. Al-Jarallah AM, Anabtawi JA, Siddiqui MAB, Aitani AM, Al-Sa'doun AW (1992) Ethylene dimerization and oligomerization to butene-1 and linear α-olefins: a review of catalytic systems and processes. Catal Today 14:1–121
5. Attridger J, Jackson R, Maddock J, Thompson T (1973) Ethylene oligomerisation with zirconium arylalkyls and alkenyls. J Chem Soc Chem Commun 1973:132–133
6. Bahuleyan BK, Ahn IY, Appukuttan V, Lee SH, Ha C-S, Kim I (2010) Ethylene oligomerization by tridentate cobalt complexes bearing pendant donor modified α-diimine ligands. Macromol Res 18:701–704
7. Belov GP, Matkovsky PE (2010) Processes for the production of higher linear α-olefins. Pet Chem 50:283–289
8. Bhagwat MS, Bhagwat SS, Sharma MM (1994) Mathematical modeling of the slurry polymerization of ethylene: gas-liquid mass transfer limitations. Ind Eng Chem Res 33:2322–2330
9. Bianchini C, Giambastiani G, Rios IG, Mantovani G, Meli A, Segarra AM (2006) Ethylene oligomerization, homopolymerization and copolymerization by iron and cobalt catalysts with 2,6-(bis-organylimino)pyridyl ligands. Coord Chem Rev 250:1391–1418
10. Bohm LL (1978) Ethylene polymerization process with a highly active Ziegler–Natta catalyst: 1. Kinet Polym (Guildf) 19:553–561
11. Bohm LL (1978) Reaction model for Ziegler–Natta polymerization processes. Polym (Guildf) 19:545–552
12. Cai T (1999) Studies of a new alkene oligomerization catalyst derived from nickel sulfate. Catal Today 51:153–160
13. De Carvalho AB, De Gloor PE, Hamielec AE (1989) A kinetic mathematical model for heterogeneous Ziegler–Natta copolymerization. Polym (Guildf) 30:280–296
14. Chandran D, Kwak CH, Oh JM, Ahn IY, Ha C-S, Kim I (2008) Ethylene oligomerizations by sterically modulated salicylaldimine cobalt(II) complexes combined with various alkyl aluminum cocatalysts. Catal Lett 125:27–34
15. Choo H, Kevan L (2001) Catalytic study of ethylene dimerization on Ni (II) -exchanged clinoptilolite. J Phys Chem B 105:6353–6360
16. Deffieux A, Fontanille M, Ribeiro MR, Portela MF (1996) Kinetic investigation of parameters governing the high-temperature polymerization of ethylene initiated by supported VCl3 catalytic systems. Eur Polym J 32:811–819
17. Dixon JT, Green MJ, Hess FM, Morgan DH (2004) Advances in selective ethylene trimerisation—a critical overview. J Organomet Chem 689:3641–3668
18. Dong-bing Liu DL (1998) Catalytic oligomerization of ethylene to lower α-olefins by the catalyst system DPA(Na)/NiC12.6H2O/Zn. Appl Catal A Gen 166:L255–L258
19. Dube MA, Soares JBP, Penlidis A, Hamielec AE (1997) Mathematical modeling of multicomponent chain-growth polymerizations in batch, semibatch, and continuous reactors: a review. Ind Eng Chem Res 36:966–1015
20. Fateen S-EK, Bonilla-Petriciolet A (2013) Intelligent firefly algorithm for global optimization. In: X. Yang (ed) Springer, Germany
21. Fernandes FAN, Lona LMF (2002) Heterogeneous modeling of fluidized bed polymerization reactors. Influence of mass diffusion into the polymer particle. Chem Eng Sci 26:841–848
22. Fernandes FAN, Lona LMF (2001) Heterogeneous modeling for fluidized-bed polymerization reactor. Chem Eng Sci 56:963–969

23. Hang W, Weidong Y, Tao J, Binbin L, Wenqing X, Jianjiang M, Youliang H (2002) Ethylene oligomerization by novel iron (II) diimine complexes/MAO. Chin Sci Bull 47:1616–1618
24. Hatzantonis H, Yiannoulakis H, Yiagopoulos A, Kiparissides C (2000) Recent developments in modeling gas-phase catalyzed olefin polymerization fluidized-bed reactors: the effect of bubble size variation on the reactor's performance. Chem Eng Sci 55:3237–3259
25. Huang J, Rempel GL (1995) Ziegler–Natta catalysts for olefin polymerization: mechanistic insights from metallocene systems. Prog Polym Sci 20:459–526
26. Ibrehem AS, Hussain MA, Ghasem NM (2009) Modified mathematical model for gas phase olefin polymerization in fluidized-bed catalytic reactor. Chem Eng J 149:353–362
27. Ibrehem PS, Hussain MA, Ghasem NM (2008) Mathematical model and advanced control for gas-phase olefin polymerization in fluidized-bed catalytic reactors. Chin J Chem Eng 16:84–89
28. Jiang T, Chen H, Ning Y, Chen W (2006) Preparation of 1-octene by ethylene tetramerization with high selectivity. Chin Sci Bull 51:521–523
29. Jin X, Hu P, Dong S, Yuan P (2004) steady-state modeling of ethylene polymerization. 5Ih World Congr. Intell Control Autom China pp 3493–3497
30. Jones D, Cavell K, Keim W (1999) Zirconium complexes as catalysts for the oligomerisation of ethylene: the role of chelate ligands and the Lewis acid cocatalyst in the generation of the active species. J Mol Catal A Chem 138:37–52
31. Asua José M (2007) Polymer reaction engineering. Blackwell Publishing Ltd, UK
32. Kazuo Soga TS (1997) Ziegler–Natta catalysts polymerizations. Prog Polym Sci 22:1503–1546
33. Khanmetov AA, Azizov AG, Ibragimova MD, Kuliev BV, Alieva RV, Kalbalieva ES, Bagirova SR, Mamedova RZ (2007) The multiple-site nature of ethylene polymerization catalysts based on zirconyl carboxylates. Pet Chem 47:176–183
34. Khanmetov AA, Azizov AG, Piraliev AG (2006) Ethylene oligomerization in the presence of catalyst systems based on mixed-ligand β-diketonatozirconium chlorides. Pet Chem 46:338–342
35. Kim I, Kwak CH, Kim JS, Ha C-S (2005) Ethylene oligomerizations to low-carbon linear α-olefins by structure modulated phenoxy-imine nickel(II) complexes combined with aluminum sesquichloride. Appl Catal A Gen 287:98–107
36. Kiparissides CA (1996) Polymerization reactor modeling: a review of recent developments and future directions. Chem Eng Sci 51:1637–1659
37. Kissin V, Mink I, Company MC, October R, Manuscript R, January R (1993) Ethylene oligomerization and chain growth mechanisms with Ziegler–Natta catalysts. Macromolecules 26:2151–2158
38. Fan-Hua Kong, Ming-He Xie (1999) Catalysts for 1-hexene Synthesis from Ethylene Oligomerization. J Nat Gas Chem 8:256–263
39. Meyer T, Keurentjes J (2005) Handbook of polymer reaction engineering, 1st edn. Wiley–VCH
40. Moustafa TM (2005) Unpublished data
41. Nelkenbaum E, Kapon M, Eisen MS (2005) Synthesis and molecular structures of neutral nickel complexes. Catalytic activity of the oligomerization of ethylene, and the dimerization of propylene. Organometallics 24:2645–2659
42. Oouchi K, Mitani M, Hayakawa M, Yamada T (1996) Ethylene oligomerization catalyzed with dichlorobis- (P-diketonato) zirconium/organoaluminium chloride systems. Macromol Chem Phys 1551:1545–1551
43. Shiraki Y, Nakamoto Y, Souma Y (2002) ZrCl 4 -TEA-EASC three-component catalyst for the oligomerization of ethylene: the role of organoaluminum co-catalysts and additives. J Mol Catal A Chem 187:283–294
44. Siedle AR, Lamanna WM, Newmark RA, Schroepfer JN (1998) Mechanism of olefin polymerization by a soluble zirconium. J Mol Catal A Chem 128:257–271
45. Skupiňsk J (1991) Oligomerization of α-olefins to higher oligomers. Chem Rev 91:613–648
46. Soares BP (2001) Mathematical modelling of the microstructure of polyoleÿns made by coordination polymerization: a review. Chem Eng Sci 56:4131–4153
47. de Souza RF, Bernardo-Gusmão K, Cunha GA, Loup C, Leca F, Réau R (2004) Ethylene dimerization into 1-butene using 2-pyridylphosphole nickel catalysts. J Catal 226:235–239
48. Su B, Feng G (2010) Influence of the metal centers of 2,6-bis(imino)pyridyl transition metal complexes on ethylene polymerization/oligomerization catalytic activities. Polym Int 59:1058–1063
49. Tembe GL, Ravindranathan M (1991) Oligomerization of ethylene to linear a-olefins by a titanium aryl oxide-alkylaluminum catalyst. Ind Eng Chem Res 30:2247–2252
50. Touloupides V, Kanellopoulos V, Pladis P, Kiparissides C, Mignon D, Van-Grambezen P (2010) Modeling and simulation of an industrial slurry-phase catalytic olefin polymerization reactor series. Chem Eng Sci 65:3208–3222
51. Turner AH (1983) Purity aspects of higher alpha olefins. J Am Oil Chem Soc 60:623–627
52. Valencia FP, Soares JBP (2007) Steady state simulation of ethylene polymerization using multiple-site coordination catalysts. Macromol Symp 259:110–115
53. Wang M, Li R, Qian M, Yu X, He R (2000) The effect of co-catalysts on the oligomerization and cyclization of ethylene catalyzed by zirconocene complexes. J Catal A Chem 160:337–341
54. Yang P, Yang Y, Zhang C, Yang X-J, Hu H-M, Gao Y, Wu B (2009) Synthesis, structure, and catalytic ethylene oligomerization of nickel(II) and cobalt(II) complexes with symmetrical and unsymmetrical 2,9-diaryl-1,10-phenanthroline ligands. Inorganica Chim Acta 362:89–96
55. Yang XS (2008) Firefly algorithm. Nature-inspired metaheuristic algorithms. Luniver Press, Frome, pp 79–90
56. Yoon WJ, Kim YS, Kim IS, Choi KY (2004) Recent advances in polymer reaction engineering: modeling and control of polymer properties. Korean J Chem Eng 21:147–167
57. Zacca JJ, Ray H (1993) Modelling of the liquid phase polymerization of olefins in loop reactors. Chem Eng Sci 48:3743–3765
58. Zaidman AV, Kayumov RR, Belov GP, Pervova IG, Lipunov IN, Kharlampidi KE (2010) Ethylene oligomerization in the presence of catalytic systems based on nickel(II) formazanates. Pet Chem 50:450–454
59. Zhang B, Wang Y, Wang G, Cao J, Sun S, Xing L, Sun Y, Han Y (2007) Oligomerization and polymerization of ethylene initiated by a novel Ni (II) -based acetyliminopyridine complexes as single-site catalysts. J Nat Gas Chem 16:64–69
60. Zhang J, Fan H, Li B-G, Zhu S (2008) Modeling and kinetics of tandem polymerization of ethylene catalyzed by bis(2-dodecylsulfanyl-ethyl)amine- and. Chem Eng Sci 63:2057–2065
61. Zhang Z, Chen S, Zhang X, Li H, Ke Y, Lu Y, Hu Y (2005) A series of novel 2,6-bis(imino)pyridyl iron catalysts: synthesis, characterization and ethylene oligomerization. J Mol Catal A Chem 230:1–8
62. Zhang Z, Zou J, Cui N, Ke Y, Hu Y (2004) Ethylene oligomerization catalyzed by a novel iron complex containing fluoro and methyl substituents. J Mol Catal A Chem 219:249–254
63. Zhukov VI, Val'kovich GV, Skorik IN, Petrov YM, Belov GP (2007) Ethylene oligomerization in the presence of ZrO(O-COR)2-Al(C2H5)2Cl-Modifier catalytic system. Pet Chem 47:49–54

6

Functionalized regio-regular linear polyethylenes from the ROMP of 3-substituted cyclooctenes

Henry Martinez · Jihua Zhang · Shingo Kobayashi · Yuewen Xu · Louis M. Pitet · Megan E. Matta · Marc A. Hillmyer

Abstract We demonstrated that the ring-opening metathesis polymerization (ROMP) of 3-substituted cyclooctenes bearing polar substituents allows the synthesis of highly regio- and stereo-regular polymers. A series of polyalkenamers with 90–99 % head-to-tail/*trans* configuration were synthetized in good yields (33–87 % yield). Upon saturation of the backbone using diimide, these polymers represent a class of linear polyethylene derivatives where the polar side chain is located on every eighth carbon. The thermal properties of both the saturated and unsaturated polymers depend strongly on the size and polarity of the functional side groups. The results presented here demonstrate that the 3-substituted cyclooctenes can be used not only for the synthesis of precisely functionalized polyethylene derivatives, but also to participate in the ring-opening metathesis copolymerization with unfunctionalized cyclic olefins to generate copolymers with tunable properties.

Keywords ROMP · Polyethylene · Substituted polyolefin · Metathesis

H. Martinez · J. Zhang · S. Kobayashi · Y. Xu ·
L. M. Pitet · M. E. Matta · M. A. Hillmyer
Department of Chemistry, University of Minnesota, 207 Pleasant St. SE, Minneapolis, MN 55455-0431, USA
e-mail: hillmyer@umn.edu

Introduction

Polyethylene (PE) is one of the most common plastics, the derivatives of which have found themselves useful in a variety of applications [1–3]. Incorporation of side chains, especially polar functional side chains, is desirable and often necessary to give the polymer desired properties and better performance. Due to the physical and chemical stability of PE, direct functionalization of PE is challenging [4]. Therefore, PE derivatives bearing functional side chains are typically produced by radical or coordination copolymerization of ethylene and functionalized olefin(s). These polymerizations, however, frequently suffer from dramatically different comonomer reaction rates, intolerance of catalyst to polar functional groups and lack of control of polymer microstructure. As a consequence, controlling the scope, content and distribution of functional side chains in the polymer is undermined. A number of methods have been developed to overcome these problems, and notable examples include development of compatible catalysts [5, 6] and metathesis polymerization of functionalized monomers. The latter method is interesting because of catalyst compatibility to polar functional groups and, more attractively, potential to precisely control the polymer structure. Wagener and coworkers have utilized acyclic diene metathesis polymerization (ADMET) [7] of functionalized α,ω-dienes to prepare linear PE derivatives bearing a broad range of polar functional groups [8–29]. Many of the ADMET-derived polymers feature precisely placed polar side chains when a symmetric monomer is used. As opposed to the condensation polymerization nature of ADMET, ring-opening metathesis polymerization (ROMP) is an addition polymerization which requires less demanding conditions and allows better control of the polymer molar mass [30]. The use of functionalized cyclic

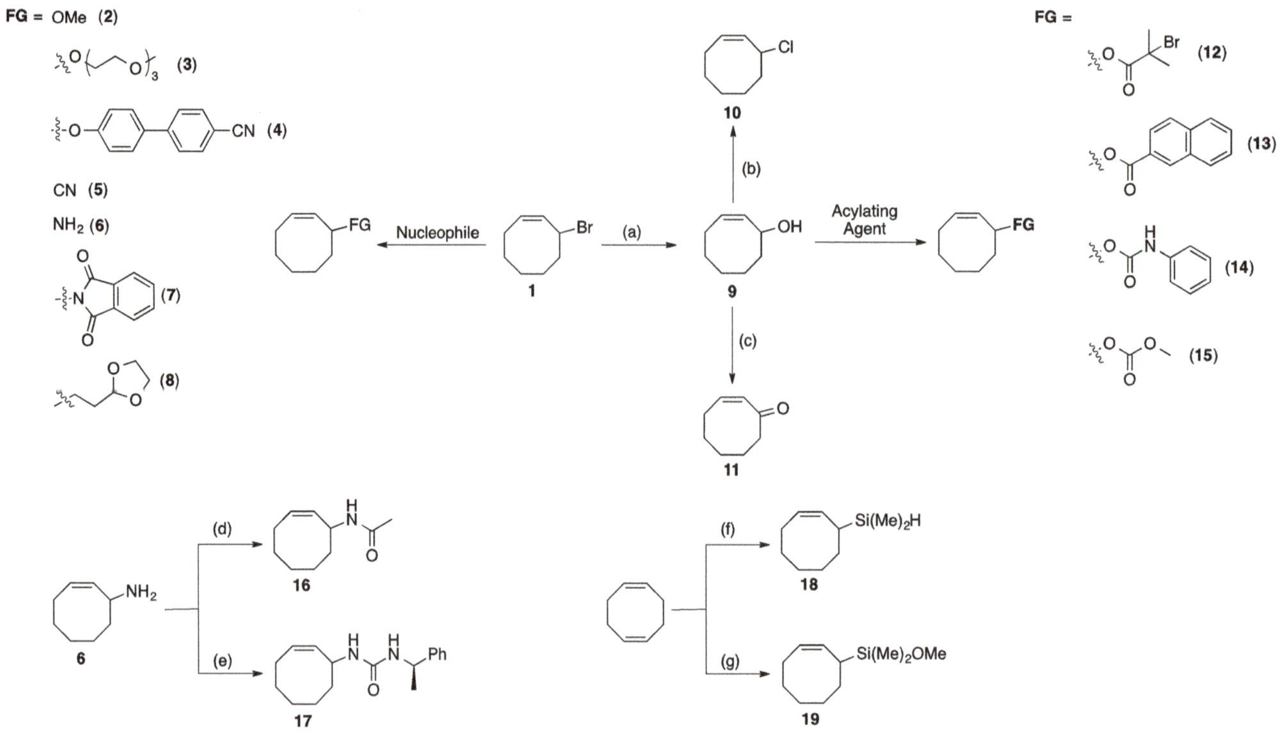

Scheme 1 Synthesis of various 3-substituted COEs. (*a*) Acetone/water, $NaHCO_3$, reflux 1 h. (*b*) $SOCl_2$, Py. (*c*) Pyridinium chlorochromate, CH_2Cl_2. (*d*) Et_3N/AcCOCl, CH_2Cl_2. (*e*) DMAP, isocyanate. (*f*) 1. $H_2PtCl_6.6H_2O$, $ClSiMe_2H$, Cy, 2. LAH, THF. (*g*) 1. $H_2PtCl_6.6H_2O$, $ClSiMe_2H$, Cy, 2. MeOH, Py

olefins in ROMP has also been explored by many research groups to access to linear PE derivatives with diverse polar side chains, including boranes, thioethers, halides, nitrile, hydroxyl, carbonyl, esters, amides, urethanes and liquid crystal moieties [31–54]. However, excluding those in which the monomer is symmetrically substituted, only few examples resulted in regio-regular products [46, 48].

We recently published a series of reports [55–59] demonstrating that single substitution at the 3-position (i.e., the allylic position) of *cis*-cyclooctene (COE) was an effective way to synthesize regio- and stereo-regular polyalkenamers by ROMP using the well-defined Grubbs second (**G2**) or third (**G3**) generation catalysts. Such substituents we have explored included hydrocarbon groups (3**R**COE, R = methyl, ethyl, hexyl or phenyl) [55, 56] and the acetoxy group [57–59]. All the reported polyalkenamers possess high levels (>95 %) of head-to-tail (HT) repeating unit connectivity and *E* double bond configuration. Hydrogenation of these polymers afforded PE derivatives with precisely positioned branches, which exhibit distinct physical properties compared to their random counterparts. More importantly, these reports suggest that ROMP of any COE substituted with a sufficiently bulky group at the 3-position could proceed in a regio- and stereo-selective manner, which in combination with the exceptional functional group tolerance of Ru-based catalysts allows us to conveniently synthesize diversely functionalized precision PE derivatives.

In the present study, we explore the scope of polar functional groups that, when placed at the 3-position (3**FG**COE, Scheme 1), are amenable to controlled ROMP with Ru-based catalysts. The polymer regio-regularity [head-to-tail (HT), head-to-head (HH) and tail-to-tail (TT)] and stereo-regularity (*cis* and *trans*) were analyzed in this study. These polymers, after the backbone is saturated, represent a class of linear PE derivatives where the polar side chain is located on every eighth carbon. The precision microstructure further enables investigation of the effect of side chain size and polarity on the polymer properties.

Results and discussion

Monomer synthesis

3-bromocyclooctene (**1**) was chosen as the principal starting material because of its scalable and highly efficient preparation. Several synthetic pathways were explored as shown in Scheme 1. The bromo group exhibited a wide range of reactivity towards the hydroxyl group in alcohols and phenol derivatives (compounds **2**–**4**). Stirring compound **1** in anhydrous methanol at room temperature afforded 3-methoxy-COE (**2**) in 24 h; and elevated temperature and extended reaction time was required to obtain compound **3** from its respective alcohol. However, no reaction of **1** with more hindered alcohols (e.g., secondary alcohols) or their deprotonated forms was observed, even after months. The phenoxide anion derived from phenol 4′-hydroxy-4-biphenylcarbonitrile, which is a liquid crystal building block, reacts readily with **1** in DMF to give 3-substituted COE with an aryl ether group (compound **4**). Other types of nucleophiles, including cyanide, ammonia, phthalimide and a Grignard reagent, reacted with **1** to generate the corresponding 3**FG**COEs (compounds **5–7**). 3-amino-COE (**6**) was used to synthesize 3**FG**COE having amide (**16**) or urea (**17**) groups substituted at the 3-position (Scheme 1).

Treatment of 3-hydroxy-COE (**9**) [59] with thionyl chloride in the presence of pyridine gave 3-chloro-COE (**10**). The hydroxyl group in **9** was also oxidized to give 2-cyclooctenone (**11**). Furthermore, a variety of ester or urethane groups were introduced through the hydroxyl group by acylation or reaction with isocyanate (3**FG**COE **12–15**).

Two silicon-containing monomers (**18** and **19**) were synthetized from hydrosilylation of *cis*-1,5-cyclooctadiene followed by nucleophilic attack of hydride or methanol to replace the chloride atom. The hydrosilylation at the 3 position was expected if taken into account that the isomerization of cyclic dienes from non-conjugated to conjugated dienes has been previously reported using H_2PtCl_6 [60]. Based on NMR <5 % of 5-substituted COE monomer was obtained using this method.

Polymer synthesis and characterization

The compatibility of these 3**FG**COEs with **G2** and the capability of these functional groups to give regio- and stereo-selective polymerizations were tested in ROMP under various conditions (Scheme 2) and the results are summarized in Table 1. To have a fair basis to compare the reactivity of these monomers, all polymerizations were carried out at the same temperature (60 °C) with comparable initial monomer concentration (1–2 M). Some 3**FG**COEs with highly polar functional groups (**4**, **7**, **16** and **17**) were essentially insoluble in $CHCl_3$ at the prescribed concentration; a co-solvent was employed in these cases to give a homogeneous solution. We attempted to use *cis*-4-octene as a chain transfer agent (CTA) to regulate the polymer molar mass as demonstrated in previous studies [55]. However, this effort only succeeded with some monomers while the others required higher catalyst loading to give polymers, suggesting these particular polar functional groups may inhibit the activity of **G2**.

Polymerization of ether-functionalized monomers **2** and **3** and carbonate monomer **15** under standard conditions resulted in polyalkenamers with the highest yields (87, 85 and 88 %, respectively) among all the monomers in this study. Polymerization and isolation of these polymers were straightforward. In contrast, polymerization of monomers **8** and **13** under standard conditions resulted in reaction mixtures containing considerable quantities of residual monomer, even after 24 h of reaction, suggesting a lower reactivity of such monomers under the chosen conditions. Initial attempts to polymerize monomer **11** yielded insoluble gels upon precipitation in methanol. However, the polymerization proceeded successfully in the presence of butylated hydroxytoluene (BHT). Additional BHT was added to a solution of the dissolved polymer immediately after precipitation to avoid subsequent cross-linking. Monomers bearing a cyclic imide (**7**), urethane (**14**) or aryl-cyanide group (**4**) were successfully polymerized, but required a relatively high catalyst loading.

Three N-containing monomers, **6**, **16** and **17**, as well as nitrile monomer **5**, did not generate any polymer using **G2** or the more reactive **G3**. In all cases, the reaction mixture turned a brown color within a few minutes, suggesting deactivation of the Ru catalyst. The failure of **5** or **6** was not surprising, as compounds containing nitrile groups or primary amines are known to deactivate the Ru catalysts [33, 36, 37, 61, 62]. However, the monomers bearing acetamide (**16**) and urea (**17**) functional groups did not polymerize either. Some studies suggested that amides and ureas were compatible with the Ru catalysts in both polymerizations [49–51] and small molecule reactions [63]. The reason for the inactivity in our trials is unknown.

The two Si-containing monomers, **18** and **19**, did not polymerize even at high catalyst loading. In both cases, the reaction mixture remained a yellow color even after 24 h, suggesting that the catalyst was active. Addition of COE after 24 h to both reaction mixtures yielded poly(COE) only in the solution that contained monomer **19**, which not

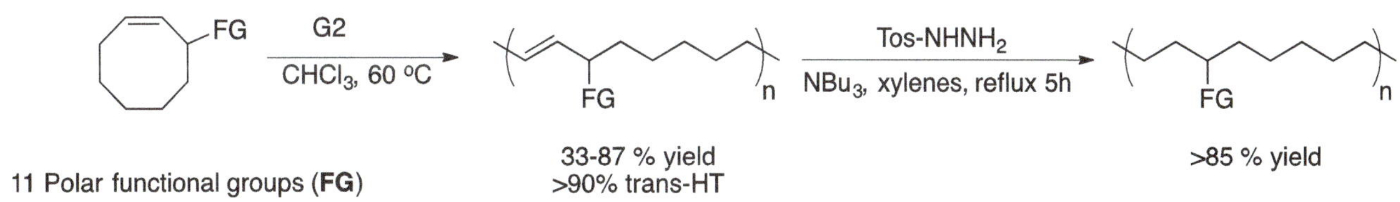

Scheme 2 Synthesis of linear polyethylene derivatives via ROMP followed by hydrogenation

Table 1 Polymers from ROMP of various 3**FG**COEs bearing polar functional groups using **G2**[a]

3**FG**COE	[G2][a] (mol %)	[CTA][a] (mol %)	Conv.[b] (%)	Yield[c] (%)	M_n (kg/mol)		*Đ*[e]	*E*-HT[f] (%)	T_g^g (°C)
					Cal.[d]	SEC[e]			
1[h]	0.20	N/A	–	69	–	–	–	–	–
2	0.04	1.0	>99	87	13.5	29.0	1.8	95	−54
3	0.10	1.0	>99	85	21.0	33.6	2.5	90	−69
4[i]	0.10	N/A	89	33	270	258	2.0	96	32
7[j]	0.10	N/A	98	71	250	240	1.4	>99	58
8	0.025	1.0	91	64	18.7	26.0	1.5	98	−34
10[h]	0.2	N/A	–	81	–	–	–	–	–
11[k]	0.025	1.0	92	73	12.1	27.6	1.7	>99	−26
13	0.025	1.0	66	42	18.1	15.3	2.0	>99	2
14	0.2	N/A	82	66	101	121	1.7	92	37
15	0.025	1.0	>99	88	18.0	38.3	1.9	97	−28

All polymerizations conducted at 60 °C with initial monomer concentration of 1–2 M in $CHCl_3$ if not otherwise stated

[a] Relative to monomer concentration, CTA = *cis*-4-octene

[b] Determined by ^1H NMR analysis of the reaction mixture

[c] Isolated yield

[d] M_n (cal) = MW (monomer) × Conv. × $[M]_0$/([G2] + [CTA])

[e] Determined by SEC in $CHCl_3$ at 30 °C versus polystyrene standards

[f] Estimated by ^1H NMR of the isolated polymer in $CDCl_3$

[g] Determined by DSC (2nd heating cycle) at 10 °C min^{-1}

[h] At room temperature

[i] 1:1 (v/v) $CHCl_3$:toluene

[j] 3:1 (v/v) $CHCl_3$:THF

[k] This polymer showed a T_m at 106 °C (ΔH = 64 J g^{-1})

only suggested that the catalyst was active, but also that the $SiMe_2OMe$ group in position 3 of the COE is bulky enough to inhibit the catalyst approach to the double bond. The fact that COE did not polymerize in the solution that contained monomer **18** might be due to oxidative addition of the Si–H group to the Ru center. Similar reactions have been used to perform hydrosilylation on terminal alkynes using Grubbs' first generation catalyst (**G1**) or **G2** catalyst [64, 65].

Both 3-hydroxy-COE (**9**) and 3-α-bromoisobutyrate-COE (**12**) [66] monomers quickly gave insoluble gels, indicating the formation of physically cross-linked networks possibly through strong hydrogen bonding (**9**) or covalent bonds (**12**) among polymer chains.

Polymers obtained from the halide-containing monomers (**1** and **10**) decomposed easily during the polymerization, generating dark insoluble rubbery materials. This result was not observed during the ROMP of 5-chloro or 5-bromo substituted COE [33–35], which suggests polymers of **1** and **10** are prone to halide elimination to generate conjugated dienes, possibly facilitated by the Ru catalyst.

Qualitatively, different rates of polymerization were observed with these 3**FG**COEs, most of which gave polymers in high or quantitative conversion. However, significantly lower conversions were obtained with 3**FG**COEs bearing aryl groups (**4** and **13**). This result is consistent with that reported on 3-phenyl COE, suggesting that bulky and rigid substituents result in lower polymerization rate. The polymer number average molar mass (M_n) determined by size exclusion chromatography (SEC) compared with polystyrene standards and the corresponding theoretical value were only in good agreement for the polymers containing aryl groups. This phenomenon can be attributed to the systematic difference of hydrodynamic radius between the polymer and the polystyrene standard, which is likely smaller in structurally similar polymers. The molar mass dispersity (*Đ*) ranged from 1.4 to 2.5, in accordance with the typical values for products from ROMP with extensive cross-metathesis.

The regio- and stereo-regularity of poly(3**FG**COE) samples was characterized by ^1H and ^{13}C NMR spectroscopy (see spectra in SI). Results comparable to those reported for our original poly(3**R**COE) [55, 59] samples were observed for every poly(3**FG**COE), namely high content of head-to-tail (HT) structure and *E* double bond configuration (Table 1). This result confirms our hypothesis that sufficiently bulky substituent at the 3-position gives rise to regio- and stereo-selective polymerization [55–59]. Thermal characterization using differential scanning calorimetry

Table 2 Characterization of hydrogenated polymers

	Yield[a] (%)	M_n^b (kg mol^{-1})	$Đ^b$	T_g^c (°C)	T_m^c (°C)	ΔH_m^c (J g^{-1})
PH2	80	36.8	1.6	−70	15[d]	25
PH3	85	34.1	1.9	−74	−5	25
PH7	90	38.3	4.1	37	–	–
PH8	92	24.9	1.6	−49	–	–
PH13	60	17.0	1.8	18	–	–
PH14	77	57.6	3.4	51	–	–
PH15	83	49.4	2.1	−32	49	0.4

[a] Isolated yield

[b] Determined by SEC in $CHCl_3$ at 30 °C versus polystyrene standards

[c] Determined by DSC (2nd heating cycle) at 10 °C min^{-1}

[d] Cold crystallization at −38 °C (ΔH = 25 J g^{-1})

(DSC) suggested nearly all poly(3**FG**COE) are amorphous with glass transition temperatures (T_g) in the range of −69 to 58 °C (Table 1). The T_g of these polyalkenamers vary substantially (from −69 to 58 °C) and are clearly dependent on the functional group identity. Poly(3**FG**COE) with relatively small and less polar functional groups (e.g., **P2**) consistently exhibit lower T_g than those with relatively bulky and more polar functional groups (e.g., **P7**). Interestingly, **P11** showed a T_m at 106 °C (ΔH = 64 J g^{-1}). When compared with the regio-irregular poly(cyclooct-4-enone) reported by Hillmyer et al. [33] in 1995 via ROMP using a Ru-based catalyst, **P11** has a higher T_g (−58 vs. −26 °C) and T_m (34 vs. 106 °C). This is consistent with the fact that at higher regio- and stereo-regularity both T_g and T_m increases [55, 59]. Similarly, the polymer microstructure has been shown to have profound influence on the thermal properties in precision polyethylenes prepared by ADMET [67].

Chemical hydrogenation of polyalkenamers using diimide has shown to be an effective way to generate saturated backbone [68]. We applied this method to hydrogenate poly(3**FG**COE), which resulted in a polymer equivalent to a copolymer of one part functionalized alpha olefin and three parts ethylene. With the exception of **P4** and **P11**, the corresponding hydrogenated polymers were obtained in decent yields (Table 2). **P4** has poor solubility in xylenes, while **P11** cross-linked while heating the solution of xylenes (even in the presence of large amounts of BHT) resulting in an insoluble gel. NMR analysis revealed the presence of the expected functional group in the expected content in the hydrogenated polymers (see SI), suggesting the reaction conditions are compatible with these functional groups. The clean and well-defined ^{13}C NMR spectra further support the regio-regular structure in the hydrogenated polymers as well as in their unsaturated precursors. With the exception of **P7** and **P14**, SEC in $CHCl_3$ versus polystyrene standard showed slight increase of molar mass and comparable dispersity for all hydrogenated polymers. Hydrogenated polymers of **P7** and **P14** (PH**7** and PH**14**, respectively) exhibited a large decrease in apparent molar mass and much broader dispersity compared to the unsaturated precursors. Considering that no noticeable change of functional group was observed in NMR spectra, we assume that the change of molar mass and dispersity may be due solely to conformational changes (i.e., hydrodynamic radius) of the hydrogenated polymers in $CHCl_3$. However, we cannot rule out some level of chain scission.

Table 2 summarizes the thermal characterization using DSC of the hydrogenated polymers. PH**7**, PH**8**, PH**13** and PH**14**, where the functional group is bulky, remained amorphous. The T_g of the saturated polymers range from −74 to 51 °C, and dependent on the functional group. Similarly to the unsaturated polymers, small substituents exhibit a much lower T_g than those with relatively bulky substituents. In addition to a glass transition, both PH**2** and PH**3** displayed well-defined melting transition transitions and PH**15** showed a small but unambiguous melting peak, indicating that these polymers are semi-crystalline. PH**3** with the larger substituent (PEG) showed a lower melting point than that in PH**2** or PH**15** with the smaller substituents (OMe and OCOMe, respectively). The only report of a 5-methoxy-COE polymerized via ROMP with a Ru-based catalyst does not report thermal properties for such polymer [69]. We previously reported that the saturated polymer of 3-methyl-COE (poly(3MeCOE)) exhibited a T_g at −59 °C, a T_m − 2 °C and ΔH_m − 30 J g^{-1} [55]. When the T_m of poly(3MeCOE) is compared with that of PH**2** and PH**15**, the more polar polymers (PH**2** and PH**15**) show a higher T_m despite being larger groups; however, the ΔH_m was lower in both cases. The more polar PH**3**, despite having the largest of all substituents, shows a T_m with equal enthalpy of melting to that of PH**2**. These results indicate that the polarity of the substituent and that of the polymer has a strong influence in the final thermal properties of the polymer. The polarity appears to have greater effect on crystallization than does the size; this could possibly be due to the polar functional group strengthening inter-chain interaction and thus favoring crystallization.

Conclusions

In summary, we demonstrated an efficient pathway to rapidly expand the scope of 3-substituted COE bearing diverse polar functional groups (3**FG**COEs). Many of these 3**FG**COEs allow ROMP using the Grubbs Ru-based catalyst (**G2**) to afford functionalized polyalkenamers, which possess HT regio-regularity and *E*-stereo-regularity. Both poly(3**FG**COE) and the saturated poly(3**FG**COE) exhibit thermal properties dependent on the size and polarity of the functional groups. This methodology, complementary to the ADMET approach, can be utilized to prepare PE derivatives with diverse polar side chains, including but not limited to those described in this study, which are difficult to be synthesized through radical or coordination copolymerization. Some examples demonstrated here further enable the incorporation of functional groups that are problematic with typical ROMP catalysts and conditions. For instance, PH**7** can be properly treated to release free amine side or PH**8** can be use to generate free aldehyde groups. It is envisioned that 3**FG**COEs would be used not only to synthesize precision functionalized PE derivatives but also to participate in the ring-opening metathesis copolymerization with unfunctionalized cyclic olefins. These advances are a step forward to new graft copolymers [66, 70] and complex supramolecular structures with PE backbone.

Acknowledgments We thank the Abu Dhabi-Minnesota Institute for Research Excellence (ADMIRE) and Dow Chemical Company for support.

References

1. Peacock A (2000) Handbook of polyethylene: structures: properties, and applications, 1st edn. CRC Press, FL
2. Fink JK (2010) Handbook of engineering and specialty thermoplastics: polyolefins and styrenics. Scrivener Publishing LLC, USA
3. Cordeiro CF, Petrocelli FP (2004) Encyclopedia of polymer science and technology. Wiley-VCH, Weinheim
4. Boaen NK, Hillmyer MA (2005) Post-polymerization functionalization of polyolefins. Chem Soc Rev 34:267–275
5. Boffa LS, Novak BM (2000) Copolymerization of polar monomers with olefins using transition-metal complexes. Chem Rev 100:1479–1494
6. Nakamura A, Ito S, Nozaki K (2009) Coordination–insertion copolymerization of fundamental polar monomers. Chem Rev 109:5215–5244
7. Baughman TW, Wagener KB (2005) Recent advances in ADMET polymerization. Adv Polym Sci 176:1–42
8. Valenti DJ, Wagener KB (1998) Direct synthesis of well-defined alcohol-functionalized polymers via acyclic diene metathesis (ADMET) polymerization. Macromolecules 31:2764–2773
9. Watson MD, Wagener KB (2000) Ethylene/vinyl acetate copolymers via acyclic diene metathesis polymerization. Examining the effect of "long" precise ethylene run lengths. Macromolecules 33:5411–5417
10. Watson MD, Wagener KB (2000) Functionalized polyethylene via acyclic diene metathesis polymerization: effect of precise placement of functional groups. Macromolecules 33:8963–8970
11. Hopkins TE, Pawlow JH, Koren DL, Deters KS, Solivan SM, Davis JA, Gomez FJ, Wagener KB (2001) Chiral polyolefins bearing amino acids. Macromolecules 34:7920–7922
12. Hopkins TE, Wagener KB (2003) Amino acid and dipeptide functionalized polyolefins. Macromolecules 36:2206–2214
13. Baughman TW, van der Aa E, Lehman SE, Wagener KB (2005) Circumventing the reactivity ratio dilemma: synthesis of ethylene-co-methyl vinyl ether copolymer. Macromolecules 38:2550–2551
14. Baughman TW, van der Aa E, Wagener KB (2006) Linear ethylene–vinyl ether copolymers: synthesis and thermal characterization. Macromolecules 39:7015–7021
15. Boz E, Wagener KB, Ghosal A, Fu RQ, Alamo RG (2006) Synthesis and crystallization of precision ADMET polyolefins containing halogens. Macromolecules 39:4437–4447
16. Baughman TW, Chan CD, Winey KI, Wagener KB (2007) Synthesis and morphology of well-defined poly(ethylene-co-acrylic acid) copolymers. Macromolecules 40:6564–6571
17. Berda EB, Lande RE, Wagener KB (2007) Precisely defined amphiphilic graft copolymers. Macromolecules 40:8547–8552
18. Boz E, Wagener KB (2007) Progress in the development of well-defined ethylene–vinyl halide polymers. Polym Rev 47:511–541
19. Berda EB, Wagener KB (2008) Inducing pendant group interactions in precision polyolefins: synthesis and thermal behavior. Macromolecules 41:5116–5122
20. Boz E, Ghiviriga I, Nemeth AJ, Jeon K, Alamo RG, Wagener KB (2008) Random, defect-free ethylene/vinyl halide model copolymers via condensation polymerization. Macromolecules 41:25–30
21. Opper KL, Fassbender B, Brunklaus G, Spiess HW, Wagener KB (2009) Polyethylene functionalized with precisely spaced phosphonic acid groups. Macromolecules 42:4407–4409
22. Opper KL, Wagener KB (2009) Precision sulfonic acid ester copolymers. Macromol Rapid Commun 30:915–919
23. Mei JG, Aitken BS, Graham KR, Wagener KB, Reynolds JR (2010) Regioregular electroactive polyolefins with precisely sequenced π-conjugated chromophores. Macromolecules 43:5909–5913
24. Aitken BS, Lee M, Hunley MT, Gibson HW, Wagener KB (2010) Synthesis of precision ionic polyolefins derived from ionic liquids. Macromolecules 43:1699–1701
25. Opper KL, Markova D, Klapper M, Mullen K, Wagener KB (2010) Precision phosphonic acid functionalized polyolefin architectures. Macromolecules 43:3690–3698
26. Aitken BS, Wieruszewski PM, Graham KR, Reynolds JR, Wagener KB (2012) Control of charge-carrier mobility via in-chain spacer length variation in sequenced triarylamine functionalized polyolefins. ACS Macro Lett 1:324–327
27. Aitken BS, Wieruszewski PM, Graham KR, Reynolds JR, Wagener KB (2012) Perfectly regioregular electroactive polyolefins: impact of inter-chromophore distance on PLED EQE. Macromolecules 45:705–712
28. Aitken BS, Buitrago CF, Heffley JD, Lee M, Gibson HW, Winey KI, Wagener KB (2012) Precision ionomers: synthesis and thermal/mechanical characterization. Macromolecules 45:681–687
29. Leonard JK, Wei YY, Wagener KB (2012) Synthesis and thermal characterization of precision poly(ethylene-co-vinyl amine) copolymers. Macromolecules 45:671–680
30. Bielawski CW, Grubbs RH (2007) Living ring-opening metathesis polymerization. Prog Polym Sci 32:1–29
31. Ramakrishnan S, Chung TC (1990) Poly(5-hydroxyoctenylene) and its derivatives: synthesis via metathesis polymerization of an organoborane monomer. Macromolecules 23:4519–4524
32. Couturier JL, Tanaka K, Leconte M, Basset JM, Ollivier J (1993) Metathesis of sulfur-containing olefins with a metallacyclic

aryloxo(chloro)neopentylidenetungsten complex. Angew Chem Int Ed 32:112–115

33. Hillmyer MA, Laredo WR, Grubbs RH (1995) Ring-opening metathesis polymerization of functionalized cyclooctenes by a ruthenium-based metathesis catalyst. Macromolecules 28:6311–6316
34. Yang HL, Islam M, Budde C, Rowan SJ (2003) Ring-opening metathesis polymerization as a route to controlled copolymers of ethylene and polar monomers: synthesis of ethylene–vinyl chloride-like copolymers. J Polym Sci Part A Polym Chem 41:2107–2116
35. Xu GJ, Wang DR, Buchmeiser MR (2012) Functional polyolefins: poly(ethylene)-graft-poly(tert-butyl acrylate) via atom transfer radical polymerization from a polybrominated alkane. Macromol Rapid Commun 33:75–79
36. Schneider MF, Gantner C, Obrecht W, Nuyken O (2010) Ring-opening metathesis polymerization of nitrile substituted cis-cyclooctenes. Macromol Rapid Commun 31:1731–1735
37. Schneider MF, Gantner C, Obrecht W, Nuyken O (2011) Ring-opening metathesis polymerization of cis-cyanocyclooct-4-ene: search for active catalysts, variation of monomer to catalyst ratios and monomer concentrations. J Polym Sci Part A Polym Chem 49:879–885
38. Scherman OA, Kim HM, Grubbs RH (2002) Synthesis of well-defined poly((vinyl alcohol)2-alt-methylene) via ring-opening metathesis polymerization. Macromolecules 35:5366–5371
39. Scherman OA, Walker R, Grubbs RH (2005) Synthesis and characterization of stereoregular ethylene–vinyl alcohol copolymers made by ring-opening metathesis polymerization. Macromolecules 38:9009–9014
40. Hejl A, Scherman OA, Grubbs RH (2005) Ring-opening metathesis polymerization of functionalized low-strain monomers with ruthenium-based catalysts. Macromolecules 38:7214–7218
41. Maughon BR, Grubbs RH (1996) Synthesis and controlled cross-linking of polymers derived from ring-opening metathesis polymerization (ROMP). Macromolecules 29:5765–5769
42. Breitenkamp K, Simeone J, Jin E, Emrick T (2002) Novel amphiphilic graft copolymers prepared by ring-opening metathesis polymerization of poly(ethylene glycol)-substituted cyclooctene macromonomers. Macromolecules 35:9249–9252
43. Breitenkamp K, Emrick T (2003) Novel polymer capsules from amphiphilic graft copolymers and cross-metathesis. J Am Chem Soc 125:12070–12071
44. Lehman SE, Wagener KB, Baugh LS, Rucker SP, Schulz DN, Varma-Nair M, Berluche E (2007) Linear copolymers of ethylene and polar vinyl monomers via olefin metathesis–hydrogenation: synthesis, characterization, and comparison to branched analogues. Macromolecules 40:2643–2656
45. Han HJ, Chen FX, Yu JH, Dang JY, Ma Z, Zhang YQ, Xie MR (2007) Ring-opening metathesis polymerization of functionalized cyclooctene by a ruthenium-based catalyst in ionic liquid. J Polym Sci Part A Polym Chem 45:3986–3993
46. Song AR, Parker KA, Sampson NS (2009) Synthesis of copolymers by alternating ROMP (AROMP). J Am Chem Soc 131:3444–3445
47. Kratz K, Breitenkamp K, Hule R, Pochan D, Emrick T (2009) PC-polyolefins: synthesis and assembly behavior in water. Macromolecules 42:3227–3229
48. Song AR, Lee JC, Parker KA, Sampson NS (2010) Scope of the ring-opening metathesis polymerization (ROMP) reaction of 1-substituted cyclobutenes. J Am Chem Soc 132:10513–10520
49. Breitenkamp RB, Ou Z, Breitenkamp K, Muthukumar M, Emrick T (2007) Synthesis and characterization of polyolefin-graft-oligopeptide polyelectrolytes. Macromolecules 40:7617–7624
50. Revanur R, McCloskey B, Breitenkamp K, Freeman BD, Emrick T (2007) Reactive amphiphilic graft copolymer coatings applied to poly(vinylidene fluoride) ultrafiltration membranes. Macromolecules 40:3624–3630
51. Kobayashi S, Kim H, Macosko CW, Hillmyer MA (2013) Functionalized linear low-density polyethylene by ring-opening metathesis polymerization. Polym Chem 4:1193
52. Shi HC, Shi DA, Yin LG, Luan SF, Zhao J, Yin JH (2010) Synthesis of amphiphilic polycyclooctene-graft-poly(ethylene glycol) copolymers by ring-opening metathesis polymerization. React Funct Polym 70:449–455
53. Winkler B, Rehab A, Ungerank M, Stelzer F (1997) A novel side-chain liquid crystal polymer of 5-substituted cis-cyclooctene via ring-opening metathesis polymerization. Macromol Chem Phys 198:1417–1425
54. Xia Y, Verduzco R, Grubbs RH, Kornfield JA (2008) Well-defined liquid crystal gels from telechelic polymers. J Am Chem Soc 130:1735–1740
55. Kobayashi S, Pitet LM, Hillmyer MA (2011) Regio- and stereoselective ring-opening metathesis polymerization of 3-substituted cyclooctenes. J Am Chem Soc 133:5794–5797
56. Martinez H, Miró P, Charbonneau P, Hillmyer MA, Cramer CJ (2012) Selectivity in ring-opening metathesis polymerization of Z-cyclooctenes catalyzed by a second-generation Grubbs catalyst. ACS Catal 2:2547–2556
57. Zhang J, Matta ME, Hillmyer MA (2012) Synthesis of sequence-specific vinyl copolymers by regioselective ROMP of multiply substituted cyclooctenes. ACS Macro Lett 1:1383–1387
58. Pitet LM, Zhang J, Hillmyer MA (2013) Sequential ROMP of cyclooctenes as a route to linear polyethylene block copolymers. Dalton Trans 42:9079–9088
59. Zhang J, Matta ME, Martinez H, Hillmyer MA (2013) Precision vinyl acetate/ethylene (VAE) Copolymers by ROMP of acetoxy-substituted cyclic alkenes. Macromolecules 46:2535–2543
60. Benkeser RA, Mozdzen EC, Muench WC, Roche RT, Siklosi MP (1979) Organic chemistry of dichlorosilane. Additions to conjugated and unconjugated diene systems followed by intramolecular cyclizations. J Org Chem 44:1370–1376
61. Slugovc C (2004) The ring opening metathesis polymerisation toolbox. Macromol Rapid Commun 25:1283–1297
62. Leitgeb A, Wappel J, Slugovc C (2010) The ROMP toolbox upgraded. Polymer 51:2927–2946
63. Formentin P, Gimeno N, Steinke JHG, Vilar R (2005) Reactivity of Grubbs' catalysts with urea- and amide-substituted olefins. Metathesis and isomerization. J Org Chem 70:8235–8238
64. Menozzi C, Dalko PI, Cossy J (2005) Hydrosilylation of terminal alkynes with alkylidene ruthenium complexes and silanes. J Org Chem 70:10717–10719
65. Polshettiwar V, Varma RS (2008) Olefin ring closing metathesis and hydrosilylation reaction in aqueous medium by Grubbs second generation ruthenium catalyst. J Org Chem 73:7417–7419
66. Xu Y, Thurber CM, Lodge TP, Hillmyer MA (2012) Synthesis and remarkable efficacy of model polyethylene-graft-poly(methyl methacrylate) copolymers as compatibilizers in polyethylene/poly(methyl methacrylate) blends. Macromolecules 45:9604–9610
67. Rojas G, Berda EB, Wagener KB (2008) Precision polyolefin structure: modeling polyethylene containing alkyl branches. Polymer 49:2985–2995
68. Hahn SF (1992) An improved method for the diimide hydrogenation of butadiene and isoprene containing polymers. J Polym Sci Part A Polym Chem 30:397–408
69. Demonceau A, Stumpf AW, Saive E, Noels AF (1997) Novel ruthenium-based catalyst systems for the ring-opening metathesis polymerization of low-strain cyclic olefins. Macromolecules 30:3127
70. Chung TC (2002) Synthesis of functional polyolefin copolymers with graft and block structures. Prog Polym Sci 27:39–85

7

Nitrides as ammonia synthesis catalysts and as potential nitrogen transfer reagents

J. S. J. Hargreaves

Abstract In this article, an overview of the application of selected metal nitrides as ammonia synthesis catalysts is presented. The potential development of some systems into nitrogen transfer reagents is also described.

Keywords Nitrogen · Hydrogen · Catalysis · Ammonia · Nitrides

Introduction

The successful development of the Haber–Bosch Process for the production of ammonia was a paramount achievement of the twentieth Century. Through the provision of an accessible route to synthetic fertiliser, it has been pivotal in sustaining a large proportion of the global population [1]. The process involves the direct combination of pure H_2 and N_2 reactants over promoted iron catalyst at moderate temperature (typically ca 400 °C) and high pressure (typically 100–200 atmospheres):

$$0.5N_{2(g)} + 1.5H_{2(g)} \leftrightarrow NH_{3(g)} \Delta H^{\circ} = -46 \text{ kJ mol}^{-1}$$

The reaction is equilibrium limited and, being exothermic, is thermodynamically favoured at lower reaction temperatures. However, the temperature of operation is dictated by the requirement to achieve acceptable process kinetics. Whilst the Haber–Bosch Process can, to some extent, be viewed as a mature technology, the development of more active (or even more poison tolerant) catalysts which would allow a reduction in the stringent conditions of operation is of significant interest. In this context, it is salient to note that, taken in its entirety, including the provision and purification of reactant feedstreams, on a global scale, the operation of the Haber Process currently consumes 1–2 % of the world's energy demand, e.g. [2, 3]. Therefore, any improvements can potentially yield massive rewards both in terms of economic and environmental benefits. The identification and application of more active catalysts could lead to lowering of the process temperatures and/or pressures of operation necessary to achieve a desired space time yield of product. Indeed, lowered reaction temperatures result in improved obtainable equilibrium yields offering the tantalising prospect of a possible win–win situation, providing sufficiently favourable kinetics can be achieved. This is, however, a very significant challenge.

Iron is cheap and abundant and, when promoted, very effective under current Haber–Bosch Process conditions. Therefore, any replacement catalyst must afford the opportunity of significantly higher activity leading to lower temperature and/or pressure of application. Indeed, a number alternative systems based upon ruthenium have been identified in the literature, e.g. [4–7]. One such system involving promoted ruthenium supported on carbon is 10–20 times more active than conventional iron systems while operating at lower reaction pressure and forms the basis of an alternative ammonia synthesis process, the Kellogg Advanced Ammonia Process [8, 9].

The focus of this overview is a description of some of the relatively recent literature relating to the activity of metal nitrides as ammonia synthesis catalysts. The potential of nitrides as possible nitrogen transfer reagents will also be considered. In some respects, this relates to historical work on the development of the Haber–Bosch Process where some early studies by Mittasch and coworkers [10] at BASF considered routes to nitrogen

J. S. J. Hargreaves
WestCHEM, School of Chemistry, University of Glasgow, Joseph Black Building, Glasgow G12 8QQ, UK

fixation involving the formation of nitrides and also demonstrated some nitrides, such as molybdenum nitride, to be intermediates in the stepwise synthesis of ammonia via nitridation of metals with N_2 and subsequent reduction by H_2.

Metal nitrides as ammonia synthesis catalysts

Binary metal nitrides

In the early studies, Mittasch [10] drew attention to molybdenum metal which was stated to catalyse ammonia synthesis differently from iron in that it formed easily detectable amounts of nitride during catalysis. It was said to resemble tungsten, manganese and uranium in this respect. It is therefore to be expected that nitrides of such metals may be active ammonia synthesis catalysts. However, mention should also be made that under industrial conditions bulk iron nitridation occurs resulting in the distorted structure associated with catalytic activity [11].

A number of binary nitrides have been identified as active ammonia synthesis catalysts in the literature, including those of molybdenum [10, 12–14], uranium [15, 16], vanadium [17, 18], rhenium [19, 20] and cerium [21].

In the case of molybdenum, arguably the most well-studied binary nitride, structure sensitivity has been reported to be operative over γ-Mo_2N in studies involving systematic variation of particle size [12]. In these studies, it was proposed that the fraction of active sites decreased with decreasing particle size. However, the effect of morphology was found to be limited in a study comparing plate like γ-Mo_2N and nanorod form of γ-Mo_2N prepared by the temperature programmed ammonolysis of plate and nanorod MoO_3, respectively [14]. Such ammonolysis is known to yield γ-Mo_2N which is pseudomorphic with respect to the MoO_3 precursor [22]. Comparison of different phases of molybdenum nitride showed β-$Mo_2N_{0.78}$ to possess similar activity to its γ-Mo_2N counterpart, whereas δ-MoN was found to be inactive (although potential deleterious effects resulting from residual sulphur left over from the preparation of this phase by ammonolysis of MoS_2 could not be excluded) [14]. The β nitride phase can be prepared by reaction of MoO_3 with N_2/H_2 mixtures [23] and the δ phase by ammonolysis of either MoS_2 or $MoCl_5$ [24]. Applying ^{15}N labelling to study the mechanism of ammonia synthesis over molybdenum nitride, Aika and Ozaki [25] deduced rate determining N_2 activation leading to intermediate $N_{(a)}$. It was argued that a proportion of the ^{15}N was lost into bulk nitride, with the following scheme being presented:

$$\begin{array}{l} N_2 \longrightarrow N_{(a)} \leftrightarrow NH_3 \\ \qquad\quad \updownarrow \\ \qquad Mo_2N \end{array}$$

and similarities with iron nitride, previously investigated by Logan et al. [26], were drawn. Contrary to this, Hillis et al. [27] had reported the hydrogenation of NH_x species to be rate determining and the rate of hydrogenation of bulk nitride to be around 20–50 times slower than ammonia synthesis in this system. Some studies have been undertaken on other binary nitride systems. For example, based upon unusual transient behaviour, the hydrogenation of bulk nitrogen in uranium nitride has been proposed with the suggestion being made of a possible dynamic process between nitrogen species not immediately hydrogenated and loosely bonded subsurface nitrogen species which subsequently diffuse to the surface where they are hydrogenated to form ammonia [15]. This is akin to some oxidation processes which occur by the Mars–van Krevelen process where lattice oxygen species are those active in catalytic oxidation reactions [28]. Phase stability may be a consideration in other systems, for example, in rhenium nitride, the active form of the catalyst is proposed to comprise a mixture of Re_3N and Re [19].

Ternary metal nitrides

In recent years, there have been reports that Co_3Mo_3N-related catalysts have high activity that have attracted interest [29–35]. In particular, it has been reported that when suitably promoted, the activity of this phase can exceed that of conventional iron based Haber–Bosch catalysts. Table 1 presents some activity data taken from a study by Aika and Kojima [29]. Cs^+ has been found to be a promoter up to a certain limit, beyond which phase decomposition occurs [29]. Preparation of the phase generally involves ammonolysis of hydrated cobalt molybdate precursor—perhaps surprisingly use of anhydrous cobalt

Table 1 Ammonia synthesis activities of various catalysts (adapted from [29])

Catalyst	Rate (μmol h^{-1} g^{-1})	Surface area (m^2 g^{-1})	Specific activity (μmol h^{-1} m^{-2})
Fe-K_2O-Al_2O_3	330	14	24
Co_3Mo_3N	652	21	31
Co_3Mo_3N-K5	869	17	51
Co_3Mo_3N-K30	364	8	46
Co_3Mo_3N-Cs2	986	16	62
Co_3Mo_3N-Cs10	586	10	59

Reaction conditions 400 °C and 0.1 MPa

molybdate is less effective. Ammonolysis may present a problem in large-scale application since, for example, heat transfer may be an issue as outlined in the case of the binary molybdenum nitride system [36, 37]. However, straight application of the hydrated molybdate precursor followed by in situ nitriding with 3/1 H_2/N_2 at 700 °C for 2 h produced a less active catalyst which, although containing some Co_3Mo_3N, comprised additional phases [38].

The activity of the Co_3Mo_3N phase has been rationalised in terms of the N_2 binding energy by Norskov and co-workers [32] as shown in Fig. 1 which presents a comparison of the calculated turnover frequency for a range of materials as a function of N_2 binding energy. The significance of N_2 binding energy as an important parameter for a reaction process in which N_2 dissociation is generally considered rate limiting has been discussed and justified [39]. In terms of the plot presented in Fig. 1, the high activity of Ru and Fe, the basis for the two major industrial processes, as discussed in the introduction, can be seen. Osmium can also be seen to be a highly active catalyst—a fact recognised in the early work as outlined by Mittasch [10] in which uranium was also highlighted as having high activity. The high activity of Fe, Ru and Os for ammonia synthesis represents an interesting group trend within the d block of the periodic table, although it is seldom, if ever, highlighted as so. In addition, the initial discovery of the activity of Os represents an early example of a currently popular aspect for consideration—scarcity of elements. At that stage, the global reserves of Os were only 100 kg and Os was also susceptible to oxidation to yield OsO_4 and so its replacement in the process was of paramount importance, which led to the discovery of the promoted Fe system. In Norskov's work, the high activity of the Co_3Mo_3N phase has been explained as an average of the N_2 binding energies for Co (which is too weakly binding) and Mo (which is too strongly binding) arriving at an optimum in the Sabatier curve shown in Fig. 1, with the Co_3Mo_3N catalyst reportedly being developed along this principle. It was argued that both components have to be in close proximity in order to be effective and that the (111) surface plane exhibits mixed Co and Mo sites. N was stated to not participate in the reaction but rather induce the required ordering. Support for this rationalisation was provided in previous microkinetic studies by the authors for which the activation energy was found to be intermediate between Co and Mo [32]. The volcano curve can also be used to rationalise the relatively high ammonia synthesis activity of the Ni_2Mo_3N phase [29, 33]. Fe_3Mo_3N, which is isostructural with Co_3Mo_3N, is also a high activity ternary nitride for the reaction [29, 33]. A perhaps somewhat related approach to the Sabatier curve are the experimental surface science studies by Attard and King [40] aimed towards the production of hybrid materials combining a

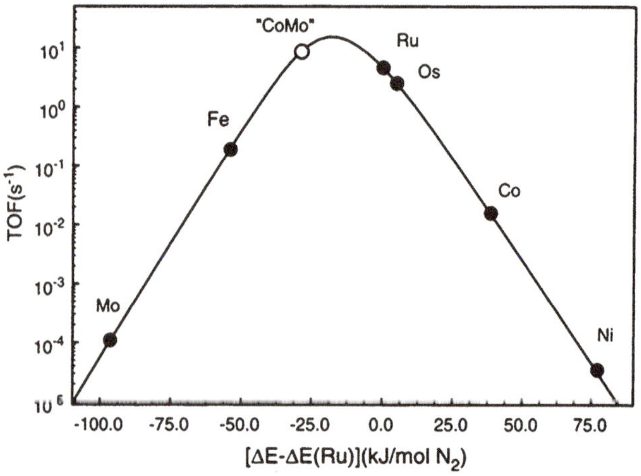

Fig. 1 Calculated turnover frequencies for ammonia synthesis as a function of the adsorption energy of nitrogen. The synthesis conditions are 400 °C, 50 bar, gas composition H_2/N_2 3/1 containing 5 % NH_3. Reproduced with permission from [32]. Copyright (2001) American Chemical Society

component with a marked ability to dissociate N_2 with a high sticking probability with a component which bonds weakly to nitrogen in order to produce a nitrogen–metal bond of similar strength to Fe–N. The former component was W, whilst Cu, Ag and Au were examples of the latter component. It was found that N_2 would not adsorb at sites consisting of both W and Cu atoms and hence it was predicted that bimetallics of W with Group IB metals would act as poor ammonia synthesis catalysts. However, if N_2 could be dissociated by another route, the possibility of copper–tungsten bimetallics functioning as effective ammonia synthesis catalysts was raised.

In the context of high ammonia synthesis activity, it is interesting to draw attention to the reported high activity of cobalt–rhenium phases [19, 41]. Whilst these do not seem to have achieved the prominence of the Co_3Mo_3N related systems, literature studies have indicated that the activity exceeds that of them. A Co/Re material ratio of 1/4 prepared by ammonolysis at 700 °C was found to be particularly effective and it was proposed that catalytic activity may relate to the presence of some form of rhenium nitride phase. However, the scarcity of rhenium and its relative cost may be a consideration in large-scale application, although it should be acknowledged that Pt–Re reforming catalysts are applied on an industrial scale and also that Co–Re catalysts for Fischer–Tropsch synthesis have been a focus of interest, e.g. [42]. Whilst many formulations were screened by Mittasch and colleagues [10] in the early development of ammonia synthesis catalysts, it is an interesting thought to note that rhenium as such would not have formed part of the matrix, since its discovery was not documented until 1925. Indeed, in these studies Co was even documented as a promoter of Mo which is fascinating

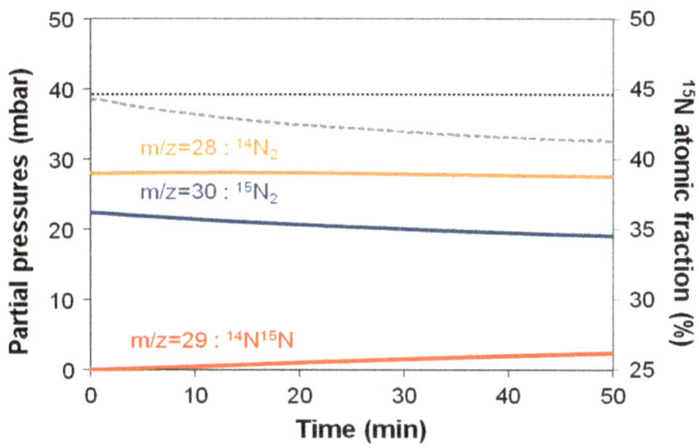

Fig. 2 Evolution of the nitrogen isotopomer partial pressures and the gas-phase ^{15}N atomic fraction during homomolecular isotopic nitrogen exchange at 450 °C over Co_3Mo_3N. Reproduced with permission from [43]. Copyright (2013) American Chemical Society

in the context of the development of the Co_3Mo_3N catalytic system at a very much later stage and illustrative of the wide ranging nature of the early work conducted at BASF.

Whilst it was stated that the lattice N of the Co_3Mo_3N catalyst did not participate in the NH_3 synthesis reaction, a nitrogen isotopic exchange reaction undertaken by Hunter et al. has shown it to be exchangeable [43]. Figure 2 presents the homomolecular isotopic exchange profile in which a mixture of $^{14}N_2$ and $^{15}N_2$ is scrambled over Co_3Mo_3N at 450 °C. Most pertinent is the observation that the atomic fraction of ^{15}N in the gas phase decreases over the course of the reaction which is a clear indication of the operation of the heterolytic exchange pathway in which exchange of gas-phase species with the lattice N (which is, of course, predominantly ^{14}N reflecting the natural distribution of nitrogen isotopes) occurs and that the dissociative activation of N_2 does not occur. Isotope exchange studies of the materials with $^{15}N_{2(g)}$ only were undertaken at higher reaction temperature and the results and the reaction and pre-treatment conditions are presented in Table 2. From this data, a clear influence of the nature of pre-treatment upon the exchangeability of lattice nitrogen is observed—in these studies, the influence of adsorbed N species was discounted and it was found that ca. 25 % of lattice N species could be exchanged at 600 °C following N_2 pre-treatment at 700 °C as compared to 6 % for the corresponding pre-treatment in Ar. Whilst these reaction temperatures are significantly above that for ammonia synthesis, the presence of H_2 in that case could accentuate the lattice N reactivity. These results were taken to demonstrate the possibility that the lattice nitrogen may indeed be reactive under reaction conditions and hence that a Mars–van Krevelen type mechanism, in which NH_3 formation occurs by direct hydrogenation of lattice species which are then replenished from gas-phase N_2, could be operative and the significant effects of pre-treatment upon the extent of exchange demonstrate the sensitivity of the system, possibly involving surface restructuring effects.

Table 2 Nitrogen isotopic exchange activity of Co_3Mo_3N as a function of pre-treatment (adapted from [43])

Pre-treatment	H_2/N_2 (3/1) 700 °C/Ar 700 °C (30 min)	H_2/N_2 (3/1) 700 °C/N_2 700 °C (30 min)	H_2/N_2 (3/1) 700 °C/ Ar 700 °C (30 min)/ N_2 700 °C (30 min)
Temperature of exchange (°C)	600	600	600
$^{15}N_2$ pressure introduced (mbar)	55.6	55.1	55.0
Total pressure at 40 min	51.4	48.5	52.3
$^{15}N_{exchanged}/N_{total331}$ at 40 min	6	28	25

Potential of nitrides as nitrogen transfer reagents

The possible existence of Mars–van Krevelen type reaction mechanisms in both the binary and ternary nitride systems raises the interesting possibility of the development of nitrides as reagents for novel nitrogen transfer pathways. The prospect of such a "mechanism" leads to the idea of a "process".

The mechanism implies two concurrent general stages to ammonia synthesis:

(i) the hydrogenation of lattice N species to yield NH_3, and
(ii) the replenishment of lattice N vacancies from gas-phase N_2

In terms of a "process" stages (i) and (ii) could be conducted under different conditions (e.g. at different temperatures) leading to potential kinetic and/or thermodynamic advantages. In oxidation catalysis there is a precedent for the distinction between "mechanism" and "process" as illustrated by the example of butane oxidation catalysed by vanadium phosphate. On the basis that this reaction occurred by a Mars–van Krevelen mechanism it has proved possible to separate the butane oxidation and catalyst re-oxidation steps into separate stages [44, 45]. In this way, the desired partial oxidation product, maleic anhydride, does not come into direct contact with O_2 in the presence of the catalyst.

Applying this general concept, it may prove possible to develop new nitrogen transfer reagents by-passing the use of NH_3 as a nitrogen transfer reagent. The potential value of this idea can be illustrated by consideration of the large-scale production of aniline via the selective reduction of nitrobenzene. Considering this process from the viewpoint of N_2 as the initial source of nitrogen, the process steps can be considered as:

$$N_2 + 3H_2 \leftrightarrow 2NH_3$$
$$2NH_3 + 4O_2 \rightarrow 2HNO_3 + 2H_2O$$
$$HNO_3 + 2H_2SO_4 \rightarrow NO_2^+ + H_3O^+ + 2HSO_4^-$$
$$C_6H_6 + NO_2^+ \rightarrow C_6H_5NO_2 + H^+$$
$$C_6H_5NO_2^+ + 3H_2 \rightarrow C_6H_5NH_2 + 2H_2O$$

As detailed above, the production of ammonia by Haber–Bosch Process is extremely energy intensive. Having applied this energy to form intermediate ammonia, the second equation (a simplified representation of the Ostwald Process) involves oxidation of some of the hydrogen atoms bonded to each ammonia molecule added in the previous process step. Taken together, these initial two stages represent an indirect way of oxidising hydrogen to yield water! The nitric acid product is then used to form the nitronium cation in corrosive mixture of concentrated acids. Having obtained the nitronium cation benzene is nitrated and the nitrobenzene product selectively reduced to yield aniline—in another stage indirectly involving the reaction between oxygen and hydrogen to yield water!

It might be imagined that the direct amination of benzene:

$$C_6H_6 + NH_3 \leftrightarrow C_6H_5NH_2 + H_2$$

is a potential alternative route, but it is thermodynamically limited and approaches to overcome this limitation by use of a cataloreactant [46] or coupled hydrogen consuming reactions [47] have proved unsuccessful.

If the lattice nitrogen of a nitride could be used in an alternative aniline forming two-stage process such as:

$$C_6H_6 + MN + 0.5H_2 \rightarrow C_6H_5NH_2 + M$$

and

$$M + \text{``N''} \rightarrow MN$$

(where MN represents a metal nitride, M the corresponding metal and "N" a nitridation agent such as NH_3 or, ideally, N_2) although a formidable challenge, the potential interest in the development of such a route is readily apparent. Indeed the nitrogen transfer and renitridation stages could be conducted under different reaction conditions. However, non-selective reduction could be a major potential issue. Aniline production is just one of a number of major "nitrogen incorporation" processes for which a similar potential approach may be imagined.

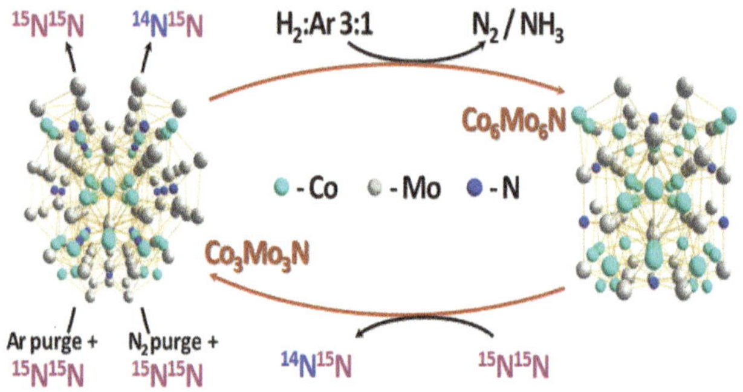

Fig. 3 Schematic of the phase transformations and isotopic exchange pathways over the Co_3Mo_3N–Co_6Mo_6N system. Reproduced with permission from [43]. Copyright (2013) American Chemical Society

Whilst not widespread, reports of the reactivity of lattice N are not without precedent, some of which were already mentioned in the section relating to binary nitride catalysts applied to ammonia synthesis. In 1948, Goodeve and Jack [48] published an investigation relating to the denitridation of iron nitride. In this study, hydrogen was shown to enhance the loss of lattice nitrogen leading to its complete conversion to ammonia in the 250–450 °C range. Interestingly, when CO was used, 25 % of the nitrogen lost formed cyanogen with isomorphous substitution of lattice N by C. Lattice nitrogen reactivity has also been implicated in NH_3 decomposition over zirconium oxynitride [49] and the ammoxidation of propane over VAlON [50].

With the Mars–van Krevelen process concept in mind, the reactivity of a number of metal nitrides towards H_2 has been investigated [14, 38, 51, 52]. Co_3Mo_3N proved a particularly interesting case in this regard—at 700 °C the previously unprecedented Co_6Mo_6N phase was formed through loss of 50 % of the lattice nitrogen [51, 52]. Despite the fact that the majority of N eliminated was in the form of N_2, with NH_3 being a minor co-product, H_2 was necessary to drive the process. In Co_3Mo_3N all the lattice nitrogen is crystallographically equivalent and it might be imagined that it would be totally lost upon reaction with H_2. However, the residual 50 % relocates to a different crystallographic site (migration from the 16c to the 8a Wyckoff site) as confirmed by powder neutron diffraction [52]. Restoration of the Co_3Mo_3N phase, which would be a necessary step for its application as a nitrogen transfer reagent, was found to be relatively facile when using a N_2/H_2 mix and achievable at higher temperature when using N_2 alone [53]. It was demonstrated that the Co_3Mo_3N/Co_6Mo_6N system can be cycled back and forth. It has tentatively been postulated that the H_2 necessary to accomplish the Co_3Mo_3N to Co_6Mo_6N phase change may function by inducing a phase change through reduction of surface stoichiometry—in terms of diffraction analysis, the transformation occurs between the two end members with no intermediate stoichiometries being formed. Isotopic nitrogen exchange over the Co_6Mo_6N phase has demonstrated that simultaneously, along with renitridation occurring leading to regeneration of the Co_3Mo_3N phase, a degree of exchange occurs with both processes ceasing simultaneously [43]. A schematic of the entire Co_3Mo_3N–Co_6Mo_6N interconversion and isotopic exchange processes is presented in Fig. 3.

Whilst Fe_3Mo_3N is iso-structural with Co_3Mo_3N, its lattice nitrogen is much less reactive. In terms of Co_3Mo_3N, the elevated temperature necessary to conduct the loss of 50 % of the lattice nitrogen would lead to carbidation upon contact with hydrocarbon reactants and so it is necessary to either find a means for enhancement of the lattice nitrogen reactivity or to develop different systems of enhanced reactivity. This could involve experimental based studies or, perhaps more excitingly, computational modelling techniques could be applied to design systems with desired lattice nitrogen reactivity. In the Co_3Mo_3N–Co_6Mo_6N system, it is probable that the renitridation step is enhanced by the maintenance of 50 % of the original nitrogen in the system. In other nitrides, such as Cu_3N, where N loss occurs at temperatures of greater potential interest for nitrogen transfer avoiding carbidation, the total loss of nitrogen which is difficult to suppress and the concomitant structural transformation have rendered renitridation extremely difficult [54].

In principle, routes to nitrogen-containing organic products such as amines could be envisaged based upon the application of carbonitrides as reagents involving hydrogenation. This could open up the exciting prospect of indirect routes from N_2 and CH_4 in three stage procedures involving nitridation, carbidation and reduction separately. In practice, each step could be conducted under different reaction conditions and the sequence of nitridation and carbidation could be reversed. This prospect is a significant challenge but arguably any demonstration of the indirect reaction of methane and nitrogen to yield any organic nitrogen-containing product, and even HCN, would be of major interest. To this end, preliminary investigations of the reduction molybdenum carbonitride systems prepared from single source routes have been undertaken [55]. In these studies, the preferential reactivity of lattice N over lattice C has been observed, indicating that it would be necessary to develop materials with more closely matched reactivity of the lattice N and C components. Again, computational modelling in which target phases are identified could play a significant role in this regard.

Related alternative strategies for ammonia synthesis

The observations outlined above indicate that it could be possible to apply Co_3Mo_3N as a material for the two-stage synthesis of ammonia in which hydrogenation of the lattice nitrogen and its subsequent replenishment from N_2 are conducted in separate stages. However, the temperatures involved in this would be high. Also, the mass fraction of N in the system is very low—an issue compounded when it is considered that only 50 % of the lattice nitrogen is "active"—and only a relatively fraction of it is hydrogenated to yield NH_3 with the major proportion being lost in the form of N_2.

The design of new, improved systems with greater efficiency of N utilisation and enhanced reactivity of N based along these lines represents an exciting opportunity for combined computational modelling and experimental studies. Cyclic operation of ammonia synthesis has been proposed previously in the literature by Amariglio and co-workers as a means of overcoming the inhibition of reaction rate by H_2 for Ru [56] and Os [57] systems. It was suggested that reaction rates of possible commercial interest could be achieved at atmospheric pressure using such an approach in these systems. For example, by applying the cyclic two-step procedure—saturation under pure N_2, followed by hydrogenation under pure H_2—extrapolation indicated that rates comparable to the industrial performance of iron catalysts could be obtained over Ru at 300 °C and atmospheric pressure.

Routes applying solar energy towards nitrogen fixation and the production of ammonia have been described recently by Pfromm and co-workers [58–60]. In these routes, metal nitrides are prepared by various means from N_2 and NH_3 is generated in a subsequent hydrolysis step. The necessary high temperatures are attained via a solar furnace. The use of a solar furnace as an environmentally friendly means to attain high temperatures for application to nitrogen fixation is an interesting concept. Historically, in the cyanamide process which immediately pre-dated the Haber–Bosch Process, nitrogen fixation was achieved by direct reaction of N_2 with calcium dicarbide producing calcium cyanamide ($CaCN_2$). Although this process required high temperatures, it was exothermic and self-sustaining once initiated [61]. However, the production of calcium carbide from CaO and C necessary for the process did involve the application of very high temperatures. $CaCN_2$ could be directly applied as a fertiliser itself or hydrolysed to liberate ammonia.

Conclusion

In this overview, selected examples of the applications of metal nitrides as catalysts for the synthesis of ammonia have been described. The potential of this class of material for the development of novel nitrogen transfer reagents has also been outlined. It can be anticipated that advances in nitride synthesis coupled with the application of more complex materials, such as quaternary systems, may result in exciting developments in these areas. It is clear that nitride materials represent an interesting opportunity. They are of increasing interest in other areas of heterogeneous catalysis.

References

1. Smil V (2002) Nitrogen and food production: proteins for human diets. AMBIO J Hum Env 31:126–131
2. Tanabe Y, Nishibayashi Y (2012) Developing more sustainable processes for ammonia synthesis. Coord Chem Rev 257:2551–2564
3. Bielawa H, Hinrichsen O, Birkner A, Muhler M (2001) The ammonia synthesis catalyst of the next generation: barium promoted oxide supported ruthenium. Angew Chemie Int Edn 40:1061
4. Jacobsen CJH (2001) A novel support for ruthenium-based ammonia synthesis catalysts. J Catal 200:1–3
5. Urabe K, Aika K, Ozaki A (1976) Activation of nitrogen by alkali metal promoted transition metal. 6. Hydrogen effect on isotopic equilibration of nitrogen and rate-determining step of ammonia-synthesis on potassium promoted ruthenium catalysts. J Catal 42:197–204
6. Rao KSR, Masthan SK, Prasad PSS, Rao PK (1991) Effect of barium addition on the ammonia synthesis activity of a cesium promoted ruthenium catalyst supported on carbon covered alumina (CAA). Appl Catal 73:L1–L5
7. Honkala K, Hellman A, Remediakis IN, Logadottir A, Carlosson A, Dahl S, Christensen CH, Norskov JK (2005) Ammonia synthesis from first principles calculations. Science 307:555–558
8. Rhodes AK (1996) New ammonia process, catalyst proven in Canadian plant. Oil Gas J 94(47):37–41
9. Liu H (2012) Ammonia synthesis catalysts—innovation and practice. World Scientific Press. ISBN: 978-981-4355-77-3
10. Mittasch A (1950) Early studies of multicomponent catalysts. In: Frankenburg WG, Rideal EK, Komorewsky VI (eds) Advances in catalysis, vol II. Academic Press, New York, pp 81–104
11. Schlogl R (2003) Catalytic synthesis of ammonia—a "never ending story?" Angew Chemie Int Edn 42:2004–2008
12. Volpe L, Boudart M (1986) Ammonia synthesis on molybdenum nitride. J Phys Chem 90:4874–4877
13. Kojima R, Aika K (2001) Molybdenum nitride and carbide catalysts for ammonia synthesis. Appl Catal A Gen 219:141–147
14. Mckay D, Hargreaves JSJ, Rico JL, Rivera JL, Sun XL (2008) The influence of phase and morphology of molybdenum nitrides on ammonia synthesis activity and reduction characteristics. J Solid State Chem 161:325–333
15. Segal N, Sebba F (1967) Ammonia synthesis catalyzed by uranium nitride. 1. Reaction mechanism. J Catal 8:105–112
16. Segal N, Sebba F (1967) Ammonia synthesis catalyzed by uranium nitride. 2. Transient behaviour. J Catal 8:113–119
17. King DA, Sebba F (1965) Catalytic synthesis of ammonia over vanadium nitride containing oxygen. 1. Reaction mechanism. J Catal 4:253–259
18. King DA, Sebba F (1965) Catalytic synthesis of ammonia over vanadium nitride containing oxygen. 2. Order-disorder transition revealed by catalytic behavior. J Catal 4:430–439
19. Kojima R, Aika K (2001) Rhenium containing binary catalysts for ammonia synthesis. Appl Catal A Gen 209:317–325
20. Kojima R, Enomoto H, Muhler M, Aika K (2003) Cesium promoted rhenium catalysts supported on alumina for ammonia synthesis. Appl Catal A Gen 245:311–322
21. Panov GI, Kharitonov AS (1985) Catalytic properties of nitrides in ammonia synthesis. React Kinet Catal Lett 29:267–274
22. Volpe L, Boudart M (1985) Compounds of molybdenum and tungsten with high surface area. 1. Nitrides. J Solid State Chem 59:332–347
23. Gong SW, Chen HK, Li W, Li BQ (2005) Synthesis of beta-$Mo_2N_{0.78}$ hydrodesulfurization catalyst in mixtures of nitrogen and hydrogen. Appl Catal A Gen 279:257–261
24. Ganin AY, Kienle L, Vajenine GV (2006) Synthesis and characterisation of hexagonal molybdenum nitrides. J Solid State Chem 179:2339–2348
25. Aika KI, Ozaki A (1969) Mechanism and isotope effect in ammonia synthesis over molybdenum nitride. J Catal 14:311–321
26. Logan SR, Moss RL, Kemball KC (1958) The catalytic decomposition of ammonia over evaporated iron films. Trans Faraday Soc 54:922–930
27. Hillis MR, Kemball C, Roberts MW (1966) Synthesis of ammonia and related processes on reduced molybdenum dioxide. Trans Faraday Soc 62:3570–3585
28. Doornkamp C, Ponec V (2000) The universal character of the Mars and van Krevelen mechanism. J Mol Chem A Chem 162:19–32
29. Kojima R, Aika K (2001) Cobalt molybdenum bimetallic nitride catalysts for ammonia synthesis—Part 1. Preparation and characterization. Appl Catal A Gen 215:149–160
30. Kojima R, Aika K (2001) Cobalt molybdenum bimetallic nitride catalysts for ammonia synthesis—Part 2. Kinetic study. Appl Catal A Gen 218:121–128
31. Kojima R, Aika K (2001) Cobalt molybdenum bimetallic catalysts for ammonia synthesis—Part 3. Reactant gas treatment. Appl Catal A Gen 219:157–170
32. Jacobsen CJH, Dahl S, Clausen BS, Bahn S, Logadottir A, Norskov JK (2001) Catalyst design by interpolation in the periodic table: bimetallic ammonia synthesis catalysts. J Am Chem Soc 123:8404–8405
33. Jacobsen CJH (2000) Novel class of ammonia synthesis catalysts. Chem Commun 12:1057–1058
34. Boisen A, Dahl S, Jacobsen CJH (2002) Promotion of binary nitride catalysts: isothermal N_2 adsorption, microkinetic model, and catalytic ammonia synthesis activity. J Catal 208:180–186
35. Kojima R, Aika K (2000) Cobalt molybdenum bimetallic catalysts for ammonia synthesis. Chem Lett 514–515
36. Wise RS, Markel EJ (1994) Catalytic NH_3 decomposition by topotactic molybdenum oxides and nitrides—effect on temperature programmed gamma-Mo_2N synthesis. J Catal 145:335–343
37. Wise RS, Markel EJ (1994) Synthesis of high surface area molybdenum nitride in mixtures of nitrogen and hydrogen. J Catal 145:344–355
38. Hargreaves JSJ, Mckay D (2009) A comparison of the reactivity of lattice nitrogen in Co_3Mo_3N and Ni_2Mo_3N catalysts. J Mol Catal A Chem 305:125–129
39. Dahl S, Logadottir A, Jacobsen CJH, Norskov J (2001) Electronic factors in catalysis: the volcano curve and the effect of promotion in catalytic ammonia synthesis. Appl Catal A Gen 222:19–29
40. Attard GA, King DA (1990) Co-adsorption of nitrogen with Cu, Ag and Au on W(100)—the role of metal adatoms in controlling surface reactivity. J Chem Soc Faraday Trans 86:2735–2741

41. Alexander A-M, Hargreaves JSJ, Mitchell C (2013) The denitridation of nitrides of iron, cobalt and rhenium under hydrogen. Top Catal 56:1963–1969
42. Iida H, Sakamoto K, Takeuchi M, Igarashi A (2013) Fischer–Tropsch synthesis over Co/SiO_2 and Co-M (M:Ru, Re)/SiO_2 catalysts prepared by a high temperature supercritical drying method. Appl Catal A Gen 466:256–263
43. Hunter SM, Gregory DH, Hargreaves JSJ, Richard M, Duprez D, Bion N (2013) A study of $^{15}N/^{14}N$ isotopic exchange over cobalt molybdenum nitrides. ACS Catal 3:1719–1725
44. Contractor RM (1999) Dupont's CBT technology for maleic anhydride. Chem Eng Sci 54:5627–5632
45. Bordes E, Contractor RM (1996) Adaptation of the microscopic properties of redox catalysts to the type of gas-solid reactor. Top Catal 3:365–375
46. Desrosiers P, Guan SH, Hagemeyer A, Lowe DM, Lugmair C, Poojary DM, Turner H, Weinberg H, Zhou XP, Armbrust R, Fengler G, Notheis U (2003) Application of combinatorial catalysis for the direct amination of benzene to aniline. Catal Today 81:319–328
47. Becker J, Holderich WF (1998) Amination of benzene in the presence of ammonia using a Group VIII metal supported on a carrier as catalyst. Catal Lett 54:125–128
48. Goodeve C, Jack KH (1948) Kinetics of nitrogen evolution from an iron-nitrogen interstitial alloy. Disc Faraday Soc 4:82–91
49. Soerijanto H, Rodel C, Wild U, Lerch M, Schomaker R, Schlogly R, Ressler T (2007) The impact of nitrogen mobility of zirconium oxynitride catalysts for ammonia decomposition. J Catal 250:19–24
50. Olea M, Florea M, Sack I, Silvy PR, Gaigneaux EM, Marin GB, Grange P (2005) Evidence for the participation of lattice nitrogen from vanadium aluminium oxynitrides in propane ammoxidation. J Catal 232:152–160
51. Mckay D, Gregory DH, Hargreaves JSJ, Hunter SM, Sun XL (2007) Towards nitrogen transfer catalysis: reactive lattice nitrogen in cobalt molybdenum nitride. Chem Commun 29:3051–3053
52. Hunter SM, Mckay D, Smith RI, Hargreaves JSJ, Gregory DH (2010) Topotactic nitrogen transfer: structural transformation in cobalt molybdenum nitrides. Chem Mater 22:2898–2907
53. Gregory DH, Hargreaves JSJ, Hunter SM (2011) On the regeneration of Co_3Mo_3N from Co_6Mo_6N and N_2. Catal Lett 141:22–26
54. Alexander A-M, Hargreaves JSJ, Mitchell C (2012) The reduction of various nitrides under hydrogen: Ni_3N, Cu_3N, Zn_3N_2 and Ta_3N_5. Top Catal 55:1046–1053
55. AlShalwi M, Hargreaves JSJ, Liggat JJ, Todd D (2012) The reactivity of lattice nitrogen species in molybdenum (oxy)carbonitrides prepared by single-source routes. Mater Res Bull 47:1251–1256
56. Rambeau G, Amariglio H (1981) Improvement of the catalytic performance of a ruthenium powder in ammonia synthesis by use of a cyclic procedure. Appl Catal 1:291–302
57. Rambeau G, Jorti A, Amariglio H (1982) Improvement of the catalytic performance of an osmium powder in ammonia synthesis by the use of a cyclic procedure. Appl Catal 3:273–282
58. Michalsky R, Pfromm PH (2011) Chromium as reactant for solar thermochemical synthesis of ammonia from steam, nitrogen and biomass at atmospheric pressure. Sol Energy 85:2642–2654
59. Michalksy R, Parman BJ, Amanor-Boadu V, Pfromm PH (2012) Solar thermochemical production of ammonia from water, air and sunlight: thermodynamic and economic analyses. Energy 42:251–260
60. Michalsky R, Pfromm PH (2012) An ionicity rationale to design solid phase metal nitride reactants for solar ammonia production. J Phys Chem C 116:23243–23251
61. Topham SA (1985) The history of the catalytic synthesis of ammonia. In: Anderson JR, Boudart M (eds) Catalysis science and technology, vol. 7, Chapter 1. Springer, Berlin

8

Carbon dioxide capture and utilization in petrochemical industry: potentials and challenges

Maryam Takht Ravanchi · Saeed Sahebdelfar

Abstract The reduction of greenhouse gas emissions is an ever-increasing challenge for production units and power plants in view of the global warming concerns. Carbon dioxide capturing from petrochemical process streams and flares has been recognized as one of the several strategies necessary for mitigating the emission of greenhouse gases into the atmosphere. To keep greenhouse gases at manageable levels, large decrease in CO_2 emissions through separation and utilization will be required. Furthermore, carbon dioxide emission potential will become an important factor in technology selection when cost of carbon dioxide emissions is set to be included in the processing cost. This article reviews the potential CO_2 separation technologies and its recycling via chemical fixation as bulk chemical products in petrochemical industry. Various separation techniques, such as absorption, adsorption, membrane separation and cryogenic, and utilization processes, such as conversion to carbon monoxide, oxidative dehydrogenation, hydrogenation and polymerization are thoroughly discussed. The technological challenges and recent developments and achievements are also presented.

Keywords Carbon dioxide · Greenhouse gas (GHG) · Carbon capture and utilization (CCU) · Catalytic conversion · C_1 chemistry · Petrochemical industry

M. Takht Ravanchi· S. Sahebdelfar
Catalyst Research Group, Petrochemical Research and Technology Company, National Petrochemical Company, No. 27, Sarv Alley, Shirazi-south, Mollasadra, P.O. Box 14358-84711, Tehran, Iran

Introduction

Environmental issues due to emissions of greenhouse gases (GHGs) have become worldwide problems. Studies have shown that increased GHG levels in atmosphere cause global warming. Carbon dioxide (CO_2), methane (CH_4), nitrous oxide (N_2O) and fluorinated gases (HFC, PFC, SF_6) are greenhouse gases among which carbon dioxide makes up a high proportion with respect to its amount present in the atmosphere (Fig. 1). Going by the prediction of Intergovernmental Panel on Climate Change (IPCC), by year 2100, the atmosphere may contain up to 570 ppm of CO_2, causing a rise in mean global temperature of around 1.9 °C and an increase in mean sea level of 3.8 m [1].

The proportion of greenhouse gases has increased significantly since the Industrial Revolution (about 1750), and global carbon dioxide (CO_2) concentration has risen by 37 %.

At present, it is accepted that the increase of GHG concentration in atmosphere is attributed to human activities. Greenhouse gas emissions are produced as the by-products of various non-energy-related industrial activities. Industrial processes can chemically transform raw materials, which often release waste gases such as CO_2. The processes include production of iron, steel, metallurgical coke, cement, lime, carbonate consuming processes (e.g., flux stone, flue gas desulfurization, and glass manufacturing), ammonia and urea, petrochemical, aluminum, soda ash, titanium dioxide, ferroalloy and glass. The production of some petrochemicals results in the release of small amounts of CO_2 emissions. Petrochemicals are chemicals isolated or derived from petroleum or natural gas. The chemical and petrochemical industry accounts for 30 % of global industrial energy use and 16 % of direct CO_2

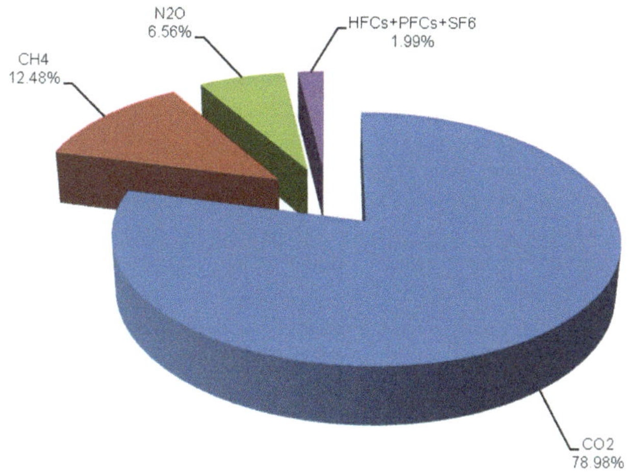

Fig. 1 Greenhouse gas emission in 2011 [1]

emissions. More than half of the energy demand is for feedstock use, which cannot be reduced through energy efficiency measures [2, 3].

The public concern about the climate change due to the greenhouse gas emissions to atmosphere led to the United Nation Framework Convention on Climate Change (UNFCCC) in 1992. The main objective of this meeting was to discuss methods for stabilization of GHG concentrations in the atmosphere at a level that the anthropogenic emissions could not interfere with the climate system. The stabilization of GHG concentrations in atmosphere can be achieved by three approaches:

- Reducing energy intensity
- Reducing carbon intensity (by use of carbon-free fuels)
- Enhancing the sequestration of CO_2

Carbon dioxide capture and sequestration (CCS) can significantly reduce emissions from CO_2 sources as well as certain industry types such as ethanol and natural gas processing plants. CCS methodologies for industrial and energy-related sources comprise three steps: CO_2 capture, CO_2 transportation and CO_2 utilization.

There are various petrochemical units that produce carbon dioxide with high purity as a by-product such as ethylene oxide and ammonia processes. Flaring is another main emission source of this greenhouse gas in petrochemical industry. In petrochemical industry, hydrocarbon wastes can be used as fuel and recycled. Therefore, the main waste products are water and CO_2. Carbon dioxide is produced in petrochemical process streams mostly by reactions with oxygenates (mainly oxygen as in oxidation of ethylene and water as in water–gas shift reaction). In fact, ethylene oxide and ammonia processes produce carbon dioxide with high purity as a by-product. However, most of the carbon dioxide in petrochemical industry is emitted in the flue gas as a result of burning fuel oil and fuel gas [4].

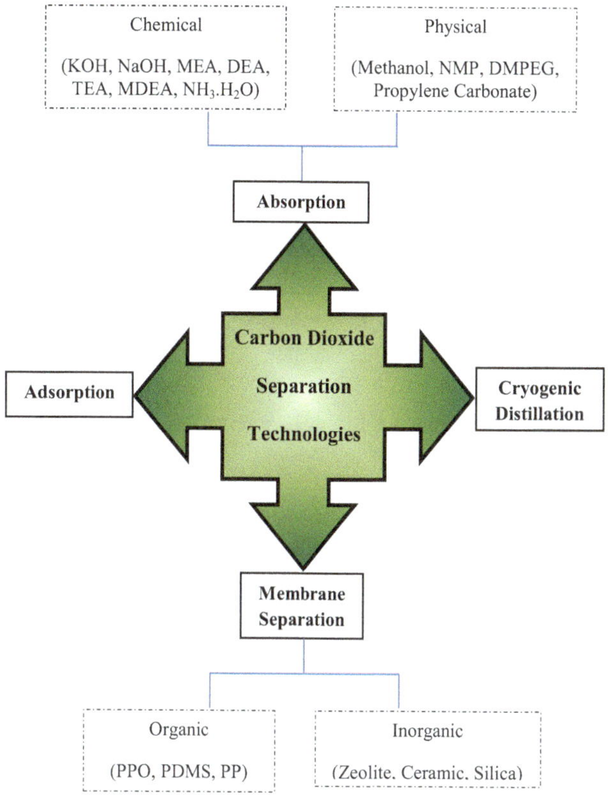

Fig. 2 Available technologies for CO_2 separation

In literature, there are some valuable reviews published in different aspects of CCS. For instance, in the field of CO_2 capture Granite and O'Brien [5]; Stewart and Hessami [6]; Olajire [7]; Pires et al. [8]; Takht Ravanchi et al. [9]; Wang et al. [10]; Kargari and Takht Ravanchi [11]; Zhao et al. [12]; Songolzadeh et al. [13]; and in the field of CO_2 utilization Wang and Zhu [14]; Song [15]; He et al. [16]; Havran et al. [17] are worth mentioning. Due to the absence of a systematic and detailed review on CO_2 separation and fixation in petrochemical industry, the present research focused on this goal.

CO_2 separation techniques

Capturing and separation of CO_2 from petrochemical gas streams are essential parameters for the carbon management for CO_2 sequestration from environment.

Generally, CO_2 separation contributes to 70–80 % of the overall CCS cost [4]. Hence, reducing separation cost is the most important issue for CCS process. In this section, common technologies of CO_2 separation from a flue gas are focused.

A wide range of technologies based on physical and chemical processes currently exist for CO_2 separation from a gas stream categorized as absorption, adsorption, membrane separation and cryogenics (Fig. 2). The choice of a suitable technology depends on the characteristics of gas stream.

Biological and hybrid processes and chemical looping combustion (CLC) are other CO_2 separation techniques which are not considered in this review. In biological processes, trees or microalgae can be used for CO_2 capture [6, 18]. Chemical looping combustion, proposed by Richter and Knoche [19], divides combustion into intermediate oxidation and reduction reactions that are performed separately with a solid oxygen carrier circulating between the separated sections. Hybrid systems can be based on membrane associated with cryogenic distillation or solvent absorption.

Absorption

Chemical absorption

The basic principle of chemical absorption is neutralization reaction in which absorbent react with CO_2 in the flue gas. The ideal chemical absorbent must have below specifications [20]:

- High reactivity with respect to CO_2—which would reduce height requirements for the absorber and/or reduce solvent circulation flow rates.
- Low regeneration cost requirements—based on a low heat of reaction with CO_2.
- High absorption capacity—which directly influences solvent circulation flow rate requirements.
- High thermal stability and reduced solvent degradation—reduced solvent waste due to thermal and chemical degradation.
- Low environmental impact.
- Low solvent costs—should be easy and cheap to produce.

Strong and weak alkaline solutions are commonly used absorbents. KOH and NaOH are examples of strong alkaline solutions, and alcohol amines (such as MEA, DEA, TEA and MDEA) and aqueous ammonia ($NH_3.H_2O$) are examples of weak alkaline solutions.

The applicability of strong alkaline solution is limited due to its strong corrosion to equipment. The alcohol amines usually suffer from the disadvantage of high energy requirement for regeneration, easy oxidation/degradation by SO_2 and O_2 in flue gas and system corrosion [21, 22]. Aqueous ammonia has the advantage of high-absorption efficiency, high-absorption capacity, low-energy requirement for absorbent regeneration and wide distribution of resources. In Table 1, advantages and disadvantages of chemical absorption by different alkaline solutions are categorized.

Table 1 Advantages and disadvantages of alkaline approaches for CO_2 capture

Process	Advantage	Disadvantage
Strong alkaline/alkaline metal solution (KOH, NaOH solution)	Fast reaction rate Large absorption capability High absorption efficiency	Cost of absorbent Strong corrosion to equipment Product treatment and disposal
Weak alkaline ($NH_3.H_2O$)	Large absorption capability and high loading capacity Low energy requirement for absorbent regeneration Utilization of products as fertilizer Wide distribution of absorbent	Easy to volatilize and leak Thermal instability of products Corrosion to equipment
Aqueous ammines (MEA, DEA, TEA, MDEA)	Less volatility Good stability of absorbent Enhancement role used as additive	Absorption efficiency High-energy consumption for regeneration Easy degradation by SO_2 and O_2 in flue gas Resulting in system corrosion

Between several chemical absorption processes, MEA would be the first to be available for immediate industrial application. Development of new solvents has achieved some progress such as ammonia [23], piperazine promoted K_2CO_3 [24], and concentrated aqueous piperazine [25].

Physical absorption

Physical absorption takes place at high CO_2 partial pressures. This process is not economical for streams with CO_2 partial pressures lower than 15 vol%.

In physical absorption, organic solvent is used to physically, rather than chemically, absorb CO_2. In this process, CO_2 removal is based on its solubility within the solvents which depends on the partial pressure and temperature of the feed gas. Higher CO_2 partial pressure and lower temperature favor the solubility of CO_2 in absorbent. The solvents are regenerated by increasing temperature or decreasing pressure. In comparison with chemical solvents, the interaction between CO_2 and absorbent is weak.

Physical absorption process and related solvents for CO_2 removal are commercially available. Selexol process is based on dimethyl ethers of polyethylene glycol (DMPEG), a Union Carbide liquid glycol-based solvent, and has been used since 1969. Glycol is effective for CO_2 capturing at high concentration. Rectisol process is based on low temperature methanol (cold methanol). Glycol carbonate is another interesting solvent because of its high selectivity for CO_2 but it has relatively low capacity. Propylene carbonate (FLUOR process) and *N*-methyl-2-pyrrolidone (NMP-Purisol) are other physical solvents [26].

Some CO_2 capture applications benefit from a mixture of physical and chemical solvents. The most commonly used examples are Sulfinol, a mixture of the physical solvent sulfolane and the amines such as diisopropyl amine (DIPA) or methyl diethanolamine (MDEA); and Amisol, a mixture of methanol and secondary amines. These hybrid solvents attempt to exploit the positive qualities of each constituent under special conditions.

Membrane separation

Membrane separation technology is a low cost process when high purity gas streams are not fundamental [27]. Membranes can separate carbon dioxide from a gas stream by size exclusion or by chemical affinity. Chemical affinity membranes are often impregnated with a scrubbing solution or chemical functional group (e.g., amine) selective for carbon dioxide. The scrubbing solutions, as well as functional groups on the surface of sorbents, are discussed in the critical review papers [28, 29]. A large body of research has been conducted on the properties of carbon dioxide selective membranes based on inorganic materials such as zeolites, alumina, carbon, and silica. Researchers proposed the separation of carbon dioxide from flue gases by multi-walled carbon nano-tube membranes [5]. Substantial advances in the selectivity, permeability, and chemical stability of inorganic membranes are needed for successful application to gas mixtures.

Membrane process has several advantages [30]:

- Higher separation energy efficiency relative to equilibrium-based processes.
- No waste streams production.
- Existence of commercial references (e.g., CO_2 separation from natural gas).
- No regeneration energy required.
- High packing density which requires small installations.
- Increasing mass transfer area in a given volume.

Its major drawbacks can be categorized as below:

- The treatment of an enormous flow rate of CO_2 containing gas requires a very large membrane area and increases the cost of this technology.
- The need of very large, expensive and energy consuming compression equipment.

Organic (polymeric) and inorganic (zeolite, ceramic and silica) membranes are of particular interest in CO_2 separation. Polymeric membranes separate gases based on solution-diffusion mechanism. Low cost, high separation performance, ease of synthesis and mechanical stability are their advantages [31, 32]. An important limitation of polymeric membrane is its application for a high-temperature gas stream as it cannot be applied without cooling of the gas stream in order not to destroy the membrane.

It was recently demonstrated that a zeolitic imidazolate framework with high thermal (up to 390 °C) and chemical stability under refluxing organic and aqueous media shows exceptional selectivity for CO_2 capture from CO_2/CO mixtures with an extraordinary capacity for storing CO_2.

Adsorption

Adsorption is also considered a feasible process for CO_2 capture at an industrial scale. It is a selective process in which molecules contained in liquid or gaseous mixtures adhere on a solid surface, the adsorbent. These molecules, even in small concentrations in the streams, can be captured by these selective materials. The properties of the adsorbed particles (molecular size, molecular weight and polarity) and the adsorbent surface (polarity, pore size and spacing) determine the adsorption quality. As adsorption is an exothermic process, the regeneration of the adsorbents through desorption can be performed by rising the temperature. However, adsorption presents lower energy requirements and avoids the shortcomings when compared to absorption [33]. Adsorption can be either physical (physisorption) or chemical (chemisorption).

Several adsorbents have been proposed for CO_2 capture, including carbon fiber monolithic adsorbents [34], activated carbon fiber-phenolic resin composites [35], melamine-formaldehyde highly porous adsorbents [36], amine immobilized adsorbents [37], red mud [38], steam-activated anthracite [39], activated carbon [40], lithium zirconate [41], lithium silicate [42, 43], alumina, metallic oxides and zeolites [44] among others [45, 46]. Sjostrom and Krutka [47] compared several adsorbents and concluded that carbon-based materials present excellent stability, while zeolites work well under dry conditions.

In the industrial context, a continuous CO_2 capture process is preferred while the technologies often studied are the cyclic processes. These adsorption processes can be grouped into two types, according to the way in which the adsorbent is exchanged between the adsorption and desorption steps: by changing the pressure or the temperature. In pressure swing adsorption (PSA), the adsorption is performed at pressures higher than atmospheric. Desorption occurs at atmospheric pressure. However, a vacuum desorption pressure can be applied and, for commercial adsorbent 13X, the costs of CO_2 capture are comparable with the cost of MEA absorption [48]. The operation with an adsorbent with higher adsorption capacity and higher selectivity to CO_2 would increase the profitability of PSA. The vacuum swing adsorption (VSA) operates at near-ambient temperature and pressures and desorption is performed at lower pressure. It is a promising technology due to the relatively low power consumption and ease of operation.

CO_2 containing gas streams to be treated by adsorption must have high CO_2 concentrations because of the low selectivity of the most available adsorbents.

Cryogenic distillation

CO_2 cryogenic distillation involves separation of gas mixture by fractional condensation and distillation at low temperature. Cryogenic temperatures are obtained by a closed-cycle operated refrigeration system consisting of a compressor, a Joule-Thompson valve (JTV), multi-stage heat exchangers and expanders.

Cryogenic distillation is a commercial process commonly used to liquefy and purify CO_2 from high purity sources (>90 %). It involves cooling the gas to a very low temperature (lower than −73.3 °C) so that CO_2 can freeze out/liquefied and separated.

The CryoCell® technology developed by Cool Energy Ltd (Australia) uses a cryogenic process for CO_2 removal. This technology eliminates water consumption, uses of chemicals and corrosion-related issues.

Advantages and constraints

In the field of carbon dioxide capture and utilization for petrochemical industry, the separation process has high importance as it should provide CO_2 within the specified purity required and with reasonable costs. Each of the above explained separation techniques has advantages and constraints. CO_2 concentration, temperature and pressure of flue gas streams are main operating parameters based on which suitable separation process can be selected.

Chemical absorption method has many obvious advantages such as high efficiency, high absorption capacity, low cost and mature technology. These solvents are usually preferred to physical ones. Physical solvents required low regeneration energy, but they have low absorption capacity and selectivity for CO_2 separation. Physical absorption is not economical for gas streams with CO_2 concentration lower than 15 vol%. Research efforts should be directed toward developing better solvents. Future progress in solvents should be:

- to reduce energy consumption;
- to avoid damage to environment and human being (when it is vented with treated gas);
- acceptable solvent degradation and corrosion to the packed column.

More important is to combine all these in one solvent. For gas streams with high temperature (about 373 K), low pressure and low CO_2 concentration (1 atm and 10–15 % mol), bulk absorption process may be the best suitable process for CO_2 separation.

Adsorption is suitable for separating CO_2 from low flow rate stream. Although, chemical adsorbents have high capacity and selectivity, but their regeneration is difficult. Physical adsorption is the most suitable for CO_2 capture at high pressures and low temperatures.

Since the pressure drop is driving force for membrane process, the gas stream must be compressed. Since compressing is very difficult and expensive, membrane separation is suitable for high pressure stream with high concentration (>10 vol%). Inorganic membranes have high thermal and chemical stability but their selectivity is lower than polymeric membranes. Polymeric membranes are very selective for CO_2 separation but they have low thermal stability. Glassy polymeric membranes have higher selectivity, while the rubbery polymeric membranes have higher thermal stability.

Membrane processes and cryogenic distillation are efficient for gas streams with high CO_2 concentration. Major disadvantage of cryogenic distillation is the large amount of energy required to provide the refrigeration.

Consequently, for capture from flue gases of petrochemical plants, adsorption and absorption appear to be the choice at least in short- and medium-term perspective.

CO_2 utilization

The disposal or utilization of the captured CO_2 is another problem for adequate reduction of CO_2 emissions. A potential and attractive route to use the captured CO_2 is its fixation as valuable petrochemical commodities or fuels. Figure 3 presents some of the current and potential utilization routes for CO_2.

Chemical fixation of carbon dioxide

Carbon dioxide, a cheap and abundant C-containing raw material, is interesting as substrate, solvent, reactant, and extracting agent. CO_2 is a potential C_1 building block for C–C chains or as competitive source of C in chemical

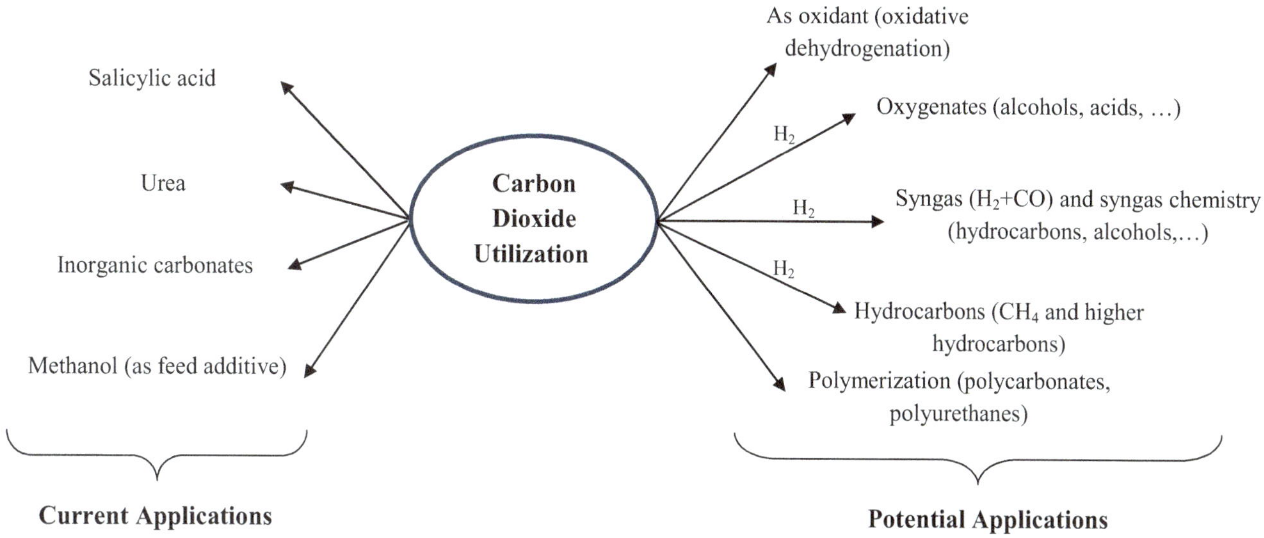

Fig. 3 Current and potential applications of CO_2 utilization

Table 2 Main uses of carbon dioxide in the chemical industry in Mtpy [55]

Process	Present	2016 forecast
Urea	114	132
Inorganic carbonates	ca. 50	70
Methanol	8	10
Total	172	217

industry and appears very attractive, although the molecule is rather inert and its reactions are energetically highly unfavorable [49]. There are several good recent review articles on CO_2 chemical fixation [17, 50–54].

The total amount of carbon dioxide used in industry is approximately 200 × 10^6 metric tons per year. Major industrial processes that utilize CO_2 as a raw material are listed in Table 2. Its utilization as a technological fluid, where carbon dioxide is not chemically converted and thus can be recovered at the end, is estimated at 30 × 10^6 metric tons per year [55]. The generation of energy used for CO_2 transformation, if based on hydrocarbon raw materials, still produces large amounts of CO_2. In addition, the resulting organic chemicals from CO_2 release carbon dioxide at the end of their use. However, despite the fact that, currently, the usage of CO_2 in the chemical industry cannot reduce significantly the global CO_2 levels, it is believed that the full potential of the fixation of CO_2 into value-added products has not yet been completely explored. The direct use of non-hydrocarbon based energy sources for reduction of CO_2 would change the above perspective and allow recycling of carbon dioxide via chemicals and liquid fuels. Thus, the further development of industrial processes, that are utilizing CO_2 for high-demand products, is of importance.

The scale of fixation for a specific product is an important parameter. Many fine chemicals or those of pharmaceutical interest produced by incorporation of CO_2 have only a limited demand worldwide (a few to a hundred ton annually). Regarding the order of magnitude of carbon dioxide emission problem (over 35 Gt annually), many of the claimed CO_2 reduction processes can be described as only a small step forward to resolve the problem and will not have a significant impact on the overall carbon emission problem. Those CO_2-fixing reactions that are feasible and show a significant bearing on overall worldwide carbon emissions are required on a very large scale. Typical examples include the conversion of CO_2 into fuels and polymeric/construction materials, which are in significant demand [54].

To date, only a few compounds have been successfully synthesized from CO_2 using heterogeneous catalytic routes, since most industrial processes utilize the more reactive but highly toxic carbon monoxide (CO) as a building block to produce higher chain hydrocarbons and liquid fuels. Processes for potential future utilization of CO_2 that have not yet been scaled up for commercialization are dependent on the development of new catalytic systems [17].

The source of energy supply should come from either renewable (solar, wind, hydro, ground heat, biomass, etc.) or integrated heat-management systems (nuclear plants or waste heat from industrial processes). Equally, the source of chemical reagents to couple with carbon dioxide should be carefully considered without any net enhancement in overall carbon emissions (to avoid a scenario that the CO_2 emission for the production of the coupling reagent is more than it captures). By taking thermodynamic limitations and the total carbon emissions of an overall process into account, many previously reported methods for carbon dioxide conversion may not be able to provide a practical way to globally reduce CO_2 emissions, although some of these reactions are of academic interest [54].

Chemistry and thermodynamic aspects

Being the most oxidized state of carbon, CO_2 as a raw material is in its lowest energy level, thus posing a major hindrance in establishing industrial processes. In other words, a large energy input is required to transform CO_2. There are four main methodologies to transform CO_2 into useful chemicals [52]:

- To use high-energy starting materials such as hydrogen, unsaturated compounds, small-membered ring compounds, and organometallics.
- To choose oxidized low-energy synthetic targets such as organic carbonates.
- To shift the equilibrium to the product side by removing a particular compound.
- To supply physical energy such as light or electricity.

Selecting appropriate reactions can lead to a negative Gibbs free energy of the reaction.

A widely accepted idea is that CO_2 is so thermodynamically and kinetically stable that it is rarely used to its fullest potential. However, due to the electron deficiency of the carbonyl carbons, CO_2 has a strong affinity toward nucleophiles and electron-donating reagents. In other words, CO_2 is an "anhydrous carbonic acid", which rapidly reacts with basic compounds [52].

Figure 4 shows the equilibrium conversion and product composition for methanation of carbon dioxide with hydrogen, obtained by constrained Gibbs free energy minimization method. This reaction is among thermodynamically favored reactions of CO_2. The equilibrium conversion is fairly high in the range of interest (300–400 °C) with a high selectivity to methane. However, as the temperature increases, the selectivity decreases and shifts to CO, indicating that the reverse reaction (dry reforming of methane) is favored. As higher organic compounds (hydrocarbons and oxygenates) are thermally unstable, lower temperatures favor higher selectivities to these compounds at the expense of lower reaction rates and CO_2 conversions.

Catalysis

Catalysts play a crucial role in CO_2 conversions, and suitable catalysts can realize interactions between CO_2 and a substrate on a transition metal center M. CO_2 has demonstrated surprising versatility by exhibiting a great variety of coordination and reaction modes in its homo- and polynuclear metal complexes. Most catalytic reactions involving CO_2 activation proceed via formal insertion of CO_2 into highly reactive M–X bonds with the formation of new C–X bonds. Coordinated CO_2 as in stable complexes is not necessarily required. Reactions are generally initiated by nucleophilic attack of X at the Lewis acidic C atom of CO_2 [49].

The conversion of carbon dioxide can be accomplished by homogeneous, heterogeneous and biological catalysts. For large-scale implementation, heterogeneous catalysis is preferred.

Homogeneous catalysis

Homogeneous catalytic processes for conversions of CO_2 are, by a large, reactions involving the incorporation of the entire CO_2 molecule into the products. These processes can be divided into two main categories: reductions of CO_2 that involve formal insertion of CO_2 into a bond in another molecule, and those that involve coupling of CO_2 with another unsaturated molecule [56].

The mild reaction conditions used in homogeneous hydrogenation of CO_2 catalyzed by transition metal complexes allow the partial hydrogenation of CO_2 to yield formic acid and derivatives:

$$CO_2 + H_2 \Leftrightarrow HCOOH \qquad \Delta H^0_{298} = -32 \text{ kJ/mol} \qquad (1)$$

Further reduction of formic acid under homogeneous conditions is more difficult and only very few examples are known reporting formation of methanol and methane catalyzed by Pd and Ru complexes in very low yields [57].

Heterogeneous catalysis

The heterogeneous catalytic conversion of CO_2 involves only hydrogenation reactions, with the exception of CO_2 reforming of CH_4, and occurs at high temperatures. Thus, the product portfolio is much more limited than for homogeneous catalysts:

$$CO_2 + H_2 \Leftrightarrow CO + H_2O \qquad \Delta H^0_{298} = +41 \text{ kJ/mol} \qquad (2)$$

$$CO_2 + CH_4 \Leftrightarrow 2CO + 2H_2 \qquad \Delta H^0_{298} = +247 \text{ kJ/mol} \qquad (3)$$

$$CO_2 + 3H_2 \Leftrightarrow CH_3OH + H_2O \qquad \Delta H^0_{298} = -49.9 \text{ kJ/mol} \qquad (4)$$

$$CO_2 + 4H_2 \Leftrightarrow CH_4 + 2H_2O \qquad \Delta H^0_{298} = -113 \text{ kJ/mol} \qquad (5)$$

Despite the limited diversity, these reactions could be potentially important for chemical utilization of CO_2 in the petrochemical industry. As these equations are equilibria, the product yields seldom reach 100 %, and, furthermore, the reactions are complicated by several equilibria occurring simultaneously at reaction temperature (see

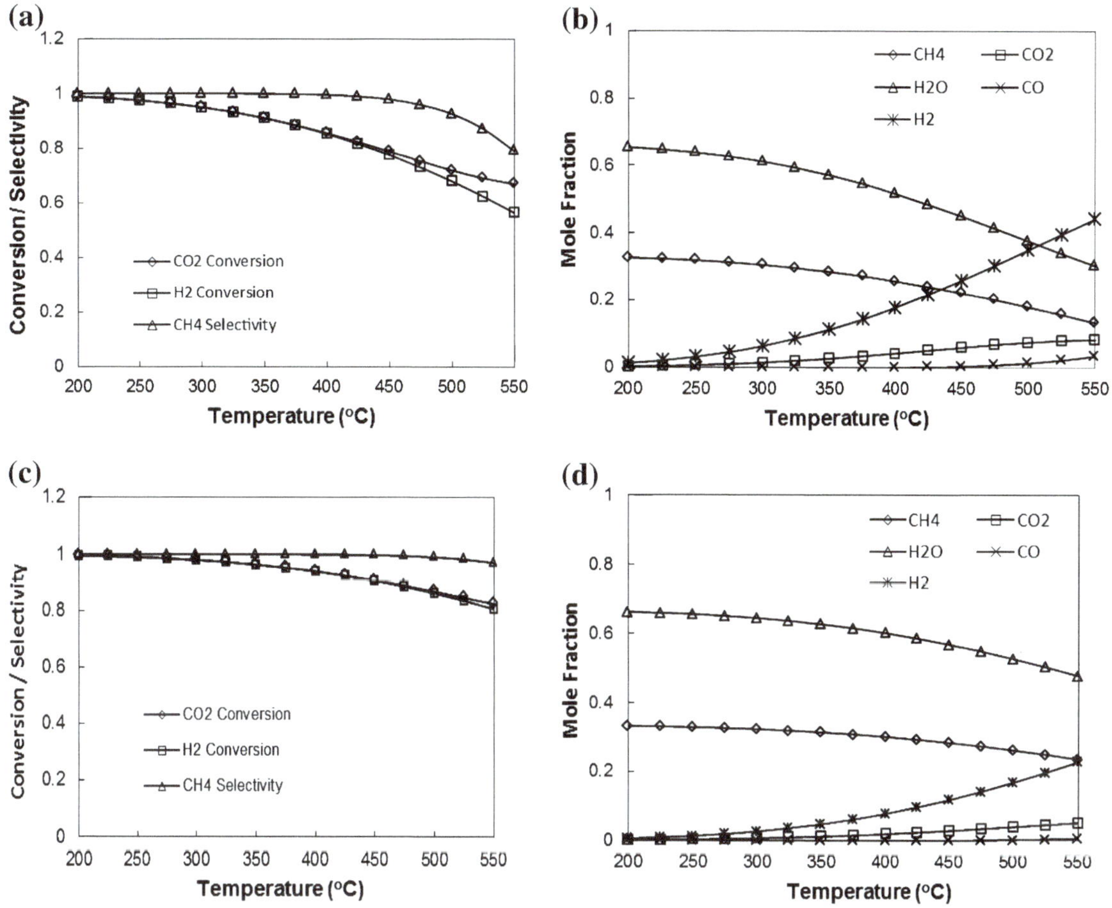

Fig. 4 The effect of temperature on conversions, selectivity and equilibrium product composition at 1 atm (**a**, **b**) and 10 atm (**c**, **d**) for stoichiometric ratio of CO_2/H_2 for methanation

Fig. 4). The art of heterogeneous catalysis is then the ability to minimize these other readily accessible reactions and steer the product formation to that desired.

Electro- and photochemical catalysis

The electro- and photochemical catalysis is identical in many respects; the major difference is the source of reducing electrons [56]. Electro- and photochemical reductions of CO_2 can also be catalyzed by metal complexes.

Methanol or related oxygenates (e.g., formic acid) can be made from CO_2 via aqueous electrocatalytic reduction in a "reversed fuel cell" without prior electrolysis of water to hydrogen [57]. The reversed fuel cell accomplishes the electrocatalytic reduction of CO_2 outside the potential range of the conventional electrolysis of water. In aqueous solutions, the selectivity depends on the electrode type. The different electrode behavior among metals can be related to their electronic configuration and can be grouped as *sp* and *d* metals. Selectivity depends on surface coverage of different reduced species. Higher surface hydrogen species concentration would lead to methane and hydrogen, while more carbon species would lead to CO and HCOOH. Copper has received much attention because it produces appreciable amounts of hydrocarbons and alcohols from CO_2 reduction, although it tends to give mixed product including hydrogen [51].

Photocatalytic reduction of CO_2 can be accomplished by suspending photosensitive semiconductor powder in aqueous solutions under irradiation, usually using UV light. Photoreduction of CO_2, however, is a completion with H_2 formation due to water decomposition, and leads to mixtures of reduced carbon products. Selectivity, therefore, is one major problem of these processes [57].

Potential processes

For effective reduction of CO_2 emissions, only reactions of high scale are of interest. Fortunately, some of heterogeneous catalytic conversions have this potential. Another need is renewable energy (hydrogen) source.

Conversion to carbon monoxide

An obvious way to use CO_2 in chemical synthesis is its reduction to CO followed by syngas (a mixture of H_2, CO and CO_2) chemistry. This can be performed by hydrogen or methane known as reverse water–gas shift (RWGS) reaction [Eq. (2)] and dry reforming of methane [Eq. (3)], respectively. Both reactions are endothermic and favored at high temperatures and susceptible to coke forming side reactions.

The commercial Fe_2O_3/Cr_2O_3, high-temperature shift catalyst might appear a logical candidate for RWGS reaction. However, despite a high initial activity, a low stability for this catalyst has been reported which was attributed to the reduction of Fe_2O_3 to Fe metal and coke formation [58]. ZnO/Cr_2O_3 catalyst, on the other hand, showed both high activity and stability with no coke formation.

The ZnO showed good specific reaction activity in the temperature range from 400 to 700 °C but this catalyst was deactivated. All the catalysts except the ZnO/Al_2O_3 (850 °C) showed low stability for the RWGS reaction and was deactivated at the reaction temperature of 600 °C. The ZnO/Al_2O_3 catalyst calcined at 850 °C was stable during 210 h under the reaction conditions of 600 °C and GHSV of 150,000 h^{-1}; showing CO selectivity of 100 % even at the pressure of 5 atm. The high stability of the ZnO/Al_2O_3 catalyst (850 °C) was attributed to the prevention of ZnO reduction by the formation of $ZnAl_2O_4$ spinel structure. The spinel structure of $ZnAl_2O_4$ phase in the ZnO/Al_2O_3 catalyst calcined at 850 °C was confirmed by XRD and electron diffraction [59].

Because of low equilibrium conversions, temperatures above 400 °C should be used. Excess hydrogen can also shift the reaction to completion. Therefore, there is need for the development of novel catalysts (e.g., Cu/ZnO, Pt/CeO_2) and reaction engineering approaches such as removal of water.

Carbon dioxide or dry reforming of methane has received much attention since 1980s. CO_2 reforming of CH_4 has great potential economic and environmental benefits. This reaction has the following important advantages:

- The formation of a suitable H_2/CO ratio for use in valuable oxygenates synthesis.
- The reduction of CO_2 and methane emissions; as both gases cause heavy greenhouse effect.
- Better use in chemical energy transmission.

The reaction is more endothermic than steam reforming. Whereas the steam reforming of methane yields syngases with a H_2/CO ratio of 3, CO_2 reforming of CH_4 results in a ratio of 1, which may be of interest for some syngas consuming reactions. The equilibrium composition is influenced by the water–gas shift (WGS) reaction. Carbon dioxide reforming needs higher temperatures (with the danger of coke formation); advantageously catalysts may be used (Ni on supports such as alumina or MgO).

Dry reforming reaction is catalyzed by Group VIII metals, such as Ni, Pt, Pd, Ru, Rh and Ir. However, there is a serious problem of catalyst deactivation due to rapid coke formation. In addition to carbon deposition, sintering, and/or metal oxidation could be the causes of catalyst deactivation. Different deactivation mechanisms are associated with different catalyst system. For example, in CO_2 reforming of CH_4 that carbon formation rather than sintering is the main cause of platinum-containing catalyst deactivation.

Precious metals are more coke resistant; however, considering the high cost and limited availability of noble metals, it is more practical in industrial standpoint to develop Ni-based catalysts with high performance and high resistance to carbon deposition. Research on the nickel catalysts used for this reaction has mainly focused on the intrinsic activity of the metal phase, stability toward carbon formation, the type of the support most suitable for improving the efficiency of the catalyst, and the reaction mechanism.

The nature of the support has also significant effects on the performances and properties of catalysts. Supports have many functions but the most important is to maintain a stable physical surface, especially at high reaction temperatures. Therefore, the support itself must resist to thermal transformation at high temperatures, which means catalyst support must have high melting points. It has been proved that type of the support and presence of modifiers greatly affect the coking tendency. The most commonly used support for dry reforming of methane is γ-Al_2O_3. Other supports such as MgO, TiO_2, SiO_2, and La_2O_3 were also used [60] and references therein). Recently, there has been interest in mixed oxides, such as $Ce_{1-x}Zr_xO_2$ [61], $LaBO_3$ [62, 63] (B = Co, Ni, Fe, Cr) and La_2NiO_4 [64, 65], either as catalysts or catalyst support. Al_2O_3 and SiO_2 are two of the most often investigated catalyst supports with high specific surface area. MgO and CaO are studied due to their high melting points which favor stable catalyst surfaces at high reaction temperatures. Also, their basicity is believed to suppress carbon formation by promoting the activation of CO_2. TiO_2 and ZrO_2 are reducible metal oxides on which metals are electron-rich and always show strong metal-support interaction (SMSI). SMSI is believed to be critical to resisting carbon formation.

Mesoporous non-crystalline zirconia has shown promising performance as a support for Ni-based catalysts. Long-term stable performance has been achieved even under conditions which favor coke formation and catalysts deactivation [66–69].

Oxidative dehydrogenation of hydrocarbons

This reaction is basically similar to the reactions discussed above and similarly carbon dioxide is not fixed. The product is an olefin and CO. Olefins are highly valued intermediates in petrochemical industry for the production of polymers and fuel additives. The simple dehydrogenation reaction is thermodynamically limited and accompanied by extensive coke formation. CO_2 is a coke removing agent as well. Bartholomew [70] reported that oxidizability in the gasification of coke was in the following order: $O_2(105) > H_2O(3) > CO_2(1) > H_2(0.003)$, where numbers in parenthesis indicated the relative ratios of the gasification with oxidants. CO_2 has been called as nontraditional, unusual or mild oxidant and oxygen transfer agent. In this sense, carbon dioxide could be proposed as an oxidant in the oxidative conversions of hydrocarbons.

The promoting effect of CO_2 is explained either by hydrogen consumption in the reverse water–gas shift reaction [Eq. (2)], which shifts the equilibrium of, for example, propane dehydrogenation [Eq. (6)] (two-step path) or its participation in direct oxidation of propane to propylene [Eq. (7)] (single-step path). Many studies considered both pathways, especially above 848 K (575 °C).

$$C_3H_8 \Leftrightarrow C_3H_6 + H_2 \qquad \Delta H^0_{298} = 124 \text{ kJ/mol} \tag{6}$$

$$C_3H_8 + CO_2 \Leftrightarrow C_3H_6 + CO + H_2O \quad \Delta H^0_{298} = 164 \text{ kJ/mol} \tag{7}$$

The oxide catalysts which are active in the non-oxidative dehydrogenation of propane are also active in the process with CO_2. Many bulk and supported oxide-based materials containing active species, such as Cr, Ga, Zn, V, In, Fe and Mn have been studied as catalysts for dehydrogenation of propane in the presence of CO_2 [71] and references therein). Among the investigated materials, those containing Cr or Ga were found to be the most active and selective one.

The single-step mechanism [direct oxidation of paraffin to olefin, Eq. (7)] has been proposed for redox oxides (Cr_2O_3, Cr_2O_5, ZnO and V_2O_5), where CO_2 can participate as an oxidizing agent in the following redox cycle [72]:

$$C_3H_8 + MeO_x \Leftrightarrow C_3H_6 + MeO_{x-1} + H_2O \tag{8}$$

$$CO_2 + MeO_{x-1} \Leftrightarrow CO + MeO_x \tag{9}$$

The two-step has been proposed for non-redox oxides such as Ga_2O_3 [73]. Production of H_2 could be an indication of two-step pathway.

The oxidative dehydrogenation of ethylbenzene to styrene over Cr or Ce catalysts is one of the most relevant of CO_2-assisted reactions [49].

Hydrogenation

The heterogeneous metal-catalyzed hydrogenation of CO_2 goes directly to methanol or methane which is the result of drastic reaction conditions applied in hydrogenations in the presence of heterogeneous catalysts. Formate, formaldehyde, and formyl species, however, have been shown by spectroscopic techniques to exist as surface intermediates during hydrogenation of carbon oxides [57].

Methanol is a versatile multi-source multi-purpose chemical. It could be used as a potentially carbon neutral fuel for automotive and fuel cell application and a feed stock for production of a variety of hydrocarbons including olefins and aromatics [74]. The methanol economy has been a suggested future economy in which methanol replaces fossil fuels as means of energy storage, fuel and raw material for synthetic hydrocarbons and their products [75].

The hydrogenation of CO_2 to methanol [Eq. (4)] is interrelated to the hydrogenation of CO [Eq. (10)] through water–gas shift reaction [Eq. (2)]. A significant disadvantage of CO_2 hydrogenation process as compared to the hydrogenation of CO is that more hydrogen is consumed because of the formation of water.

$$CO_2 + 2H_2 \Leftrightarrow CH_3OH \qquad \Delta H^0_{298} = -90.8 \text{ kJ/mol} \tag{10}$$

It is known through isotope labeling studies that in the hydrogenation of CO the primary source of methanol is CO_2 [57]. It is also known that the traditional three-component Cu catalysts (Cu–ZnO–Al_2O_3 and Cu–ZnO–Cr_2O_3) exhibiting high activity and selectivity in hydrogenation of CO to methanol show the low activity and selectivity in hydrogenation of pure CO_2. Therefore, efforts have been made to find new and more effective catalysts for direct hydrogenation. The problem is to develop catalysts with high selectivity to methanol formation but low selectivity to reverse water–gas shift reaction.

It has been widely reported that Cu–ZnO-based catalysts with additional oxides are among the most useful systems for catalytic hydrogenation of CO_2 to methanol. It has been well documented that the Cu/ZnO/ZrO_2 catalyst is active in the methanol synthesis from CO_2 and H_2 [76]. To increase its activity and stability, this basic catalyst was often modified by different oxide additives. Besides ZrO_2, other irreducible oxides such as Al_2O_3, TiO_2, and Ga_2O_3 have been investigated. An example is the high performance multicomponent Cu–ZnO–ZrO_2–Al_2O_3–Ga_2O_3 catalyst prepared by coprecipitation [77]. The addition of a small amount (0.6 wt%) of silica greatly improved the long-term stability of the catalyst [78].

As these findings indicate, ZnO is not an ideal component for CO_2 hydrogenation catalysts. As an alternative, a Cu–Ga_2O_3 catalyst was prepared using ZnO as support, and as support and a silica-supported Cu–ZnO–Ga_2O_3 sample was also investigated [79]. The latter catalyst exhibited high selectivity to methanol and the conversion to CO was very low.

Catalysts with metals other than Cu have also been prepared and used. Ru on zeolite NaY prepared by ion exchange and promoted with Co showed high activity for methanol production [80].

Methanation, that is, selective hydrogenation of CO_2 to methane [the Sabatier reaction, Eq. (4)], is potentially important transformation, although at present there is no practical reason for producing methane from carbon dioxide and hydrogen. Studies, nevertheless, have been carried out to develop suitable catalysts and find effective reaction conditions. The reaction is highly exothermic and heat removal from the reactor is a major issue.

The full reduction of CO_2 into CH_4 requires the transfer of eight electrons which leads to significant kinetic limitation. Among all catalysts studied, Ru, Rh and Ni supported on metal oxides exhibit the highest activity, and Ru has been reported to be the most selective. The supports exert a marked influence of both the adsorption and specific activity of the metals [57]. In most cases, TiO_2 is the best support material to achieve good catalyst performance, which attributed to a strong interaction between the metal and the support.

Carbon dioxide usually adsorbs dissociatively on noble metals, and hydrogen facilitates adsorption. On supported Ru, Rh and Pd catalysts, chemisorptions of CO_2 in the presence of hydrogen lead to the formation of surface carbonyl hydride [57]. Dissociation of CO_2 promoted by hydrogen, the dissociation of surface carbonyl hydride into reactive surface carbon, and hydrogenation of the latter are proposed as important steps in the hydrogenation process.

A related process is the production of ethanol and higher alcohols which have potential as gasoline additives. The reactions are thermodynamically more favorable than hydrogenation of CO_2 to methanol. The most successful catalysts are Rh-based, which is prohibitively expensive for large-scale production [81].

Polymerization

Regarding possible CO_2 mitigation by chemical fixation on a large scale, the conversion of carbon dioxide into polymeric and construction materials could be made commercially feasible.

The chemistry of polymerization catalysis is identical, often, to some of the reactions in homogeneous catalysis. Generally, however, the chemistry of copolymerization of CO_2 with organic compounds is not nearly as well developed as homogeneous catalysis. By far, the most common copolymers with CO_2 are polycarbonates, which is one of the type of biodegradable polymer [56, 82].

A promising methodology in the utilization of CO_2 is its transformation into valuable polycarbonates and/or cyclic carbonates. The synthesis of cyclic carbonates from epoxides and CO_2 (Fig. 5) is one of the major ways to transform CO_2 to valuable chemicals [51]. The first successful synthesis of cyclic carbonate using room temperature ion in liquids (ILs) of imidazolium and pyridinium salts was reported by Peng and Deng [83]. They found that 1-*n*-butyl-3-methylimidazolium tetraflouroburate was the most active catalyst with nearly 100 % selectivity, and a suitable CO_2-propylene oxide ratio was required for high conversion. The ionic liquid used as the catalyst was recyclable.

Copolymerization of monomers such as azetidines and aziridines with CO_2 for poly(urethane)s may lead to new polymers and construction materials for important uses [54]. Bayer announced to plan to market a process to produce polyurethane-based plastics using CO_2 as a raw material by 2015 [81] and references therein).

Recent developments and applications

Processes based on CO_2 are known for synthesis of hydrocarbons (via Fischer–Tropsch synthesis), energy-rich C_1 molecules (CH_4, CH_3OH, HCOOH, etc.) and fine chemicals containing functionalities such as –COO– (acids, esters, lactones), –N–COO– (carbamates), or –N–CO– (urea, amides) (Cornilis et al. 2003).

As was shown before the major drawback of dry CO_2 reforming is the high thermodynamic driving force to produce coke. The SPARG process and the CALCOR process have overcome the coking problem during reforming [81].

The sulfur passivated reforming (SPARG) process was developed to decrease the syngas H_2/CO ratio from 2.7 obtained from steam reforming to 1.8 without changing the steam reforming facility. Consequently, the SPARG process is suitable as a retrofit to the existing steam reforming plants. Because the use of CO_2 instead of steam lowers H/C ratios in the product, part of the steam was replaced by CO_2. The increased probability of carbon formation and the associated catalyst deactivation is minimized by introducing a partially sulfur-poisoned nickel catalyst. By partially sulfiding of the Ni catalyst, the sites for coke formation are blocked because of ensemble effect while sufficient sites for reforming reaction are maintained [84].

Due to lowering catalyst activity, higher operating temperatures are necessary. The process operates at 1,188–1,218 K which decreases the CH_4 slip but the

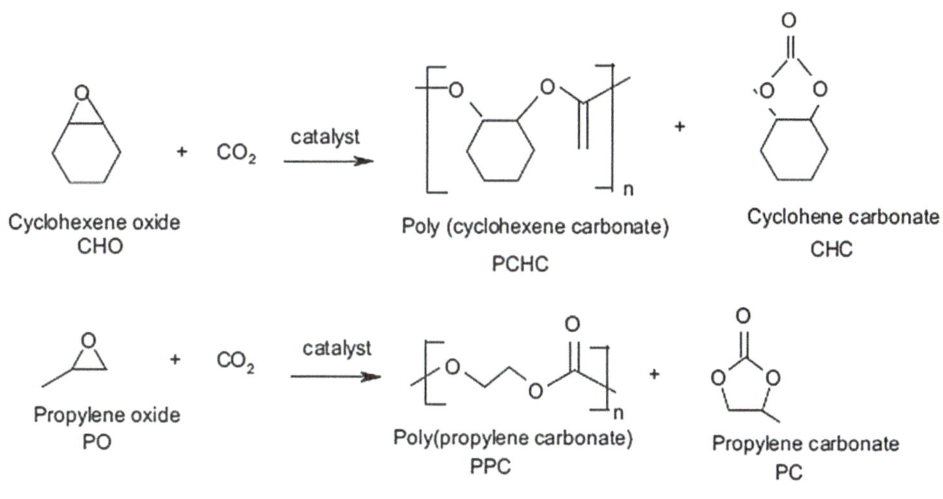

Fig. 5 Coupling of CO_2 and epoxides to afford polycarbonates and/or cyclic carbonates

product syngas still contains about 2.7 mol % methane. When CO_2/CH_4 reforming is used to produce CO for polycarbonate synthesis, the impurities in the CO (mainly H_2 and CH_4) cause inferior mechanical properties of polycarbonates made via the phosgene route.

Under the reaction conditions applied in the SPARG process, higher hydrocarbons (C_{2+}) in the feed will be thermally cracked to form pyrolytic carbon. The C_{2+} species in the feed should be removed by a pre-reforming step. Thus, traces of sulfur might be present in the syngas produced. When a pure syngas is required, an additional cleaning step will be necessary which involves an additional investment. Therefore, a catalyst which operates carbon free without the addition of sulfur is preferred.

The CALCOR process aiming at production of high purity CO from natural gas ($H_2/CO = 0.42$ in produced syngas) or LPG was developed to overcome problems involved in transportation of the toxic CO and the need for high product quality. These factors favor the production of CO on-site even for small capacities.

The CALCOR process operates under dry reforming conditions in an excess of CO_2. Details of catalysts used are proprietary. The prevention of carbon formation is claimed to be achieved using catalysts with different activities and shape and by their specific arrangement throughout the reformer tube. The syngas produced contains below 500 ppm of methane. Two options for purification of CO produced are available: membrane purification leads to CO purity of about 98.9 vol%, while the low temperature pressure swing adsorption purification can yield CO with purity higher than 99.4 vol%. The specific CO_2 emission per ton of CO product is 15 kg CO_2 for Calcor Standard and 360 kg CO_2 for Calcor Economy (emitted CO_2 minus imported CO_2) process, compared to ca. 1,800 kg steam reforming with CO_2 recycle [85].

The new Tri-reforming process uses flue gases as CO_2 source. Tri-reforming refers to simultaneous oxidation, CO_2 and steam reforming from natural gas. It is a synergic combination of endothermic CO_2 and steam reforming and exothermic partial oxidation of CH_4. CO_2 from flue gas does not need to be separated. In fact, H_2, O_2 and CO_2 in waste flue gas from fossil fuel power plants will be used to tri-reforming natural gas and produce syngas.

Tri-reforming does have a different requirement on the catalyst composition to dry and steam reforming. It can be performed with stable operation and no carbon formation and no appreciable deactivation of tri-forming conditions. Pt/ZrO_2 and Ni/ZrO_2 showed high activity for steam and dry reforming and partial oxidation of methane [86]. Ni on ZrO_2 and ZrO_2 doped with yttrium and a lanthanide and alkaline earth metal has been patented for this process [87].

In the CAMERE (carbon dioxide hydrogenation to form methanol via a reverse water–gas-shift reaction) of Korean Institute of Science and Technology, CO_2 is converted to methanol [58]. In CAMERE process, carbon dioxide and hydrogen are converted to CO and H_2O by RWGS reaction over a $ZnAl_2O_4$ catalyst in a separate reactor and then the gas mixture is fed into a methanol synthesis reactor after water has been removed.

A commercial plant constructed by Carbon Recycling International (CRI) company for CO_2 hydrogenation to methanol is currently operative in Iceland using geothermal energy source [81, 88]. Recently in Japan, the construction of a 100 tpy demonstration plant for single-step production of methanol from CO_2 was announced [89]. The CO_2 feed comes as a byproduct of Mitsui Chemical Complex at Osaka and H_2 from water using renewable (solar) energy [90].

Many petrochemical and licensing companies are working on carbon dioxide capture and use to achieve sustainability in the petrochemical industry. As reported by Reuters on Jan 14, 2014, SABIC and KACST are examples in the Middle East. Several projects at KACST contribute to the sustainability development in the petrochemical sector. Examples include power generation with available fuels at well heads with CO_2 capture and use for enhance oil recovery (EOR), CO_2 utilization to produce syngas (dry reforming project) and utilizing carbon dioxide as regenerative agent to increase the commercial potential of methane dry reforming.

Saudi Basic Industries Corp (SABIC) has hired Germany's Linde Group to build the world's largest plant for capturing and using climate-warming carbon dioxide. The United Jubail Petrochemical Company, an affiliate of SABIC, plans to capture around 1,500 tons a day of carbon dioxide from ethylene plants and purify it for use in SABIC-owned petrochemical plants in the industrial city of Jubail. The high cost of catching gas emitted by factories and power plants and storing it safely underground have deterred commercial CCS projects.

Mitsubishi Heavy Industries, Ltd. (MHI) has signed a license agreement for carbon dioxide (CO_2) recovery technology with Gulf Petrochemical Industries Company (GPIC), a manufacturer of fertilizers and petrochemicals in Bahrain. GPIC will use the technology to recover CO_2 from flue gas emitted at its existing petrochemical plant and utilize the captured CO_2 to increase urea and methanol production. The recovery units can capture 450 metric tons of CO_2 per day, one of the world's largest capacities for the chemical application. In addition to urea production, CO_2 recovery technology can be used in chemical applications such as production of methanol and dimethyl ether (DME) and, in the food and beverage industries, production of carbonated beverages and dry ice. Another important possible application is enhanced oil recovery (EOR) enabling increased crude oil production; in this case, CO_2 is injected into an oil reservoir suffering from low productivity [91].

Conclusions

There are many potential options for carbon dioxide capture from petrochemical industry and its utilization as a C_1 feedstock for the production of petrochemical intermediates and/or products. However, as the capture and the transformation of CO_2 by itself requires energy input and CO_2 gives no enthalpy contribution to the final product, the contribution to net CO_2 emission could not be significant unless more effective techniques and renewable energy (or hydrogen) source are available. Consequently, the product demand, the source of energy and overall net carbon emissions have to be taken into consideration. Large impact of carbon dioxide capture and fixation on global CO_2 emissions could be realized only if bulk chemicals are the target of conversion. Several potential utilization routes are possible for the petrochemical industry. Based on the above discussions on CO_2 separation methods, it can be concluded that flue gases properties (mainly concentration of CO_2, temperature and pressure) are the most effective factors for selection of suitable process for CO_2 separation. Catalytic CO_2 hydrogenation seems to have the highest potential for implementation in petrochemical industry. Further R&D is needed on catalysis and hydrogen (energy) source. Electrochemical or photo-electrochemical CO_2 conversion is promising, but technology is still in its infancy.

References

1. United Nations Framework Convention on Climate Change (UNFCCC). http://unfccc.int/ghg_data/ghg_data_unfccc/ghg_profiles/items/4625.php. Accessed January 2014
2. Sahebdelfar S, Tahriri Zangeneh F, Takht Ravanchi M (2009) Chemical recycling of carbon dioxide to valuable petrochemicals. In: The 7th APCSEET conference, Qingdao, China
3. Takht Ravanchi M, Sahebdelfar S, Tahriri Zangeneh F (2009a) Carbon dioxide sequestration in petrochemical industries with the aim of reduction in greenhouse gas emissions. In: The 7th APCSEET conference, Qingdao, China
4. Seddon D (2010) Petrochemical economics technology selection in a carbon constrained world. Imperial College, London
5. Granite EJ, O'Brien T (2005) Review of novel methods for carbon dioxide separation from flue and fuel gases. Fuel Proc Technol 86:1423–1434
6. Stewart C, Hessami MA (2005) A study of methods of carbon dioxide capture and sequestration–the sustainability of a photosynthetic bioreactor approach. Energ Convers Manage 46:403–420
7. Olajire AA (2010) CO_2 capture and separation technologies for end-of-pipe applications—a review. Energy 35:2610–2628
8. Pires JCM, Martins FG, Alvim-Ferraz MCM, Simes M (2011) Recent developments on carbon capture and storage: an overview. Chem Eng Res Des 89:1446–1460
9. Takht Ravanchi M, Sahebdelfar S, Tahriri Zangeneh F (2011) Carbon dioxide sequestration in petrochemical industries with the aim of reduction in greenhouse gas emissions. Front Chem Sci Eng 5:173–178
10. Wang M, Lawal A, Stephenson P, Sidders J, Ramshaw C (2011) Post-combustion CO_2 capture with chemical absorption: a state-of-the-art review. Chem Eng Res Des 89:1609–1624
11. Kargari A, Takht Ravanchi M (2012) Carbon dioxide: capturing and utilization. In: Liu G (ed) Greenhouse Gases INTECH, pp 1–30
12. Zhao B, Su Y, Tao W, Li L, Peng Y (2012) Post-combustion CO_2 capture by aqueous ammonia: a state-of-the-art review. Int J Greenh Gas Control 9:355–371
13. Songolzadeh M, Soleimani M, Takht Ravanchi M, Songolzadeh R (2014) Carbon dioxide separation from flue gases; a technological review emphasizing on reduction in greenhouse gas emissions. Sci World J. doi:
14. Wang Sh, Zhu ZH (2004) Catalytic conversion of alkanes to olefins by carbon dioxide oxidative dehydrogenation—a review. Energ Fuel 18:1126–1139
15. Song Ch (2006) Global challenges and strategies for control, conversion and utilization of CO_2 for sustainable development

involving energy, catalysis, adsorption and chemical processing. Catal Today 115:2–32
16. He LN, Wang JQ, Wang JL (2009) Carbon dioxide chemistry: examples and challenges in chemical utilization of carbon dioxide. Pure Appl Chem 81:2069–2080
17. Havran V, Dudukovic MP, Lo CS (2011) Conversion of methane and carbon dioxide to higher value products. Ind Eng Chem Res 50:7089–7100
18. Skjånes K, Lindblad P, Muller J (2007) BioCO_2—a multidisciplinary, biological approach using solar energy to capture CO_2 while producing H_2 and high value products. Biomol Eng 24:405–413
19. Richter HJ, Knoche K (1983) Reversibility of combustion processes, efficiency and costing- second law analysis of processes. ACS Symp Ser 235:71–85
20. Davidson RM (2007) Post-combustion carbon capture from coal fired plants—solvent scrubbing. IEA Clean Coal Centre, CCC/125
21. Alie C, Backham L, Croiset E, Douglas PL (2005) Simulation of CO_2 capture using MEA scrubbing: a flowsheet decomposition method. Energy Conv Manag 46:475–487
22. Mamun S, Svendsen HF, Haff KA, Juliussen O (2007) Selection of new absorbents for carbon dioxide capture. Energy Conver Manage 48:251–258
23. Darde V, Thomsen K, van Well WJM, Stenby EH (2010) Chilled ammonia process for CO_2 capture. Int J Greenh Gas Control 4:131–136
24. Cullinane JT, Rochelle GT (2005) Thermodynamics of aqueous potassium carbonate, piperazine, and carbon dioxide. Fluid Phase Equilib 227:197–213
25. Freeman SA, Dugas R, Van Wagener DH, Nguyen T, Rochelle GT (2010) Carbon dioxide capture with concentrated, aqueous piperazine. Int J Greenh Gas Control 4:119–124
26. Kovvali AS, Sirkar KK (2002) Carbon dioxide separation with novel solvents as liquid membranes. Ind Eng Chem Res 41:2287–2295
27. Takht Ravanchi M, Kaghazchi T, Kargari A (2009) Application of membrane separation processes in petrochemical industry: a review. Desal 235:199–244
28. Chow JC, Watson JG, Herzog A, Benson SM, Hidy GM, Gunter WD, Penkala SJ, White CM (2003) Critical review discussion-separation and capture of CO_2 from large stationary sources and sequestration in geological formations. J Air Waste Manage Assoc 53:1172–1182
29. White CM, Strazisar BR, Granite EJ, Hoffman JS, Pennline HW (2003) Separation and capture of CO_2 from large stationary sources and sequestration in geological formations—coal beds and deep saline aquifers. J Air Waste Manage Assoc 53:645–715
30. Bounaceur R, Lape N, Roizard D, Vallieres C, Favre E (2006) Membrane processes for post-combustion carbon dioxide capture: a parametric study. Energy 31:2556–2570
31. Scholes CA, Kentish SE, Stevens GW (2009) The effect of condensable minor components on the gas separation performance of polymeric membranes for carbon dioxide capture. Energy Procedia 1:311–317
32. Xomeritakis G, Tsai CY, Jiang YB, Brinker CJ (2009) Tubular ceramic-supported sol-gel silica-based membranes for flue gas carbon dioxide capture and sequestration. J Membr Sci 341:30–36
33. Drage TC, Smith KM, Pevida C, Arenillas A, Snape CE (2009) Development of adsorbent technologies for post-combustion CO_2 capture. Energy Procedia 1:881–884
34. Thiruvenkatachari R, Su S, An H, Yu XX (2009) Post combustion CO_2 capture by carbon fibre monolithic adsorbents. Prog Energ Combust 35:438–455
35. An H, Feng B, Su S (2009) CO_2 capture capacities of activated carbon fibre-phenolic resin composites. Carbon 47:2396–2405
36. Pevida C, Snape CE, Drage TC (2009) Templated polymeric materials as adsorbents for the postcombustion capture of CO_2. Energy Procedia 1:869–874
37. Chatti R, Bansiwal AK, Thote JA, Kumar V, Jadhav P, Lokhande SK, Biniwale RB, Labhsetwar NK, Rayalu SS (2009) Amine loaded zeolites for carbon dioxide capture: amine loading and adsorption studies. Micropor Mesopor Mater 121:84–89
38. Yadav VS, Prasad M, Khan J, Amritphale SS, Singh M, Raju CB (2010) Sequestration of carbon dioxide (CO_2) using red mud. J Hazard Mater 176:1044–1050
39. Maroto-Valer MM, Tang Z, Zhang Y (2005) CO_2 capture by activated and impregnated anthracites. Fuel Process Technol 86:1487–1502
40. Pevida C, Plaza MG, Arias B, Fermoso J, Rubiera F, Pis JJ (2008) Surface modification of activated carbons for CO_2 capture. Appl Surf Sci 254:7165–7172
41. Fauth DJ, Frommell EA, Hoffman JS, Reasbeck RP, Pennline HW (2005) Eutectic salt promoted lithium zirconate: novel high temperature sorbent for CO_2 capture. Fuel Process Technol 86:1503–1521
42. Essaki K, Nakagawa K, Kato M, Uemoto H (2004) CO_2 absorption by lithium silicate at room temperature. J Chem Eng Jpn 37:772–777
43. Kato M, Nakagawa K, Essaki K, Maezawa Y, Takeda S, Kogo R (2005) Novel CO_2 absorbents using lithium-containing oxide. Int J Appl Ceramic Technol 2:467–475
44. Zhao Z, Cui X, Ma J, Li R (2007) Adsorption of carbon dioxide on alkali-modified zeolite 13× adsorbents. Int J Greenh Gas Control 1:355–359
45. Drage TC, Arenillas A, Smith KM, Pevida C, Piippo S, Snape CE (2007) Preparation of carbon dioxide adsorbents from the chemical activation of urea-formaldehyde and melamine-formaldehyde resins. Fuel 86:22–31
46. Singh R, Reddy MKR, Wilson S, Joshi K, Diniz da Costa JC, Webley P (2009) High temperature materials for CO_2 capture. Energy Procedia 1:623–630
47. Sjostrom S, Krutka H (2010) Evaluation of solid sorbents as a retrofit technology for CO_2 capture. Fuel 89:1298–1306
48. Ho MT, Allinson GW, Wiley DE (2008) Reducing the cost of CO_2 capture from flue gases using pressure swing adsorption. Ind Eng Chem Res 47:4883–4890
49. Cornils B, Herrmann WA, Schlogl R, Wong C-H (2003) Catalysis from A to Z: A Concise Encyclopedia. Wiley, Weinheim
50. Hua B, Guild C, Suib SL (2013) Thermal, electrochemical, and photochemical conversion of CO_2 to fuels and value-added products. J CO_2 Util 1:18–27
51. Olajire AA (2013) Valorization of greenhouse carbon dioxide emissions into value-added products by catalytic processes. J CO_2 Util 3–4:74–92
52. Sakakura T, Choi J-C, Yasuda H (2007) Transformation of carbon dioxide. Chem Rev 107:2365–2387
53. Tahriri Zangeneh F, Sahebdelfar S, Takht Ravanchi M (2011) Conversion of carbon dioxide to valuable petrochemicals: an approach to clean development mechanism. J Nat Gas Chem 20:219–231
54. Yu KMK, Curcic I, Gabriel J, Tsang SCE (2008) Recent advances in CO_2 capture and utilization. ChemSusChem 1:893–899
55. Aresta M, Dibenedetto A, Angelini A (2013) The changing paradigm in CO_2 utilization. J CO_2 Util 3–4:65–73
56. Heyn RH (2003) Carbon dioxide conversion. In: Horvath IT (ed) Encyclopedia of catalysis. Wiley, New Jersey, pp 19–71
57. Olah GA, Molnar A (2003) Hydrocarbon chemistry. Wiley, New Jersey

58. Park SW, Joo OS, Jung KD, Kim H, Han SH (2000) ZrO_2/Cr_2O_3 catalyst for reverse-water-gas-shift reaction of CAMERE process. Korean J Chem Eng 17:719–722
59. Joo O-S, Jung K-D (2003) Stability of $ZnAl_2O_4$ catalyst for reverse-water-gas-shift reaction (RWGSR). Bull Korean Chem Soc 24:86–90
60. Guo J, Lou H, Zhao H, Chai D, Zheng X (2004) Dry reforming of methane over nickel catalysts supported on magnesium aluminate spinels. Appl Catal A Gen 273:75–82
61. Montoya JA, Romero-Pascual E, Gimon C, Del Angel P, Monzon A (2000) Methane reforming with CO_2 over Ni/ZrO_2-CeO_2 catalysts prepared by sol-gel. Catal Today 63:71–85
62. Guo JJ, Lou H, Zhu YH, Zheng XM (2003) CO_2 reforming of CH_4 over nickel and cobalt catalysts prepared from La-based perovskite precursors. J Nat Gas Chem 12:17–22
63. Wu YY, Kawaguchi O, Matsuda T (1998) Catalytic reforming of methane with carbon dioxide on $LaBO_3$ (B = Co, Ni, Fe, Cr) catalysts. Bull Chem Soc Jpn 71:563–572
64. Guo JJ, Lou H, Zhu YH, Zheng XM (2003) La-based perovskite preparations, properties, and its catalytic action for CO_2 reforming of CH_4. Mater Lett 57:4450–4456
65. Nam JW, Chae H, Lee SH, Jung H, Lee KY (1998) Natural gas conversion. Stud Surf Sci Catal 119:843–847
66. Rezaei M, Alavi SM, Sahebdelfar S, Yan ZF (2006) Nanocrystalline zirconia as support for nickel catalyst in methane reforming with CO_2. Energy Fuels 20:923–929
67. Rezaei M, Alavi SM, Sahebdelfar S, Liu XM, Qian L, Yan ZF (2007) CO_2-CH_4 reforming over nickel catalysts supported on mesoporous nanocrystalline zirconia with high surface area. Energy Fuels 21:581–589
68. Rezaei M, Alavi SM, Sahebdelfar S, Yan ZF (2008) Effect of CO_2 concentration on the activity and stability of nickel catalyst supported on mesoporous nanocrystalline zirconia. J Nat Gas Chem 17:278–288
69. Rezaei M, Alavi SM, Sahebdelfar S, Yan ZF (2009) A highly stable catalyst in methane reforming with carbon dioxide. Scripta Mater 61:173–176
70. Bartholomew CH (1984) Catalyst deactivation. Chem Eng 91:96–112
71. Michorczyk P, Ogonowski J, Zenczak K (2011) Activity of chromium oxide deposited on different silica supports in the dehydrogenation of propane with CO_2—a comparative study. J Mol Catal A Chem 349:1–12
72. Michorczyk P, Ogonowski J, Kustrowski P, Chmielarz L (2008) Chromium oxide supported on MCM-41 as a highly active and selective catalyst for dehydrogenation of propane with CO_2. Appl Catal A Gen 349:62–69
73. Xu B, Zheng B, Hua W, Yue Y, Gao Z (2006) Support effect in dehydrogenation of propane in the presence of CO_2 over supported gallium oxide catalysts. J Catal 239:470–477
74. Olah GA (2005) Beyond oil and gas: the methanol economy. Angew Chem Int Ed 44:2636–2639
75. Olah GA, Goeppert A, Prakash GKS (2006) Beyond oil and gas: the methanol economy. Wiley-VCH, Weinheim
76. Słoczynski J, Grabowski R, Olszewski P, Kozłowska A, Stoch J, Lachowska M, Skrzypek J (2006) Effect of metal oxide additives on the activity and stability of $Cu/ZnO/ZrO_2$ catalysts in the synthesis of methanol from CO_2 and H_2. App Catal A Gen 310:127–137
77. Saito M, Fujitani T, Takeuchi M, Watanabe T (1996) Development of copper/zinc oxide-based multicomponent catalysts for methanol synthesis from carbon dioxide and hydrogen. Appl Catal A 138:311–318
78. Saito M, Takeuchi M, Fujitani T, Toyir J, Lou S, Wu J, Mabuse H, Ushikoshi K, Mori K, Watanabe T (2000) Advances in joint research between NIRE and RITE for developing a novel technology for methanol synthesis from CO_2 and H_2. Appl Organomet Chem 14:763–772
79. Toyir J, de la Piscina RR, Fierro JLG, Homs N (2001) Highly effective conversion of CO_2 to methanol over supported and promoted copper-based catalyst: influence of support and promoter. App Catal B 29:207–215
80. Bando KK, Arakawa H, Ichikuni N (1999) CO_2 hydrogenation over micro- and mesoporous oxides supported catalysts. Catal Lett 60:125–132
81. Homs N, Toyir J, de la Piscia PR (2013) Catalytic processes for activation of CO_2. In: Suib SL (ed) New and future developments in catalysis: activation of carbon dioxide, 1st edn. Elsevier, Amsterdam, pp 1–23
82. Miao CX, Wang JQ, He LN (2008) Catalytic processes for chemical conversion of carbon dioxide into cyclic carbonates and polycarbonates. Open Organic Chem J 2:68–82
83. Peng JJ, Deng YQ (2001) Cycloaddition of carbon dioxide to propylene oxide by ionic liquids. New J Chem 25:639–641
84. Verykios XE (2003) Catalytic dry reforming of natural gas for production of chemicals and hydrogen. Int J Hydrog Energy 28:1045–1063
85. Teuner STC, Neuman P, Linde FV (2001) The calcor standard and calcor economy processes. Oil Gas Euro Mag 3:44–46
86. Hegarty HES, Ross JRH, O'Conner AM (1998) Syngas production from natural gas using ZrO_2-supported metals. Catal Today 42:225–232
87. Moon DJ, Kang JS, Nho WS, Kim DH, Lee SD, Lee BG (2008) Ni-based catalyst for tri-reforming of methane and its catalytic application for the production of syngas. US 20080260628 A
88. Shulenberger AM, Jonsson FR, Ingolfsson O, Tran KC (2007) US patent application 0244208 A1
89. Trembaly JF (2008) CO_2 as a feedstock, Mitsui will make methanol from greenhouse gas. Chem Eng News 86:13
90. Olah GA, Goeppert Prakash GKS (2009) Chemical recycling of carbon dioxide to methanol and dimethyl ether: from Greenhouse Gas to renewable environmentally carbon neutral and synthetic hydrocarbons. J Org Chem 74:487–498
91. Bahrain: Betting on Industry. http://www.bahiw.com/pdf/ecnomicmarch0109.pdf. Accessed January 2014

9

Ethanol photoreactions over Au–Pd/TiO_2

A. K. Wahab · S. Bashir · Y. Al-Salik · H. Idriss

Abstract A prototype Au–Pd/TiO_2 catalyst was prepared, characterized and tested for the photoreaction of ethanol. XPS Au4f and Pd3d indicated that the as-prepared material is composed of metallic Au, metallic Pd as well oxidized Pd (Pd^{2+}). Ar ion sputtering (5 min) of the catalyst surface resulted in almost total reduction of Pd^{2+} to metallic Pd in addition to considerable reduction of surface Ti cations to Ti^{3+} and Ti^{2+} cations; XPS Au4f lines were not affected. Transmission electron microscopic studies indicated that Au particles have a mean particle size of about 3.5 nm while Pd particles are smaller 1–1.5 nm in size. UV excitation of the catalyst in ultrahigh vacuum (UHV) conditions resulted in the formation of acetaldehyde and hydrogen in addition to photodesorption of the reactant ethanol. Hydrogen production, representing ca. 30 % of the desorbing products, was delayed compared to acetaldehyde desorption. This was interpreted as due to kinetic effect whereby initially most electrons transferred to the conduction band are trapped by surface hydroxyls as well inevitable presence of oxygen in the powder accelerating acetaldehyde formation (dehydrogenation). Only once most oxygen-containing species have reacted and molecular hydrogen was formed. To our knowledge, this is the first UHV in situ study of hydrogen production from ethanol photocatalytically over M/TiO_2 system.

Keywords Hydrogen production · Ethanol photoreaction · Gold plasmon · Synergism · Ethanol dehydrogenation

A. K. Wahab · S. Bashir · Y. Al-Salik · H. Idriss
SABIC-Corporate Research and Innovation (CRI) at KAUST, Thuwal, Saudi Arabia

Introduction

Light from the sun, the most abundant source of energy on Earth, contributes by <0.05 % of the total power (15,000 GW annual) used by humans (excluding solar heating which contributes around 0.6 %). The estimated practical and convertible power[1] that the Earth surface receives is equivalent to that provided by 600,000 nuclear reactors (one nuclear power plant generates on average 1 GW power). One mode of solar energy utilization is the use of sunlight to generate energy carriers such as hydrogen from renewable sources (e.g., ethanol and water) using semiconductor photocatalysts.

The photo-assisted splitting of water into hydrogen and oxygen was first achieved by Fujishima and Honda [1], who showed that hydrogen and oxygen could be generated in an electrochemical cell containing a titania photoelectrode, provided an external bias was applied. Since that time, numerous researchers have explored ways of achieving direct water dissociation without the need for an external bias. Much work has been conducted since, a large fraction of which is discussed in recent reviews [2–6]. Among the many issues affecting direct water splitting is the need to separate hydrogen from oxygen and the relatively low hydrogen evolution rates so far achieved. These in addition to the need of using UV light (>3 eV) to excite TiO_2 and other related materials has been one of the main obstacles for practical applications. Many authors have sought modified photocatalysts which, unlike pure TiO_2, respond to visible (sunlight) excitation, with limited success to date; see some of these materials in ref. [3].

[1] The total amount of sun light reaching the earth surface is orders of magnitude higher than the quoted figure.

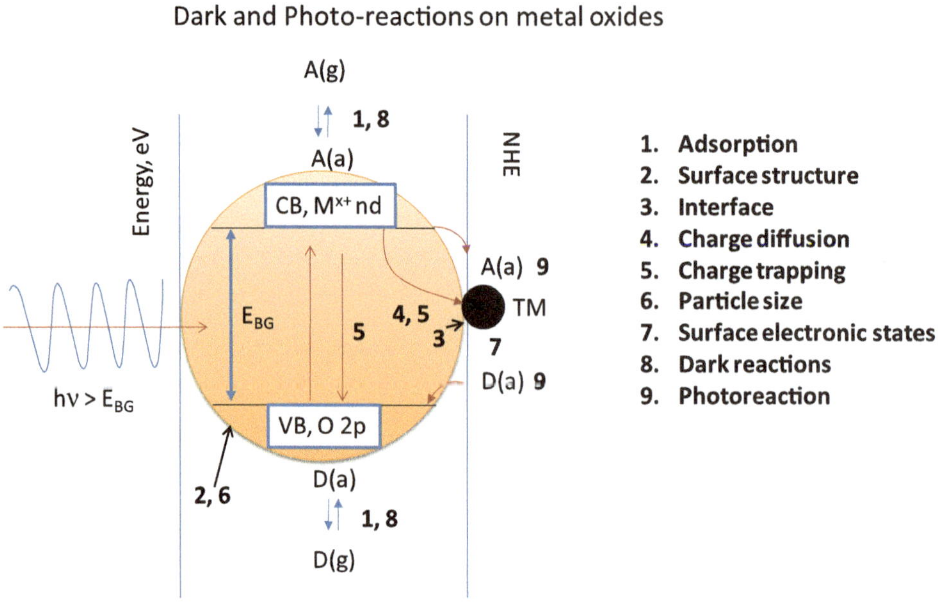

Fig. 1 Schematic representation of the main steps during a photocatalytic reaction involving a metal (*small black circle*) deposited on a metal oxide semiconductor (*large circle*). Also shown are the valence and conduction bands energy with respect to the normal hydrogen electrode (NHE) scale. Adsorbed species are represented by A (electron acceptor) and D (electron donor)

Many approaches have been conducted to design a photocatalyst that can work under direct sunlight in stable conditions. To achieve this, a few key factors need to be overcome chiefs of them are light absorption efficiency, charge carrier life time and materials stability. To enhance light absorption, a large number of photocatalysts were designed based on visible light-range band gap either by solid solutions, hybrid materials or doping of wide-band gap semiconductors. To decrease the charge carriers' life time and also hydride semiconductors, metal nanoparticles are added and sacrificial agents are used [2–6]. The scheme in Fig. 1 presents several steps and concepts involved in photoreactions composed of a semiconductor on which a metal particle is deposited. As can be seen, many factors can affect the reaction in addition to basic thermodynamic requirements and these include charge carrier (electrons and holes) diffusion (step 4) from the bulk to the surface [and interface (step 3)] both affected by bulk and surface structures (step 2) which are in turn related to particle sizes (step 6) and oxidation states (step 7). Kinetics is also important to consider in particular when sacrificial organic compounds are used, therefore, the effect of concentrations and temperatures on the surface populations (step 8) and ultimately the photoreaction rate (step 9) is not negligible [7].

In this work, we present a study on Au–Pd/TiO_2 that is found active for hydrogen production [8]. The Au/TiO_2 system has been studied by us and others in numerous works previously [9–13]. Au metal has at least two important roles in photoreaction, first it acts as sink for transferred electrons from the conduction band and second it contributes by its plasmon response (excited by visible light) in the electron transfer reaction [14–16] albeit is not a fully understood mechanism. To act as an efficient plasmon, Au particles need to be relatively large. Due to this requirement they are less dispersed on the surface of TiO_2. Pd is used to decorate the surface around Au so in essence Pd acts as efficient electron trap center, while Au induces the electric field due to it is electronic oscillation; Pd particles are easily prepared in 1 nm size on top of TiO_2. The TiO_2 used in this work is composed of anatase and rutile phases in ca. 80–20 ratio. This ratio is thought to be in the range where considerable synergism occurs between the two phases further increasing the reaction rate [17]. The catalyst has therefore a combination of properties that are important in photocatalysis: surface plasmons (Au), synergism (anatase + rutile), metal–semiconductor interface (Au/TiO_2 and Pd/TiO_2). In this work, we focus on its properties as well as a mechanistic study of ethanol reaction under UV excitation in which the channels for hydrogen desorption and acetaldehyde desorption seem to be decoupled on the powder material.

Experimental

X-ray photoelectron spectroscopy was conducted using a Thermo scientific ESCALAB 250 Xi. The base pressure of the chamber was typically in the low 10^{-10} to high 10^{-11} mbar range. Charge neutralization was used for all samples. Spectra were calibrated with respect to C1s at 285.0 eV. Au4f, Pd 3d, O1s, Ti2p, C1s and valence band energy regions were scanned for all materials. Typical acquisition conditions were as follows: pass energy = 30 eV and scan rate = 0.1 eV per 200 ms. Ar ion bombardment was performed with an EX06 ion gun at

1 kV beam energy and 10 mA emission current; sample current was typically 0.9–1.0 nA. Self-supported oxide disks of approximately 0.5 cm diameter were loaded into the chamber for analysis. UV–Vis absorbance spectra of the powdered catalysts were collected over the wavelength range of 250–900 nm on a Thermo Fisher Scientific UV–Vis spectrophotometer equipped with praying mantis diffuse reflectance accessory. Absorbance (A) and reflectance (% R) of the samples were measured. The reflectance (% R) data were used to calculate the band gap of the samples using the Tauc plot (Kubelka–Munk function). The TiO_2 exhibited the typical two band gaps of anatase and rutile at 3.2 and 3.0 eV, respectively. BET surface area was measured using Quantachrome Autosorb analyzer by N_2 adsorption. The surface area was found to be 55 m^2/g_{Catal}. Transmission electron microscopy studies were performed at 200 kV with a JEOL JEM 2010F instrument equipped with a field emission source. The microscope was operated in HAADF-STEM mode (Z-contrast). Samples were dispersed in alcohol in an ultrasonic bath and a drop of supernatant suspension was poured onto a holey carbon-coated grid. In situ study was conducted in an ultrahigh vacuum chamber equipped with an online quadrupole mass spectrometer (RGA Hiden; 300 AMU), a sputter ion gun (PHI), a dosing line for ethanol exposure. The chamber is pumped down with an ion pump (PHI), turbo pump (Pfeiffer) and a titanium sublimation pump (PHI). Base pressure was in the low to mid 10^{-9} torr. Catalyst (as a pressed pellet of about 0.5 cm in diameter; total amount about 50 mg) was loaded into an *x*, *y*, *z* R sample mount on top of a Mo sample holder. Prior to reaction the catalyst was cleaned using UV light in presence of 10^{-6} torr of O_2 for 60 min (twice). Ethanol was subjected to freeze–thaw pump to remove residual water and CO_2 and its purity is checked by the online mass spectrometer until a typical ethanol fragment is obtained [18]. Six experiments with different dosing of ethanol were conducted from 90 s to 10 min at a chamber of pressure close to 10^{-7} torr. After dosing, the chamber was pumped down to the 10^{-9} torr range. Prior to UV light form a 100 W ultraviolet lamp (H-144GC-100, Sylvania par 38) with a flux of ca. 10 mW/cm^2 at a distance of 25 cm the sample was rotated away from the mass spectrometer which was turned on to measure the back ground of the chamber. A negligible signal of *m/e* 31, the parent ion of the ethanol compared to that of residual N_2/CO (*m/e* 28) was indicative of the removal of gas-phase ethanol. The following masses were scanned at a dwell time of 50 ms for each (*m/e* 2, 12, 14, 15, 16, 18, 28, 29, 31 and 44) for each run. The catalyst pellet was then turned toward the mass spectrometer which is enveloped with a shroud having an orifice smaller than that of the pellet to minimize stray desorption, at the same time the UV shutter was removed to start the excitation. Au–Pd/TiO_2 (85 % anatase/15 % rutile) catalysts were synthesized by co-impregnation method (0.13 wt% of gold and 0.2 wt% of Pd; 1:3 molar ratio). The precursor of gold and palladium were $AuCl_4$ (dissolved in aqua regia) and $PdCl_2$ in 1 normal HCl. TiO_2 semiconductor, ca. 85 % anatase and 15 % rutile, was used as a support martial. First, TiO_2 was placed into Pyrex beaker. Then, the aqua regia solution of Au and Pd in 1 normal HCl were, respectively, poured into a certain amount of TiO_2 under magnetic stirring (170 rpm) at 80 °C for 12–24 h. The precipitate formed was dried for >4 h, at 120 °C. Finally, the material was calcined at 300 °C for 5 h; afterward it was crushed using a mortar to fine powder.

Results and discussion

Figure 2 presents XPS of the as-prepared Au–Pd/TiO_2 catalyst. Ti2p peaks at 459.2 and 464.6 eV are attributed to the $Ti2p_{3/2}$ and $Ti2p_{1/2}$, respectively. Both the narrow FWHM (1.1 eV) of the $Ti2p_{3/2}$ and the splitting of 5.4 eV are consistent with the presence of Ti^{4+} cations only. O1s peak at 530.3 eV is due to lattice oxygen while the large peak at 532.6 eV is due to irreversibly adsorbed water on the surface; surface hydroxyls at ca. 531.5 eV are masked by the large water contribution. Gold presence can be seen by their XPS Au4f at 83.8 and 87.4 eV attributed to $Au4f_{7/2}$ and $Au4f_{5/2}$, respectively. Both the peak positions and spin splitting of 3.6 eV are characteristic of metallic gold [the 0.2 eV deviation from pure metallic gold (binding energy at 84.0 eV) is slightly larger than the resolution limit and most likely not due to calibration issues]. It has been previously indicated that a charge transfer from TiO_2 to Au results in lowering the binding energy of small metallic particles of Au [19] and this might be the cause. It has also been reported that alloy of Au with Pd results in shifting the Au4f by up to 0.5 eV and at the same time increase in the binding energy of Pd3d by up to 0.8 eV [20]. Also seen in the figure is the contribution of Pd4s lines [because Pd loading is larger (3–1 molar ratio) and their particles are smaller therefore more dispersed; see TEM pictures below]. XPS Pd3d region indicates that Pd exists in two oxidation states, Pd^0 and Pd^{2+}. It is also worth noting that almost at the same binding energy position of Pd3d, Au4d lines are overlapping [21]. However, the signal of Au4d lines is about 5 times weaker than that of Au4f [22]. Given the weak signal of Au4f it is safe to neglect its contribution into the spectrum. The splitting between the XPS $3d_{5/2}$ and $3d_{3/2}$ of both oxidation states of 5.2–5.3 eV is consistent with many other works on both oxidation states [26]. The binding energy of Pd^0 is, however, slightly shifted upward (by 0.3 eV) when compared to Pd metal [20], which may also indicate the presence of some alloy (Au–Pd) (Fig. 3).

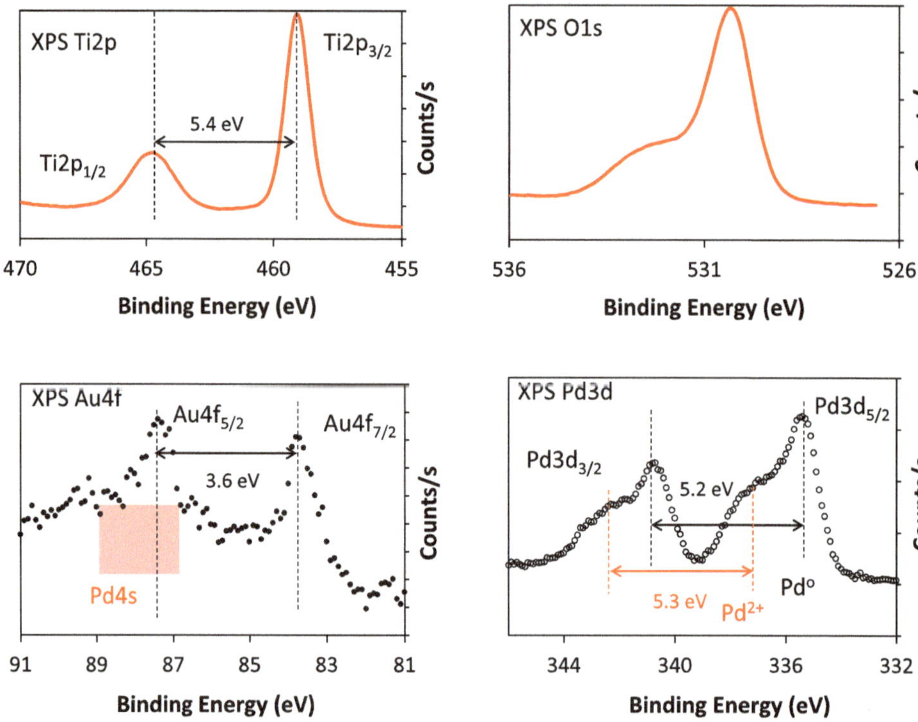

Fig. 2 XPS Au4f, Pd3d, Ti2p and O1s of as-prepared Au–Pd/TiO_2

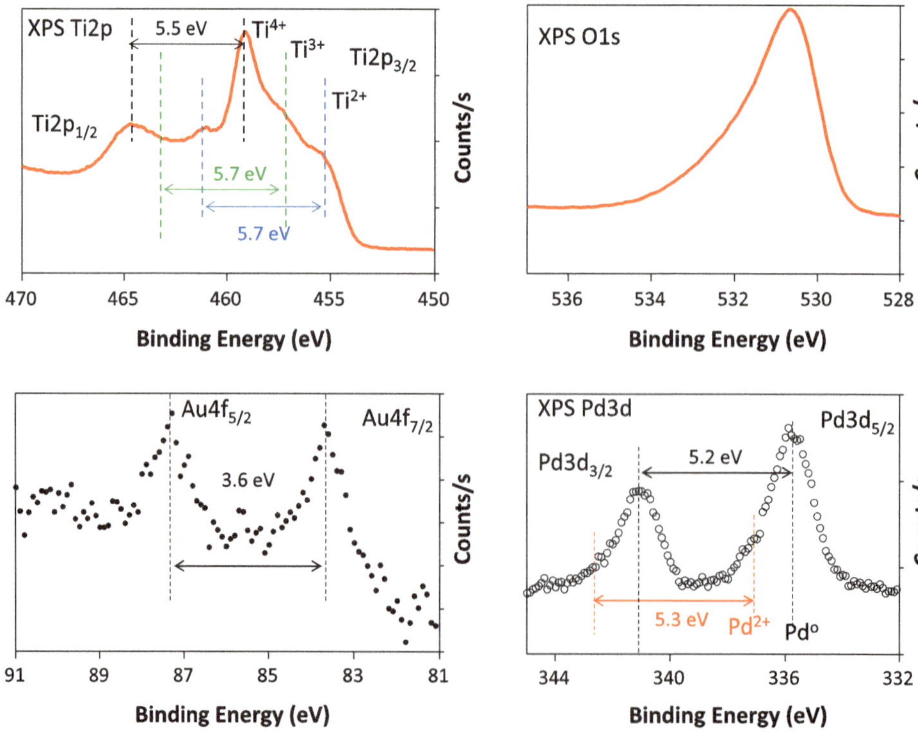

Fig. 3 XPS Au4f, Pd3d, Ti2p and O1s of Ar^+ sputtered sample (5 min) Au–Pd/TiO_2

Figure 2 presents the same lines of Fig. 1 but after Ar^+ sputtering for 5 min to reduce the materials and see for the effect of these on possible electronic shifts. The Ti2p region is that of typical reduced Ti cations (Ti^{4+}, Ti^{3+} and Ti^{2+} and possibly less reduced states with smaller contributions). The XPS $Ti2p_{3/2}$ lines attributed to Ti^{3+} and Ti^{2+} are at 457.3 and 455.3 eV, respectively [23, 24] in agreement with other works. The $Ti2p_{1/2}$ of Ti^{2+} is clear while that of Ti^{3+} is masked by the $Ti2p_{1/2}$ of Ti^{4+} (although one can see a considerable broadening of the peak at the lower binding energy side due to multiple electronic states). The splitting of the three states is very similar (5.5–5.7 eV).

Table 1 XPS surface composition of As prepared and Ar ion reduced Au-Pd/TiO_2

	Area CPS × eV	Sensitivity factor	Corrected area	Atomic (%)
As prepared				
Au4f	222.43	5.24	42.45	0.02
Pd3d	2,423.01	4.64	521.98	0.25
Ti2p	63,711.57	1.80	35,434.69	17.25
O1s	85,127.25	0.71	119,728.90	58.27
C1s	14,724.15	0.30	49,743.75	24.21
Ar^+ sputtered				
Au4f	178.04	5.24	33.98	0.02
Pd3d	1,195.65	4.64	257.57	0.14
Ti2p	76,212.77	1.80	42,387.53	22.44
O1s	64,737.00	0.71	91,050.63	48.20
C1s	16,327.96	0.30	55,162.03	29.20

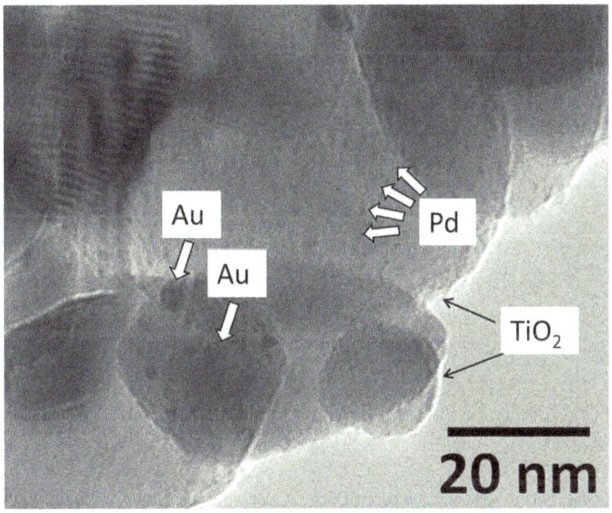

Fig. 4 Transmission electron microscopy of Au/Pd TiO_2 (anatase + rutile)

The O1s region also has the typical reduced shape as seen by Doniac broadening at the high binding energy side [25]. The XPS Au4f region has changed slightly also the background appeared to be less prominent probably because of a more homogenous electronic distribution of Pd particles (contributing by their 4s into the spectrum). The XPS Pd3d indicates that most of Pd particles are now in their reduced state. Both XPS Au4f and 3d main peak positions did not change considerably upon Ar ion sputtering. Table 1 presents the atomic % of the Au–Pd/TiO_2 before and after Ar ions sputtering.

In Fig. 4 a representative transmission electron microscopy image of the Au–Pd/TiO_2 catalyst. A large number of images were collected. The mean particle size of Au is about 3.5 nm while that of Pd is 1.5 nm (although the Pd particle size distribution was asymmetric at the low size indicating a larger contribution of small particles <1 nm); there was negligible number of Pd particles above 2 nm. The TEM images of the anatase and rutile phases of this TiO_2 have been presented elsewhere; the difference between the two phases can be made based on their local diffraction lines as well as the lattice spacing from high-resolution images.

Table 2 Photoreaction for hydrogen production from water in presence of 5 vol% of organic sacrificial agent in water (ethylene glycol in the case of Pd/TiO_2 and Au–Pd/TiO_2 and ethanol in the case of Au/TiO_2—both sacrificial agents are very efficient for the hole capturing with slight variations (ca. 10 %)

Catalyst	Rate (mol/g_{Catal} min)
0.15 at.% Pd/TiO_2	2.3×10^{-5}
0.20 at.% Au/TiO_2	1.7×10^{-4}
0.05 at.% Au–0.15 at.% Pd/TiO_2	2.8×10^{-4}

Amount of catalyst = 25 mg; reactor volume 100 mL, light Hg mercury lamp with 360 nm cutoff filter. Flux ca. 4–5mW/cm^2 (comparable to that provided from the sun in summer midday at a longitude = 25°N)

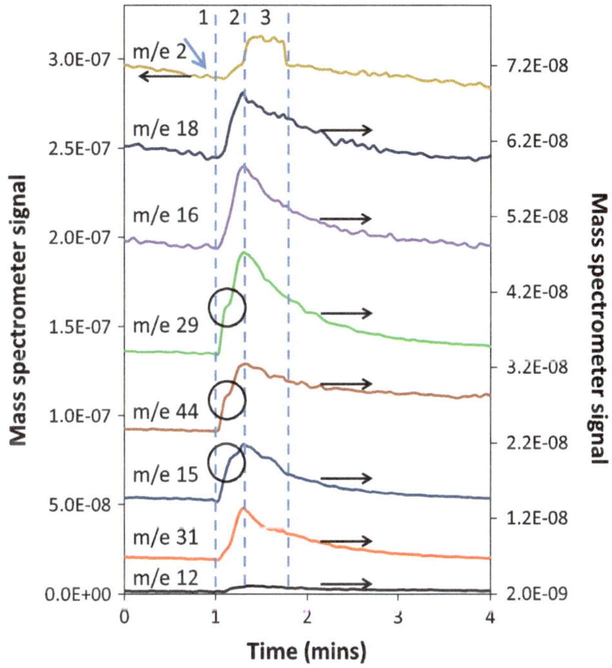

Fig. 5 Mass spectrometer fragmentation pattern of products collected upon ethanol photoreaction under UV excitation for a previously exposed Au–Pd/TiO_2 to ethanol for 4 min at 10^{-7} torr (this exposure was found to ensure surface saturation based on monitoring of desorption products)

Table 2 presents the rate of reaction for hydrogen production for Pd/TiO_2, Au/TiO_2 and Pd–Au/TiO_2. The support (85 % anatase/15 % rutile) was the same for the three catalysts. It is to be noted that variations from preparations to the other may contribute by up to 25 % of changes in the

Table 3 CF for mass spectrometer correction factor

m/e	Peak area	Product	CF	Peak area	Selectivity (%)
2	6.9×10^{-7}	Hydrogen	0.5	3.5×10^{-7}	29.4
12	1.1×10^{-8}				
15	5.9×10^{-8}				
16	1.9×10^{-7}				
18	1.7×10^{-7}	Water	1.1	1.9×10^{-7}	15.9
29	2.6×10^{-7}	Acetaldehyde	2.1	5.5×10^{-7}	46.5
31	5.4×10^{-8}	Ethanol	1.8	9.7×10^{-8}	8.3
44	1.2×10^{-7}		1.2		
Total	1.6×10^{-6}			1.2×10^{-6}	

reaction rate and that the rate is dependent on the % of the metal in a non-correlated way as often observed (see for example ref. [12]). Still it is clear that the catalyst composed of Pd–Au/TiO_2 is more active than that from Au/TiO_2 and Pd/TiO_2. A similar observation was reported on Pt–Au/TiO_2 previously [26].

To analyze the reaction products and their evolution with time upon radiation we have opted to conduct experiment in which the photocatalyst is illuminated with online mass spectrometer. Figure 5 presents the UV-excited photoreaction over Au–Pd/TiO_2 that was initially saturated with ethanol (prior to excitation). Upon illumination (indicated by the thick arrow at the top of the figure) fragments of reaction products are observed to desorb followed by a decrease in the signal due to surface depletion of the reactant. The desorption of these products can be grouped in three modes: photodesorption, photoreduction and photooxidation. Ethanol adsorption on TiO_2, as well as M/TiO_2 (M = Au, Pd or both) at room temperature gives both molecular and dissociated species. This has been seen both experimentally [27] and by DFT computation methods [28]. As indicated in Fig. 2, the surface contains (in addition to ethanol) non-negligible amounts of irreversibly adsorbed water and these ultimately give hydroxyl radicals upon photoexcitation. There are three regions in the figure denoted by lines 1, 2 and 3. In region 1–2, desorption of all products occurs (except hydrogen). In addition, this desorption is different for the set of products monitored. Ethanol (*m/e* 31), water (*m/e* 18) and oxygen (*m/e* 16; *m/e* 32 has the same desorption profile) all have similar shape and these may be considered as due to photodesorption upon UV excitation. In addition, fragments related to acetaldehyde are clearly noticed (*m/e* 29 and *m/e* 44). *m/e* 29 is largely due to acetaldehyde (CHO) fragment as ethanol contribution into this fragment is small (typically <30 % of *m/e* 31). *m/e* 44 (parent acetaldehyde molecule) has contribution from CO_2 in addition. The ratio *m/e* 29 to *m/e* 44 of about 5–2 indicates that in this region most of desorption is due to acetaldehyde [29]. *m/e* 15 (CH_3) has contribution from both acetaldehyde and ethanol. Seen in this region is a clear shoulder (indicated by a circle) composed of these three masses only. This is a clear indication of acetaldehyde desorption due to ethanol dehydrogenation. The formation of acetaldehyde from ethanol over TiO_2 upon photoreaction in presence of oxygen on TiO_2 single crystal has been reported before [27]. The reaction that involves two-electron injections into the valence band can be explained as follow.

Dissociative adsorption of ethanol and water occurs on the surface of TiO_2 in the presence or absence of light.

$$CH_3CH_2OH + Ti^{4+} - O_s^{2-} \rightarrow CH_3CH_2O - Ti^{4+} + OH(a) \quad (1)$$

$$H_2O + Ti^{4+} - O_s^{2-} \rightarrow HO - Ti^{4+} + OH(a) \quad (2)$$

s for surface, (a) for adsorbed.

Light excitation of the catalyst resulting in electron (e^-)–hole (h^+) pair formation.

$$TiO_2 + nUV \rightarrow xe^- + xh^+ \quad (3)$$

where $n > x$ due to scattering and absorption cross section.

Hole scavenging (two electrons injected per ethoxide into the VB of TiO_2) followed by acetaldehyde formation.

$$CH_3CH_2O - Ti_s^{4+} - O_s^{2-} + 2\,h^+ \rightarrow CH_3CHO\ (g) + OH(a) + Ti_s^{4+} \quad (4)$$

On average, acetaldehyde maximum separate desorption (in region 1–2) occurred after 3 s from UV excitation. The maximum desorption is, however, seen together with ethanol and water desorption at about 20 s of light excitation. In region 2–3, all products start to decrease but hydrogen desorption increased reached a short plateau before a sharp decrease. The observed profile can be rationalized kinetically and due to diffusion. Initially the surface contains large amounts of adsorbed ethanol but also adsorbed water and irreversibly adsorbed O_2 molecules (in the sample cleaning process). Upon illumination, the initial reaction is photooxidation producing acetaldehyde. The formation of acetaldehyde can, if coupled to oxygen radical species, prevent the reduction of H ions because O_2 is a faster electron scavenger. Once a large fraction of oxygen atoms and molecules have been consumed the remaining ethanol and water could farther react with the generated holes. Electrons transferred to the conduction band can then be transferred to Au and Pd particles that in turn reduce hydrogen ions (of surface hydroxyls) to molecular hydrogen; as follow [considering Eqs. (1)–(4) above].

Electron transfer from the CB of TiO_2 to H^+ (via M nanoparticles) resulting in molecular hydrogen formation, while the hole transfer occurs from one OH species of water in addition.

$$4\,OH\,(a) + 4\,e^- + 2\,h^+ \rightarrow 3O_s^{2-} + \tfrac{1}{2}\,O_2 + 2\,H_2 \qquad (5)$$

Because of the small size of hydrogen ions, intra and inter diffusion is much faster than for ethanol, acetaldehyde and water, is a plausible explanation of the abrupt decrease of the signal. The signal of all, but *m/e* 44 products return to the base line. The remaining *m/e* 44 signal is due to total oxidation of traces of ethanol to CO_2 (as it is not mirrored by *m/e* 29 and *m/e* 15; fingerprint of acetaldehyde); it is worth indicating that photooxidation of ethanol is a set of consecutive reactions studied in some details previously on TiO_2 [30] as well as on Ru/TiO_2 [31].

Table 3 presents the computed peak areas of the mass spectrometer signals of the main products. In addition, the corrected peak areas, to the mass spectrometer sensitivity factors for hydrogen, water, ethanol and acetaldehyde are given. While the results are preliminary it is, however, clear that hydrogen desorption contributes by a non-negligible fraction of the overall desorption and that acetaldehyde represents the largest fraction. In the absence of molecular oxygen one would expect that the amount of hydrogen would be at least equal to that of acetaldehyde if all hydrogen is made from ethanol and would be larger if additional hydrogen is made from adsorbed water.

Conclusions

Au–Pd/TiO_2 catalysts combining both plasmonic and synergism behavior are prepared, characterized and tested for the photoreaction of ethanol under ultrahigh vacuum condition upon excitation with UV light. The catalyst is composed of metallic Au and Pd in their reduced state. The photoreaction of ethanol results in the production of acetaldehyde and hydrogen. Hydrogen production is retarded due to the initial reaction of surface hydroxyls and adsorbed oxygen atoms and or molecules. Only once most of these species were consumed and molecular hydrogen was formed.

References

1. Fujishima A, Honda K (1972) Electrochemical photolysis of water at a semiconductor electrode. Nature 238:37
2. Kudo A, Miseki Y (2009) Heterogeneous photocatalyst materials for water splitting. Chem Soc Rev 38:253
3. Connelly KA, Idriss H (2012) The photoreaction of TiO_2 and Au/TiO_2 single crystal and powder surfaces with organic adsorbates. Emphasis on hydrogen production from renewable. Green Chem 14:260–280
4. Nadeem MA, Connelly KA, Idriss H (2012) The photoreaction of TiO_2 and Au/TiO_2 single crystal and powder with organic adsorbates. Int J Nanotechnol (special edition on nanotechnology in Scotland) 9:121–162
5. Connelly KA, Wahab AK, Idriss H (2012) Photoreaction of Au/TiO_2 for hydrogen production from renewables: a review on the synergistic effect between anatase and rutile phases of TiO_2. Mater Renew Sustain Energy 1:1–12
6. Walter MG, Warren EL, McKone JR, Boettcher SW, Mi Q, Santori EA, Lewis NS (2010) Solar water splitting cells. Chem Rev 110:6446–6473
7. Yang YZ, Chang CH, Idriss H (2006) Photo-catalytic production of hydrogen from ethanol over M/TiO_2 catalysts (M = Pd, Pt or Rh). Appl Catal B: Environ 67:217–222
8. Wahab AK, Al-Oufi M, Bashir S, Al-Salik Y, Katsiev H, Idriss H (2013) Photocatalyst, method of preparation, photolysis system. World Patent 13T&I0037-WO-PCT (2013); Gulf Cooperation Council (13T&I0037-GC-NP), Serial Number GCC/P/2013/25576 International Procedure (13T&I0037-WO-PCT), Serial Number PCT/IB13/59406
9. Jovic V, Al-Azria ZHN, Sun-Waterhouse D, Idriss H, Waterhouse GIN (2013) Photocatalytic H_2 production from bioethanol over Au/TiO_2 and Pt/TiO_2 photocatalysts under UV irradiation—a comparative study. Top Catal 56:1139–1151
10. Waterhouse GIN, Wahab AK, Al-Oufi M, Jovic V, Anjum DH, Sun-Waterhouse D, Llorca J, Idriss H (2013) Hydrogen production by tuning the photonic band gap with the electronic band gap of TiO_2. Scientific Rep 3:1–5
11. Jovic V, Chen WT, Blackford MG, Idriss H, Waterhouse GIN (2013) Effect of gold loading and TiO_2 support composition on the activity of Au/TiO_2 photocatalysts for H_2 production from ethanol–water mixtures. J Catal 305:307–317
12. Bowker M, Millard L, Greaves J, James D, Soares J (2004) Photocatalysis by Au nanoparticles: reforming of methanol. Gold Bull 37:3–4
13. Subramanian V, Wolf EE, Kamat PV (2004) Catalysis with TiO_2/gold nano composites: effect of metal particle size on the fermi level equilibration. J Am Chem Soc 126:4943–4950
14. Du L, Furube A, Yamamoto K, Hara K, Katoh R, Tachiya M (2009) Plasmon induced charge separation and recombination dynamics in gold TiO_2 nanoparticles systems: dependence on TiO_2 particle size. J Phys Chem C 113:6454–6462
15. Seh ZW, Liu S, Low M, Zhang SY, Liu Z, Mlayah A, Han MY (2012) Janus Au–TiO_2 photocatalysts with strong localization of plasmonic near-fields for efficient visible-light hydrogen generation. Adv Mater 24:2310–2314
16. Zhang Z, Zhang L, Hedhili MN, Zhang H, Wang P (2013) Plasmonic gold nanocrystals coupled with photonic crystal seamlessly on TiO_2 nanotube photoelectrodes for efficient visible light photoelectrochemical water splitting. Nano Lett 13:14–20
17. Bashir S, Wahab AK, Idriss H (2014) Synergism and photocatalytic water splitting to hydrogen over Pt/TiO_2 catalysts: effect particle size. Catal Today (submitted)
18. http://webbook.nist.gov/Massfragmentationethanol
19. Kruse N, Chenakin S (2011) XPS characterization of Au/TiO_2 catalysts: binding energy assessment and irradiation effects. Appl Catal A 391:367–376
20. Yi CW, Luo K, Wei T, Goodman DW (2005) The composition and structure of Pd–Au surfaces. J Phys Chem B 109:18535
21. Li Z, Gao F, Wang Y, Calaza F, Burkholder L, Tysoe WT (2007) Formation and characterization of Au/Pd surface alloys on Pd (111). Surf Sci 601:1898–1908

22. Moulder JF, Stickle WF, Sobol PE, Bomben KD (1992) Handbook of X-ray photoelectron spectroscopy, Perkin-Elmer Corporation
23. Idriss H, Barteau MA (1994) Reactions of p-benzoquinone on TiO_2(001) single crystal: oligomerization and polymerization by reductive coupling. Langmuir 10:3639–3700
24. Idriss H, Barteau MA (1994) Characterization of TiO_2 surfaces active for novel organic synthesis. Catal Lett 26:123–139
25. Doniach S, Sunjic M (1970) Many-electron singularity in X-ray photoemission and X-ray line spectra from metals. J Phys C Solid 3:285
26. Montini T, Marelli M, Minguzzi A, Gombac V, Psaro R, Fornasiero P, DalSanto V (2012) H_2 production by renewables photoreforming on Pt–Au/TiO_2 catalysts activated by reduction. ChemSusChem 5:1800–1811
27. Nadeem MA, Murdoch M, Waterhouse GIN, Metson JB, Keane MA, Llorca J, Idriss H (2010) Photoreaction of ethanol on Au/TiO_2 anatase. Comparing the micro to nano particle size activities of the support for hydrogen production. J PhotoChem PhotoBiol A: Chem 216:250–255
28. Muir JN, Choi YM, Idriss H (2012) DFT study of ethanol on TiO_2 (110) rutile surface. Phys Chem Chem Phys 14:11910–11919
29. Idriss H, Kim KS, Barteau MA (1993) Carbon–carbon bond formation via aldolization of acetaldehyde on single crystal and polycrystalline TiO_2 surfaces. J Catal 139:119–133
30. Reztova T, Chang C-H, Koersh J, Idriss H (1999) Dark and photoreactions of ethanol and acetaldehyde over TiO_2/carbon molecular sieve fiber. J Catal 185:223–235
31. Kundu S, Vidal AB, Nadeem A, Senanayake SD, Stacchiola D, Idriss H, Rodriguez J (2013) Ethanol photo reaction on ruthenium metal/metal oxide modified rutile TiO_2 (110) single crystal surface. J Phys Chem C 117:11149–11158

10

CO_2 recycling using microalgae for the production of fuels

M. H. Wilson · J. Groppo · A. Placido · S. Graham · S. A. Morton III · E. Santillan-Jimenez · A. Shea · M. Crocker · C. Crofcheck · R. Andrews

Abstract CO_2 capture and recycle using microalgae was demonstrated at a coal-fired power plant (Duke Energy's East Bend Station, Kentucky). Using an in-house designed closed loop, vertical tube photobioreactor, *Scenedesmus acutus* was cultured using flue gas as the CO_2 source. Algae productivity of 39 g/(m^2 day) in June–July was achieved at significant scale (18,000 L), while average daily productivity slightly in excess of 10 g/(m^2 day) was demonstrated in the month of December. A protocol for low-cost algae harvesting and dewatering was developed, and the conversion of algal lipids—extracted from the harvested biomass—to diesel-range hydrocarbons via catalytic deoxygenation was demonstrated. Assuming an amortization period of 10 years, calculations suggest that the current cost of capturing and recycling CO_2 using this approach will fall close to \$1,600/ton CO_2, the main expense corresponding to the capital cost of the photobioreactor system and the associated installation cost. From this it follows that future cost reduction measures should focus on the design of a culturing system which is less expensive to build and install. In even the most optimistic scenario, the cost of algae-based CO_2 capture is unlikely to fall below \$225/ton, corresponding to a production cost of ~\$400/ton biomass. Hence, the value of the algal biomass produced will be critical in determining the overall economics of CO_2 capture and recycle.

Keywords Microalgae · Carbon dioxide · Flue gas · Capture · Techno-economic analysis · Biofuels

M. H. Wilson · J. Groppo · A. Placido · S. Graham ·
E. Santillan-Jimenez · M. Crocker · R. Andrews
Center for Applied Energy Research, University of Kentucky, Lexington, KY 40511, USA

S. A. Morton III
Department of Engineering, James Madison University, Harrisonburg, VA 22807, USA

A. Shea · C. Crofcheck
Department of Biosystems and Agricultural Engineering, University of Kentucky, Lexington, KY 40506, USA

R. Andrews
Department of Chemical and Materials Engineering, University of Kentucky, Lexington, KY 40506, USA

Introduction

Despite concerns surrounding the contribution of fossil fuel combustion to global warming, the need for fossil fuels will remain significant for the foreseeable future. With direct replacement unlikely, strategies to reduce the emitted CO_2 are in high demand. The current array of options encompasses four main areas: (1) modifications to existing power plants to increase the efficiency of combustion, (2) improvements to the efficiency of energy use by consumers, (3) chemical carbon capture with subsequent sequestration, and (4) bio-mitigation with carbon recycling. All of these methods have promise, yet they are beset with significant technical and non-technical challenges. Alterations to the methods of use and production involve issues related to expensive plant modifications or changes to the behavioral patterns of consumers. Despite technological gains for CO_2 capture and sequestration, the costs associated with energy-intensive CO_2 concentration and compression are significant and anticipated to result in a parasitic power plant load on the order of 30–40 %. In addition, the uncertainty surrounding risk and liability issues related to long-term geologic sequestration is potentially strong enough to deter investment and adoption.

As is frequently stated, microalgae are the fastest growing photosynthetic organisms with growth rates and CO_2 bio-fixation potentials generally in excess of terrestrial plants [21]. Depending on growth conditions (light intensity, temperature, and physical nature of the environment), the levels of available CO_2, and the nutrient needs of the organism, a combination of carbohydrates, proteins, and lipids are produced. From these metabolites a range of fuel and chemical feedstocks/resources can be produced [4, 29]. It is this combination of growth rate and lipid productivity that has led to algae being touted as an ideal source of bio-derived oil [28].

The use of microalgae-based CO_2 mitigation suffers from two principal disadvantages: (1) in bio-mitigation, CO_2 is captured and subsequently recycled (effectively, the carbon is used twice); in other words, CO_2 is not permanently removed from the carbon cycle; and (2) a range of challenges exist which are primarily related to system complexity and scale-up issues that are driven more by economic constraints than technical issues. However, biological carbon capture and recycling has the potential to generate a revenue stream to offset, at least in part, the overall cost of implementation [12]. Indeed, the use of algae as a carbon dioxide bio-mitigation strategy and as a potential source of renewable fuels has long been a focus of research and development [2, 5, 7, 14, 16, 20]. A primary concern is that the cultivation strategy selected (typically a large shallow open pond) requires vast amounts of water and land. This is further complicated by the need to keep the algae cultivation system close to the carbon dioxide source, where a primary limiting factor becomes the suitability of the available land. To resolve this problem, most studies have inferred that the optimal approach would be to construct a fossil fuel power facility where land availability is not a problem for large-scale cultivation; however, this solution does not resolve the issue of carbon dioxide mitigation from existing fossil fuel energy production facilities, or, for that matter, other types of CO_2 point sources.

A few recent reports have attempted to address these challenges through the development of closed loop photobioreactors, which on an areal basis are typically more productive than open ponds [33]. Indeed, a study by Doucha et al. [10] employing flue gas from a natural gas fired boiler fed to a thin layer photobioreactor containing *Chlorella* sp. found that up to 50 % CO_2 removal could be attained. A study reported by Vunjak-Novakovic et al. [31] placed this figure as high as 82 % on sunny days for *Dunaliella* strains grown in an airlift reactor, with 50 % CO_2 removal on cloudy days. An older study by Laws and Berning [17] performed in Hawaii using the marine chlorophyte *Tetraselmis suecica* (grown in outdoor flumes) indicated that CO_2 emitted from an oil-fired electric plant could be successfully substituted for pure CO_2 for the cultivation of algae. More limited than the number of these studies are the commercially viable microalgae production processes that utilize flue gas from power plants. In fact, only Seambiotic in Israel has produced significant quantities of algae in this manner, using the flue gas from the Israel Electric Corporation's coal-fired Ashkelon power plant (Seambiotic [27]).

Against this background, we set out to determine the technical and economic feasibility of algae-based carbon capture in Kentucky. This required designing and demonstrating a process capable of utilizing flue gas through operation of a continuous microalgae culture, and evaluating the economics of CO_2 capture using this approach. This paper summarizes the results of our initial work, including the development and scale-up of component technologies, and demonstration of the integrated process at Duke Energy's East Bend Station, situated in northern Kentucky.

Experimental

Algae culturing

Scenedesmus acutus was obtained from the University of Texas Culture Collection (UTEX B72) and was used for all experiments. Cultures were grown in urea medium previously optimized for this *S.* strain (Crofcheck et al. [8]). Initial cultures were grown in 500 mL Erlenmeyer flasks under warm (Philips F32T8/TL741 Alto, 32 Watts) and cool white (Philips F32T8/TL735 Alto, 32 Watts) fluorescent lights [70 $\mu mol/(m^2\ s)$] in a 16:8 h light:dark illumination period. Flasks were bubbled with 3 % CO_2 (from gas cylinders) and kept at room temperature (22 °C). The flask cultures were eventually transferred to 7.5 L airlift photobioreactors (PBRs). These airlifts also received a constant supply of 3 % CO_2, but were grown under natural light conditions in a greenhouse. A number of airlift PBRs were used to inoculate a 650 L Varicon BioFence PBR which, in turn, was used to seed a 1,000 L PBR. Both of these large greenhouse reactors were needed to produce enough algae to inoculate the East Bend Station PBR. The larger PBRs were constantly monitored by probes for pH (Hach DPD1R1), dO_2 (Hach 5740DOB), temperature, dCO_2 (Mettler Toledo InPro 5000i), and photosynthetically active radiation (PAR, Apogee Instruments SQ-215). The large greenhouse PBRs were fed CO_2 whenever the culture pH rose above a certain set point (usually pH 7.0). The East Bend Station PBR operated the same way using flue gas as its CO_2 source.

Culture growth was monitored by means of dry mass (g/L) (Crofcheck et al. [8]) and qualitative microscopy analyses. In addition, ultraviolet–visible spectrophotometry

(Thermo Scientific Evolution 60) was used to monitor the density of algal cultures, absorbance being measured at 680 nm. Typically, one 50 mL sample was taken daily from the PBRs for analysis. In addition, ion and urea concentrations in the cultures in the large PBRs were monitored on a regular basis by ion chromatography (IC) and high performance liquid chromatography (HPLC), respectively. The concentrations of urea and specific nutrient ions were tracked to determine the rate of nutrient consumption. Elemental analysis of harvested algal biomass was conducted using inductively coupled plasma-optical emission spectroscopy (ICP-OES).

Lipid extraction and purification

The algae used in all experiments were *S. acutus* (UTEX B72) autotrophically cultured at East Bend Station using a urea-based medium (Crofcheck et al. [8]). After harvesting and dewatering, the algae (10–15 % solids) were dried in an oven at 60 °C for 24 h. The oven-dried *S.* algae were ground up in a coffee grinder until the algae particles were reduced to a size of <1 mm. After grinding and before all extractions, the algae used in the extractions were heated to 100 °C for 20 min to remove residual water (moisture content <3 wt%). Extractions were performed according to the Bligh–Dyer method [3] with one modification, namely, a biomass to total solvent ratio of 10 g/180 g was used. After removal of solvent, the crude lipid was weighed, re-dissolved in $CHCl_3$ and filtered through a plug of K10 montmorillonite (Sigma-Aldrich) to remove the chlorophyll present (which remained strongly adsorbed on the clay). Solvent was then removed under vacuum to afford the purified lipid as a colorless waxy solid.

Lipid conversion to fatty acid methyl esters

Algal lipids were converted to the corresponding methyl esters using a two-step process of esterification (to convert free fatty acids) and transesterification (to convert triacylglycerides) (Canakci and Van Gerpen [6]). 1 g of extracted, purified algal lipids was mixed with 1.5 mL anhydrous methanol containing 2 % H_2SO_4 (wt/wt) and refluxed at 65 °C for 2 h. The reaction was then cooled in an ice bath and the contents were mixed with approximately 3 mL of a 1:1 (v/v) mixture of water:cyclohexane. The organic layer was extracted, dried over sodium sulfate and the cyclohexane was removed by a rotary evaporator. The remaining lipid was then reacted with methanol containing potassium methoxide (0.5 wt% of lipid feedstock). This mixture was refluxed at 65 °C for 30 min. The reaction contents were then cooled, mixed with a water:cyclohexane mixture and the organic layer was separated, dried over sodium sulfate and the cyclohexane was removed on a rotary evaporator. FAME analysis was performed using an HP6890 GC equipped with a J&W Scientific HP-88 capillary column (30 m × 250 μm × 0.2 μm). The inlet was set to 250 °C. The split ratio was 20:1 with a constant flow of 1 mL/min. The oven began at 50 °C and was ramped at 20 °C/min to 140 °C and held for 5 min prior to a second ramp of 3 °C/min to 240 °C. The detector was held at 300 °C. The GC was calibrated using a 37-component FAME GC standard (Sigma-Aldrich). Samples were diluted 1,000:1 in cyclohexane and toluene was used as an internal standard.

Lipid conversion to hydrocarbons

Lipid deoxygenation experiments were performed in a fixed bed stainless steel tubular reactor (1/2 in. o.d.) equipped with an HPLC pump. 0.5 g of Ni–Al LDH catalyst (particle size 150–300 μm) was first reduced under H_2 at 400 °C for 3 h. Details of the catalyst preparation and characterization have been reported elsewhere [25]. After reduction of the catalyst, the system was taken to the reaction temperature (300 °C) and pressurized with H_2 to 580 psi. A 1.33 wt% solution of the algal lipids dissolved in dodecane was introduced to the system at a rate of 0.1 mL/min along with a flow of H_2 (50 mL/min). Samples were collected from a liquid/gas separator placed downstream from the catalyst bed. The liquid feed and reaction products were analyzed using an Agilent 7890A GC equipped with an Agilent J&W DB-5HT column (30 m × 250 μm × 0.1 μm), an Agilent Multimode inlet, a deactivated open ended helix liner and a flame ionization detector (FID). Data acquired using the GC-FID were processed using SimDis Expert 9 software purchased from Separation Systems, Inc. The dodecane solvent was subtracted and/or quenched from the chromatogram prior to processing the chromatographic data. Further details can be found in [25].

Results and discussion

Photobioreactor development

The cultivation of an autotrophic organism requires the provision of a controlled growth environment, which involves exposure of the organism to appropriate levels of sunlight, CO_2, and nutrients [30]. The mass cultivation of algae can be realized in either an open culture system (pond), or a closed loop system (photobioreactor). The selection of an open or closed culture system revolves around a number of system parameters: (1) the microalgae to be cultured, (2) the anticipated carbon source, (3) the accessibility to required resources, and (4) the cost of construction, operation, and maintenance of the culture system. Based upon simple mass balance calculations, an algae unit size to reduce the CO_2 output of a power plant would need to be of an enormous scale. Photobioreactors (PBRs) were chosen as the cultivation method in this study on the basis of their higher areal productivities [33] and limited water loss [30] due to evaporation. A number of prototype reactors of different configurations were first constructed in an effort to incorporate the lessons learned into larger scale reactors. Specifically, variations in construction materials, tube orientation and spacing, as well as flow patterns, were examined. The most important factor in designing photobioreactors is to allow exposure of the algal culture to sunlight to drive photosynthesis. Given that a vertical system typically enables a higher surface to footprint ratio than other configurations, a design based on a tubular photobioreactor was selected, oriented vertically, and constructed from low-cost, off-the-shelf parts.

The hydrodynamics of the reactor is another area of specific concern. Having a good understanding of the flow characteristics of the PBR is an important step toward enabling process control. Having a well-mixed system, with even flow and limited dead zones, ensures that measurements taken at a centralized point are descriptive of the entire system. In addition, any stagnant areas or zones with lower flow (with the potential to collect biomass) should be limited. If the biomass remains trapped in the reactor it can degrade and release compounds that affect culture health. Moreover, poor mixing can lead to anaerobic conditions which favor microbial denitrification, resulting in N_2O emissions [11, 15]. Given that N_2O is a potent greenhouse gas (GHG), this negatively affects the GHG balance of the system.

Different methods for circulating the algal culture and keeping it well mixed were, therefore, evaluated before developing a series flow, serpentine-style PBR (Fig. 1). This style of reactor most closely resembles a plug flow reactor, and is constructed by connecting multiple tubes in series to provide the algae access to the solar radiation needed to drive growth. Initial concerns over oxygen accumulation and low carbon dioxide levels were set aside due to the relatively low kinetic rate of photosynthesis (Grima et al. [13]). In order for detrimental concentrations of dissolved O_2 to accumulate, the liquid path, and thereby the residence time, would have to be extremely long. If care is taken in overall reactor design and operation, this issue can be resolved.

Historically, one of the main challenges in algae cultivation is CO_2 limitation. Carbon constitutes almost 50 % by weight of the elemental composition of algal biomass, with CO_2 representing the most significant nutrient requirement. One of the inherent benefits of working with coal flue gas is the high percentage of CO_2 in the gas (10–15 %) as compared to atmospheric conditions (0.04 %). Introducing CO_2-rich gas to the system can often be energy intensive, as is the case in systems requiring gas compression and bubbling. To minimize the costs associated with CO_2 entrainment, while maximizing mass transfer, a liquid driven vacuum pump (i.e., venturi or eductor) was employed in this study. An eductor uses the Bernoulli principle to entrain gas in a driven liquid flow. The extremely turbulent nature of the biphasic flow encourages good mixing and mass transfer, thereby facilitating CO_2 dissolution in the growth medium.

Photobioreactor operating strategy

Daily productivity rates of algal cultures are dependent on multiple factors, including the nature of the organism being cultured, nutrient concentrations, the concentration of dissolved carbon, temperature, light intensity (i.e., photosynthetically active radiation, PAR), and pH. In this study

Fig. 1 CAD image showing PBR design and photograph of the PBR installed in a greenhouse

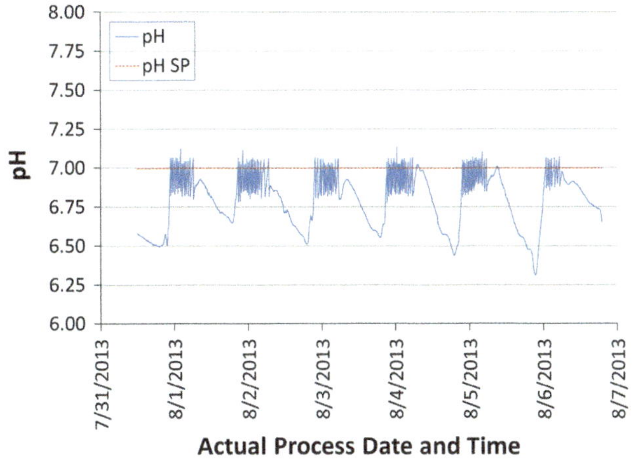

Fig. 2 Photobioreactor pH control. pH SP refers to the pH set point

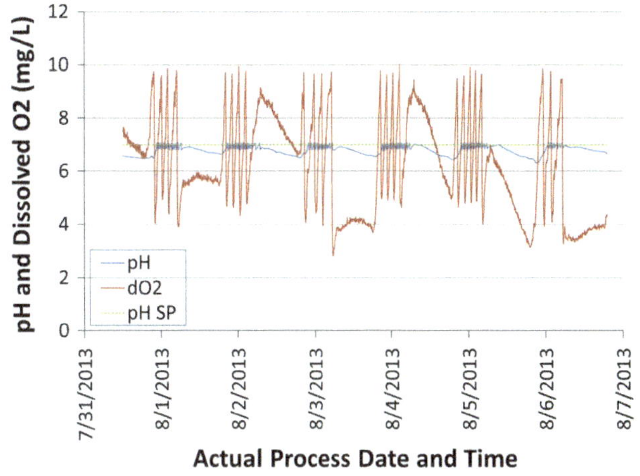

Fig. 3 O_2 production and system pH. The oscillations of the dO_2 signal are due to automated sparging of the culture with N_2 to maintain the dO_2 concentration below the set point value of 10 mg/L

S. acutus was cultured, a prior screening study having shown it to be appropriate for CO_2 capture based on its robust growth, tolerance to a wide range of pH values (Crofcheck et al. [9]), as well as ease of harvesting. Combustion flue gases can be a rich source of CO_2 for algae cultivation; however, the addition of excess flue gas must be avoided as this can result in over-acidification of the culture medium. This can be achieved by feeding CO_2 (as flue gas) on demand based upon the pH of the system. As algae grow, consuming CO_2, the pH of the solution is increased. The introduction of a CO_2-rich gas increases the concentration of carbonic acid and other dissolved carbon species, thereby lowering the pH. This approach maintains the system pH within an optimum pH range for algal growth while providing enough CO_2 to sustain growth. This is particularly important if other acidic flue gas components such as SOx and NOx are present. The dissolution of SOx in particular, which forms H_2SO_3/H_2SO_4, can result in over-acidification of the culture medium, thereby inhibiting growth (Crofcheck et al. [9]). For this reason, it is important that SOx is not added to the cultivation system faster than its dissolution products can be utilized by the algae.

Figure 2 illustrates the pH control method used to regulate CO_2 flow to the reactor during 6 days of algal cultivation in a 650 L PBR. The horizontal line indicates the pH set point of the reactor, while the trace shows the measured pH of the system. This graph also captures the occurrence of respiration, which produces CO_2, thereby lowering the pH during the night hours.

Appropriate reactor design can eliminate the buildup of O_2 in the photoactive portion of the reactor, but dissolved O_2 accumulation can still occur in a closed system over time. Elevated levels of dissolved O_2 can inhibit photosynthesis so it is important to have a method to remove excess O_2 from the system (Weissman et al. [32]). One method is to periodically sparge the main process tank with an oxygen-lean gas (such as post-combustion flue gas or nitrogen) which will remove dissolved O_2 preferentially over dissolved CO_2. Figure 3 shows the response of a PBR with N_2 sparging to remove excessive O_2 concentrations (i.e., >100 % atmospheric saturation or ~9 ppm). As anticipated, a high frequency of pH oscillation (corresponding to strong CO_2 consumption) is paired with a strong response in dissolved O_2. The effects of respiration on the pH and dissolved O_2 concentration during the night are also illustrated.

Another important variable that must be controlled is the culture density of the reactor. As a culture increases its number of cells, the increased chlorophyll concentration of the culture attenuates light much more quickly, starving some cells of required levels of solar radiation. Regular harvesting and dilution of the culture are, therefore, required to maintain a stable system capable of operating for an extended period of time.

Demonstration facility

Field testing of the system described above was conducted at Duke Power's East Bend Station (650 MW) located in Boone County, Kentucky. This single unit plant burns high sulfur coal as the fuel source and utilizes a wet limestone scrubber for SOx control and selective catalytic reduction (SCR) with ammonia injection for NOx control. Flue gas used for algae growth studies was obtained after the scrubber and SCR treatments with typical composition summarized in Table 1.

The site layout consisted of a PBR tube array located on an embankment situated approximately 7.5 m above a lower level where the 19,000 L feed tank, 5,700 L harvest tank and system control enclosure were located. The PBR assembly was constructed on a concrete pad poured above

Table 1 East Bend flue gas analysis (3/1/11–3/1/12)

	CO_2 (%)	NOx (ppm)	SO_2 (ppm)
Average:	8.9	53.4	28.0
Minimum:	7.2	14.5	6.5
Maximum:	9.6	97.2	84.3

Table 2 Analysis of source water at East Bend Station (average value of duplicate measurements ± standard deviation)

Analyte	Concentration, ppm
Chloride	3.79 ± 0.01
Nitrate–N	3.89 ± 1.17
Sulfate	25.15 ± 0.21
Phosphorus, total	<0.04
Calcium	89.45 ± 2.90
Magnesium	28.10 ± 0
Hardness by calculation	335.5 ± 12
Potassium	1.17 ± 0
Sodium	4.11 ± 0.78

Fig. 4 Photobioreactor installed at East Bend Station

a gravel drainage bed lined with a geomembrane below a French drain to collect all surface run-off and potential tube leakages. The drain flowed down the 7.5 m embankment to another concrete pad poured to provide a stable foundation for the feed and harvest tanks.

Water used to fill the PBR was drawn from several wells located on the property; typical analyses are shown in Table 2. Before water was fed into the PBR, it was passed through a UV sterilizer to minimize potential contamination by any organisms that may be present in this otherwise untreated water.

The PBR was constructed of clear PET (polyethylene terephthalate) tubes (8.9 cm diameter × 244 cm high) connected by 7.5 cm diameter schedule 40 PVC (polyvinyl chloride) pipe. Reactor tubes were arranged in 10 parallel flow paths, each consisting of 51 tubes connected in a serpentine path extending linearly for 18.3 m (Fig. 4). Feed was introduced by a centrifugal pump via a manifold where flow velocity through each tube was maintained at 16 cm/s, providing a residence time of approximately 13 min in the photosynthetically active volume of the clear tubes for each pass through the PBR. At the end of the PBR, the flow from each parallel flow path was combined in a common manifold and returned to the feed tank. As the return volume flowed back to the top of the 19,000 L feed tank, flow was directed to fall through T-pipe fittings to create suction, which was used as a means to introduce flue gas into the system (see Fig. 5 for a schematic of the PBR system).

The suction end of the return piping arrangement was connected to an air manifold with three automated control valves. When slurry pH rose above the desired set point (pH 7.0) the control valve connected to the flue gas would open, allowing flue gas to be introduced to the feed tank. When the pH dropped below the set point, the flue gas control valve would close and another control valve would open, allowing air in the head space to be recirculated, preventing CO_2 in the head space from venting from the system. The third control valve was used as a relief valve to prevent the feed tank from becoming pressurized as flue gas was added to the system. In this manner, as CO_2 was consumed by the algae, additional CO_2 was automatically fed into the system as needed to maintain the desired operating pH.

The PBR at East Bend Station was seeded on 7 December 2012 and operated continuously until 31 December 2012. During this time, flue gas was added as needed to maintain the pH at 7.0. Summary results (Fig. 6) show that productivity as high as 23 g/(m^2 day) was achieved during this period. These data also illustrate that productivity is related to available sunlight (i.e., PAR) as growth rate increased following periods of increased available PAR. The fact that the productivity data do not align perfectly with the PAR values requires comment, albeit that a clear trend is evident. This can be explained on the basis that (1) productivity is dependent on a combination of PAR and temperature (indeed, little growth was observed below 10 °C), and (2) the time at which the reactor is sampled during the day (am or pm) can introduce a lag into the data, i.e., samples taken in the early morning reflect growth the previous day (as well as night losses due to respiration), while samples taken in late afternoon reflect growth on the same day. During this time period, average daily temperature ranged from 4 to 20.5 °C. While this particular organism is not known to be particularly well suited for winter growth, reasonable growth rates were achieved, provided that sufficient PAR was available.

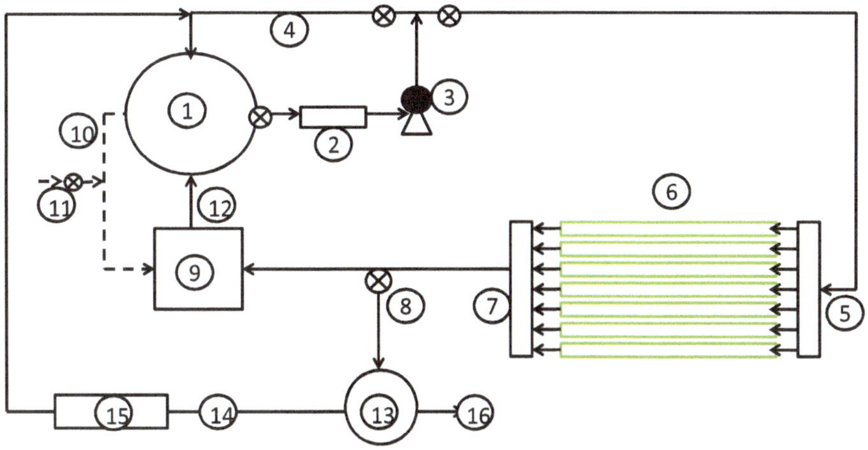

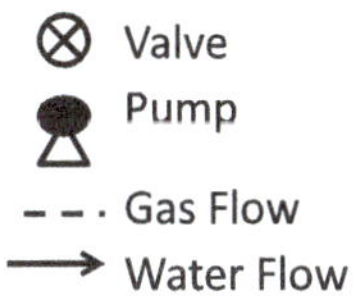

1	Main Process Tank	9	Eduction / Gas Entrainment
2	Probe Housing	10	Head Space Circulation
3	Process Pump	11	CO_2 Addition
4	Bypass Loop	12	Process Return to Tank
5	Feed Manifold	13	Harvest Tank
6	Phototube Array	14	Supernate/ Nutrient Return
7	Return Manifold	15	UV Sterilization
8	Harvest Port	16	Concentrated Biomass Removal

Fig. 5 Schematic of photobioreactor system at East Bend Station

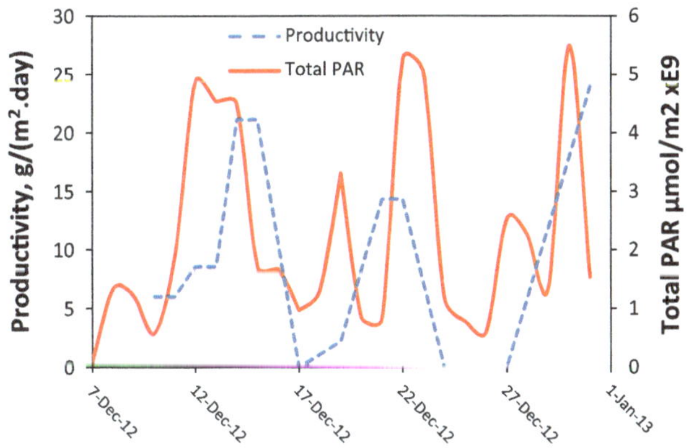

Fig. 6 Algal productivity and PAR during December growth study

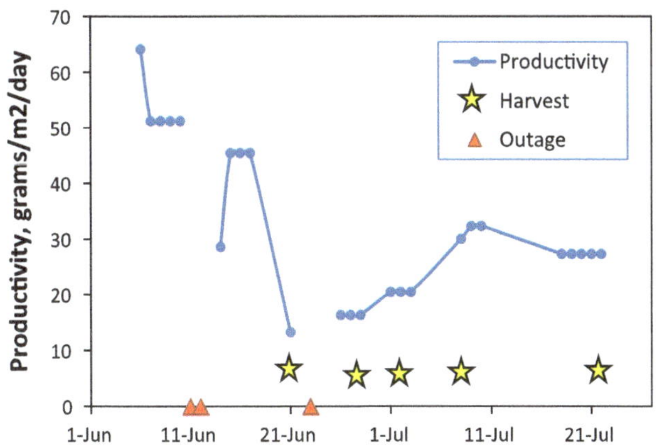

Fig. 7 Algal productivity during June–July growth study

Summer growth studies were conducted during June and July 2013 (see Fig. 7). As was observed with the winter growth study, productivity essentially followed periods of available sunlight (not shown). This study was complicated by several unforced electrical outages and an electrical failure to the feed system power supply caused by a lightning strike. Nevertheless, during periods of operation a mean growth rate of 32.9 g/(m^2 day) [with a standard deviation of 14.2 g/(m^2 day)] was recorded.

Harvesting and dewatering

During continuous PBR operation, it is necessary to periodically remove algae to control culture density to minimize self-shading and maintain culture health. Development of a suitable harvesting/dewatering strategy also addressed other system needs such as recycling clarified water to minimize water consumption and recycle unused soluble nutrients. Since the algae culture is very dilute (0.4–1.0 g dry mass/L), a cost effective, high capacity solid/liquid separation strategy was warranted. After considering a number of options, including dissolved air flotation and centrifugation, it was decided to pursue the use of sedimentation, thickening and filtration, an approach commonly used for treatment of industrial waste water.

Harvesting cycles were conducted as deemed necessary to maintain culture density, typically on a cycle of two to three times per week. A schematic diagram of the process is shown in Fig. 8. While the system continued normal

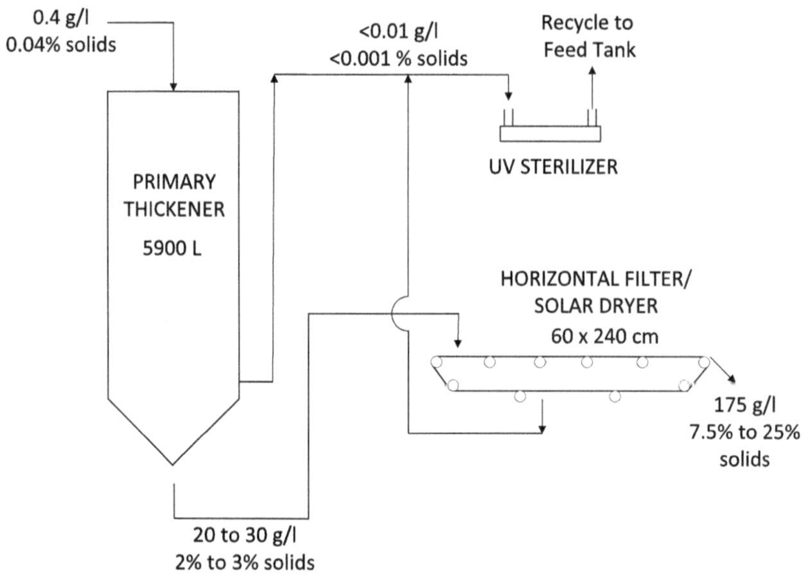

Fig. 8 Schematic diagram of harvesting/dewatering process

Table 3 Ultimate analyses of algal biomass harvested at East Bend Station (average of seven separate algae harvests ± standard deviation)

Carbon (%)	42.47 ± 4.18
Hydrogen (%)	6.50 ± 0.55
Nitrogen (%)	6.77 ± 0.70
Total sulfur (%)	0.52 ± 0.07
Oxygen (%)	24.38 ± 1.60
Ash (%)	19.36 ± 6.65
Volatile matter (%)	66.54 ± 4.25
Fixed carbon (%)	9.09 ± 2.40
As (ppm)	<0.1
Se (ppm)	<0.1
Cd (ppm)	<0.1
Hg (ppm)	<0.1

operation, approximately 4,000 L of culture was diverted into a cylindrical, cone bottomed harvesting tank. Moderate molecular weight cationic polyacrylamide flocculant was added at a dosage of 3–5 ppm, dependent upon harvest culture density, and mixed using a recirculating centrifugal pump. The flocculated system was allowed to settle for 4–8 h, after which settled biomass (20–30 g/L) was removed from the thickener and clarified water was pumped back into the feed tank to recycle water along with remaining soluble nutrients. As water was recycled, it was passed through a UV sterilizer. The concentrated algae slurry was then transferred to a horizontal gravity filter/ solar dryer and allowed to drain on a multifilament filter fabric. Clear filtrate was passed through the UV sterilizer and returned to the feed tank and dewatered biomass cake was removed from the filter/dryer. If adequate sunlight was available, the filter cake was dry ($\leq$3 % moisture) after approximately 24 h. If inadequate sunlight was available, the filter cake typically contained 7.5–25 % solids and was transferred to an oven and dried at 100 °C. After the biomass was thoroughly dried, a representative portion was characterized and the remainder stored in a freezer for utilization studies such as lipid extraction and upgrading.

Ultimate analyses of algal biomass harvested at East Bend are summarized in Table 3. The harvested biomass is characterized by an average of 42.47 % C and very high volatile matter content (66.54 %). Elemental analysis showed no detectable concentration of trace elements As, Se, Cd, and Hg within the detection limit of 0.1 ppm.

Upgrading of algal lipids to liquid fuels

Fatty acid methyl esters

To analyze the fatty acid profile of the lipids present in the harvested algae, lipids were extracted by the Bligh–Dyer method and converted to fatty acid methyl esters (FAME, more commonly referred to as biodiesel) via a sequence of esterification and transesterification. The gas chromatogram of the resulting FAME mixture is shown in Fig. 9 and the corresponding composition is summarized in Table 4. These results show that the oil consists mainly of C16:0 (palmitic), C18:1 (elaidic and oleic), C18:2 (linoleic) and C18:3 (linolenic) fatty acid chains. While C16 and C18 chain lengths are suitable for the production of diesel fuel hydrocarbons via hydrodeoxygenation or decarboxylation/ decarbonylation (vide infra), or indeed for the production of biodiesel (FAME), higher value fatty acids such as EPA (eicosapentaenoic acid) and DHA (docosahexaenoic acid) are not present. Consequently, the value of the oil for nutraceutical purposes would be low.

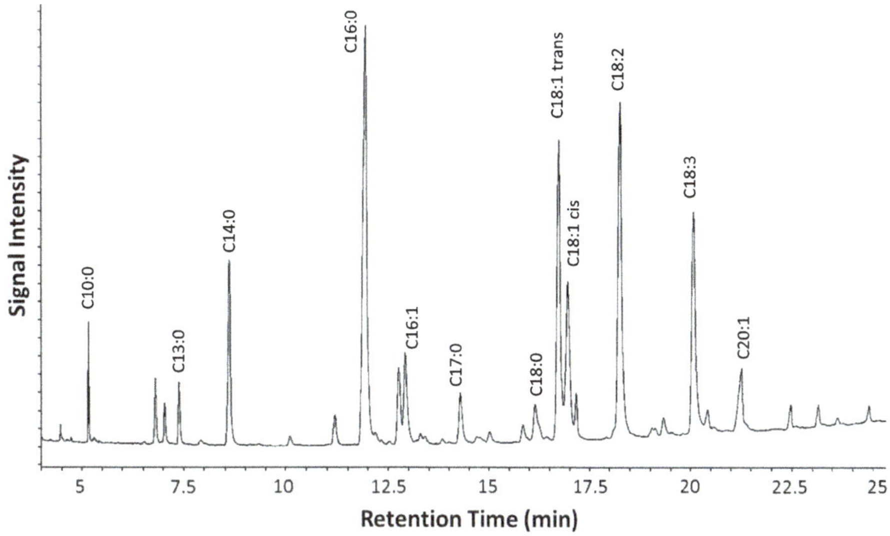

Fig. 9 Gas chromatogram of fatty acid methyl esters obtained from lipids extracted from *Scenedesmus acutus*

Table 4 Distribution of fatty acid chains in lipids extracted from *Scenedesmus acutus*

Fatty acid chain (*X*:*Y*)[a]	Algal lipid (GC area %)
Capric (10:0)	1.2
Tridecanoic (13:0)	1.3
Myristic (14:0)	5.3
Palmitic (16:0)	20.6
Palmitoleic (16:1)	3.9
Heptadecanoic (17:1)	2.0
Stearic (18:0)	2.1
Elaidic (18:1n9t)	12.1
Oleic (18:1n9c)	7.9
Linoleic (18:2)	15.3
γ-Linolenic (18:3n6)	10.5
Cis-11-eicosenoic (20:1)	3.3
Other	14.5

[a] *X*;*Y* = carbon number: number of double bonds

Diesel-range hydrocarbons

Hydrodeoxygenation (-H_2O) via hydrotreating forms the basis of a number of commercial or semi-commercial processes for the production of high quality drop-in hydrocarbon fuels from the lipids in vegetable oils and animal fats. Unfortunately, these processes require sulfided catalysts that risk contaminating the products with sulfur; in addition, they are constrained to use high pressures of H_2 that are typically only available in centralized facilities. An alternative lies in the deoxygenation of lipids via decarboxylation/decarbonylation (deCO$_x$), an approach that proceeds under considerably lower H_2 pressures and uses simple metal catalysts [24]. In recent work, we have shown that Ni-based catalysts are highly active for the upgrading of soybean oil and model triglycerides via deCO$_x$ [18, 19, 26]. Similarly, Lercher and co-workers have demonstrated that Ni-containing bifunctional catalysts can be employed to convert algal lipids to diesel-range alkanes in both batch and continuous modes (Peng et al. [22]; Peng et al. [23]; Zhao et al. [34]).

Table 5 Conversion of algae oil to diesel-range hydrocarbons

Catalyst	Conversion	Selectivity to C10–C17 (%)[a]	Selectivity to C17 (%)
Ni–Al LDH	95	73	7

Conditions: fixed bed reactor, 300 °C, 580 psi H_2, feed = 1.33 wt% algae oil in dodecane, feed rate = 6 mL/h

[a] Note that this value underestimates actual C10–C17 selectivity due to the fact that any C12 produced is not included in the calculation (given that C12 is used as the reaction solvent)

Building on the above studies, oil extracted from *Scenedesmus* microalgae harvested from the East Bend facility was subjected to upgrading via catalytic decarboxylation/decarbonylation. Prior to upgrading, the crude lipids were purified by filtration through K10 montmorillonite (an acid-treated clay) to remove chlorophyll (the presence of which might lead to the formation of deposits such as coke and Mg^{2+} on the catalyst during reaction). The purified oil was then upgraded as a solution in dodecane over a Ni/Al layered double hydroxide (LDH) catalyst [25] in fixed bed mode under H_2 (580 psi). Results are summarized in Table 5 and Figs. 10 and 11.

These results confirm that catalytic deCO$_x$ is a viable process for the conversion of microalgal lipids to diesel/jet fuel range hydrocarbons. Notably, some C18 is obtained (with 5 % selectivity), indicating that hydrodeoxygenation occurs in parallel with decarbonylation/decarboxylation, although together the latter processes constitute the major pathway given the higher selectivity of C17 observed. It is

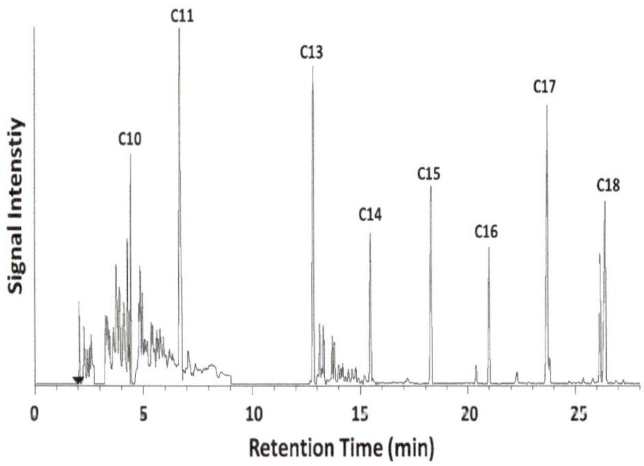

Fig. 10 Gas chromatogram of liquid product sampled after 4 h on stream during decarbonylation/decarboxylation of algal lipids. Note that the dodecane solvent (C12) has been subtracted from the chromatogram

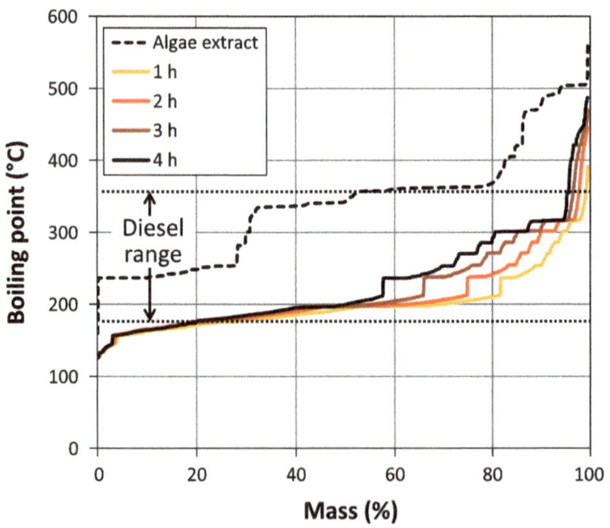

Fig. 11 Simulated-distillation boiling point distribution plots of algal lipid feed and liquid product sampled at 1, 2, 3, and 4 h during decarboxylation/decarbonylation. Note that the dodecane solvent (C12) has been removed from the plots

also evident that cracking of the unsaturated C18 fatty acid chains occurs, given the significant amounts of C10–C13 hydrocarbons obtained. Such cracking may or may not be beneficial depending on whether hydrocarbons are being targeted for jet fuel/lighter diesel-range applications or not. In other work [18, 19], we have found that less highly unsaturated fatty acid chains produce comparatively higher yields of the longer chain hydrocarbons (e.g., C15 and C17), which are well suited for diesel fuel blending.

Preliminary techno-economic analysis

The baseline scenario considered for this study was a 1,000 MW power plant, requiring 30 % CO_2 capture. Key inputs and assumptions used are collected in Table 6, while Table 7 summarizes the calculated costs associated with CO_2 capture. The techno-economic model is based on the capital and operating costs of a microalgae cultivation system sized to consume a given amount of CO_2. On a stoichiometric basis, algae consume ~1.76 tons of CO_2 to produce 1 ton of algal biomass, the exact figure depending on the elemental composition of the biomass produced [1]. The CO_2 emissions are based on the average rating of a coal-burning power plant, which relates the BTU content of the coal to the CO_2 emission. Algal productivity is expressed in grams per meters squared per day, a value of 30 g/(m^2 day) being used in this analysis. This number is derived from East Bend Station data (collected in the months of December, June and July) and data collected at the University of Kentucky (UK) over an approximately 12 month period; the East Bend data followed the trends previously observed at UK with respect to algae productivity as a function of the time of year. Relating algal

Table 6 Summary of inputs and assumptions used in the techno-economic analysis (base case)

Input	Value	Comment
Required CO_2 capture efficiency	30 %	Required capture efficiency if CO_2 emissions are to be maintained at 1,990 value
PBR cost, $/L (raw materials)	0.55	Custom design, PETG and PVC parts. Current cost, discounted by 55 % for bulk manufacture of parts
PBR installation cost, $/L	100	Assumed to be 100 % of raw material costs
PBR tube useful life, years	5	UV degradation limits tube life
Operation and maintenance costs, $/L	5 %	Labor + minor consumables, 5 % of PBR material costs
Areal productivity, g/(m^2 day) (year average)	30	Value based on data collected at East Bend Station and at the University of Kentucky
Nutrient costs, $/kg algae	0.14	Nutrient recipe utilizes bulk grade fertilizer
Nutrient recycle	97 %	Water + nutrients from algae harvesting and dewatering are recycled
Flocculant concentration, ppm	3	Commercial cationic flocculant
Flocculant cost, $/kg	4.40	Commercial cationic flocculant
Electricity cost, $/kWh	0.02	Discounted rate at utility site
Water cost, $/L	–	Water at site is free
Operating days per year	300	Estimated power plant operation (allowing for plant maintenance and unscheduled outages)
Payback period, years	10	Assumed payback time

Table 7 Summary of costs associated with CO_2 mitigation using microalgae

	Cost in $ per ton of CO_2 removed
Growing system	
PBR capital	775
PBR installation	775
PBR operation and maintenance	40
Energy	1
Nutrients	15
Growth subtotal	1,606
Dewatering	
1st stage dewatering	15
2nd stage dewatering	0.50
Dewatering subtotal	15.50
Total cost	1,621

productivity to the carbon emissions of a typical coal-fired power plant results in the total land area required for the cultivation system, i.e., an algae farm big enough to consume the requisite amount of CO_2. Based on this productivity, a cultivation system equivalent to 26,200 acres (40.9 square miles) would be required for the baseline scenario (30 % CO_2 mitigation for a 1,000 MW capacity plant). The costs associated with this process (capital, operating, dewatering) are then normalized by land area ($/m^2) to be compared with areal productivity, and thereby CO_2 consumption. As with capital costs, the energy consumption of the PBR system is extrapolated from current system values. The pressure drop associated with adding additional tubes in series is negligible, resulting in a lower energy cost per unit area (Watts/m^2) of installed and operating PBR. [Note that each tube pair has a calculated pressure drop of 0.00134 psi based on the frictional resistance; a module of 500 tubes as for the East Bend PBR has 25 tube pairs (10 rows) and would, therefore, have a pressure drop of 0.034 psi]. In this way, we are able to estimate the costs associated with consuming a ton of CO_2.

The key assumptions start with the size of the power plant, combined with the percentage of flue gas to be consumed, which together set the CO_2 emission rate of the system. In this case, we based our system on a 300 MW slip stream from a 1 GW plant, i.e., a 30 % slipstream. Operating days per year (300) and an amortization period are chosen to calculate the total amount of CO_2 that would be emitted over the lifetime of the algae-based mitigation system. The overall capital costs are estimated based on the current design of the demonstration facility at East Bend Station and are normalized based on the land area that the system would occupy. The most important assumption is areal productivity, which controls the size and thereby cost of the overall system.

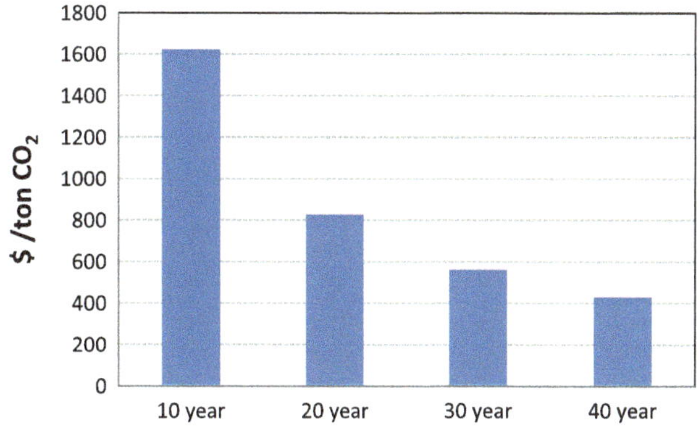

Fig. 12 Effect of amortization period on the cost of CO_2 capture

Although significant progress was made in reducing both the capital and recurring operating costs of the PBR employed, according to the analysis in Table 7, the current cost of capturing CO_2 falls close to $1,600/ton. Moreover, the capital cost and installation of the algae growth facility constitute 98 % of the overall cost of CO_2 capture. Further progress is clearly needed to reduce the overall capital cost of the system and thus reduce the cost of CO_2 capture. In addition, it should be noted that an amortization period of 10 years was used in the analysis. A less conservative approach would involve increasing this 10 year period to longer periods, bearing in mind that the contribution of operations and maintenance should be increased accordingly. Specifically, allowance has to be made for replacement of the PET tubes every 5 years, these comprising 8 % of the total capital cost; other parts are fabricated from PVC and are assumed to have a lifetime of >30 years. As shown in Fig. 12, extending the operating life of the facility improves the cost per ton of CO_2 mitigated considerably, although further cost reductions would require a decrease in the various cost elements associated with the PBR. A second option would be to increase the areal productivity although the scope for this seems limited given climatic constraints.

An important consideration is that this analysis takes no account of the value of the algal biomass produced. From this, it follows that there is a strong incentive to maximize this value, e.g., by conversion of the biomass to valuable products (nutraceuticals, animal food additive, premium organic fertilizer, etc.), to generate a revenue stream which can help to defray the costs of CO_2 capture/recycle. The size of the markets for algae-derived products is inversely related to the product price. Lower value products such as liquid fuels may be less attractive from a profitability perspective, but a utility-scale installation would produce a quantity of algal biomass that would inevitably oversupply lower volume, higher value product markets. However, there is no reason that utility-scale biomass utilization could not focus on developing both markets; lower value markets to utilize significant volumes, along with limited participation in higher value markets to maximize profits. By doing so, higher value market supply, and hence, profitability will be maintained.

Conclusions

Based on our initial work, we conclude that CO_2 capture and recycle using microalgae is feasible from a technical standpoint. By applying PBR technology, that was developed in-house, at East Bend Station, Kentucky, and using flue gas as the CO_2 source, algae productivity of routinely $\geq$30 g/(m^2 day) in the summer months was achieved at significant scale (18,000 L). These values compare favorably with values reported in the literature for both pond- and PBR-based cultivation studies. Moreover, average daily productivity slightly in excess of 10 g/(m^2 day) was demonstrated in the month of December and 39 g/(m^2 day) in June–July. To harvest and dewater the produced algal biomass, a protocol was developed based on flocculation and sedimentation, followed by filtration. Extraction of lipids from the harvested biomass was also demonstrated, followed by their conversion to diesel-range hydrocarbons via catalytic deoxygenation.

Conservative estimates suggest that the current cost of capturing and recycling CO_2 using this approach will fall close to $1,600/ton CO_2 (assuming an amortization period of 10 years). The largest sources of cost reside in the algae culturing stage of the process, corresponding mainly to the capital cost of the photobioreactor system and the associated installation cost. From this it follows that future cost reduction measures should focus on the design of a culturing system which is less expensive to build and install. Even in the most optimistic scenario, the cost of algae-based CO_2 capture is unlikely to fall below $225/ton, corresponding to a production cost of ~$400/ton biomass. Clearly, the economics of CO_2 capture and recycle can be significantly improved if the algal biomass produced can be sold. In view of the fact that the markets for algae are in their infancy, the value of algal biomass is at present hard to quantify with the exception of its fuel value. That said, the literature suggests that several large-volume markets do exist, such as animal feed and organic fertilizer, albeit that these applications require the absence of bioaccumulated heavy metals in the biomass. In this regard, the absence of heavy metals in the algae grown at East Bend, at a detection level of 0.1 ppm, is encouraging.

Acknowledgments The authors would like to thank Dr. Anne Ware, Tonya Morgan, Landon Mills, Wade Bailey and Neil McCarthy for assistance with lipid extraction and upgrading experiments, together with Nick Rhea and many undergraduate students who contributed to the construction of the photobioreactor. We would also like to acknowledge the considerable assistance of Doug Durst (Duke Energy) with the activities at East Bend Station, and Prof. James Dawson (Pittsburg State University) for preliminary experiments. The Kentucky Department of Energy Development and Independence, Duke Energy and the University of Kentucky are thanked for financial support.

References

1. Bayless DJ, Kremer G, Vis M, Stuart B, Shi L, Ono E, Cuello JL (2006) Photosynthetic CO_2 mitigation using a novel membrane-based photobioreactor. J Environ Eng Manag 16(4):209–215
2. Benemann JR, Oswald WJ (1996) Systems and economic analysis of microalgae ponds for conversion of CO_2 to biomass. Final report. DOE/PC/93204–T5; p 214
3. Bligh EG, Dyer WJ (1959) A rapid method of total lipid extraction and purification. Can J Biochem Phys 37:911–917
4. Brennan L, Owende P (2010) Biofuels from microalgae—a review of technologies for production, processing, and extractions of biofuels and co-products. Renew Sustain Energy Rev 14:557–577
5. Campbell PK, Beer T, Batten D (2009) Greenhouse gas sequestration by algae—energy and greenhouse gas life cycle studies. CSIRO, Australia
6. Canakci M, Van Gerpen J (2001) Biodiesel production from oils and fats with high free fatty acids. ASAE Trans 4(6):1429–1436
7. Copeland CH, Paul P, Samantha W, Paul E, Richard S, David B (2003) Chemical fixation of CO_2 in coal combustion products and recycling through biosystems. TVA/DE-FC26-00NT40933; p 70
8. Crofcheck C, Shea A, E X, Montross M, Crocker M, Andrews R (2012) Influence of media composition on the growth rate of *Chlorella vulgaris* and *Scenedesmus acutus* utilized for CO2 mitigation. J Biochem Technol 4(2):589–594
9. Crofcheck C, Shea A, Montross M, Crocker M, Andrews R (2013) Influence of flue gas components on the growth rate of *Chlorella vulgaris* and *Scenedesmus acutus*. Trans ASABE 56(6):1421–1429
10. Doucha J, Straka F, Livansky K (2005) Utilization of flue gas for cultivation of microalgae (*Chlorella* sp.) in an outdoor open thin-layer photobioreactor. J Appl Phycol 17(5):403–412
11. Fagerstone KD, Quinn JC, Bradley TH, De Long SK, Marchese AJ (2011) Quantitative measurement of direct nitrous oxide emissions from microalgae cultivation. Environ Sci Technol 45:9449–9456
12. Farrelly DJ, Everard CD, Fagan CC, McDonnell KP (2013) Carbon sequestration and the role of biological mitigation: a review. Renew Sustain Energy Rev 21:712–727
13. Grima EM, Fernandez A, Camacho FG, Chisti Y (1999) Photobioreactors: light regime, mass transfer, and scaleup. J Biotechnol 70:231–247
14. Johnson DA, Sprague S (1987). FY 1987 Aquatic species program: annual report. SERI/SP-231-3206; p 269
15. Harter T, Bossier P, Verreth J, Bodé S, Van der Ha D, Debeer A, Boon N, Boeckx P, Vyverman W, Nevejan N (2013) Carbon and nitrogen mass balances during flue gas treatment with *Dunaliella salina* cultures. J Appl Phycol 25:359–368
16. Kadam KL (2001) Microalgae production from power plant flue gas: environmental implications on a life cycle basis. NREL/TP-510-29417; p 55
17. Laws EA, Berning JL (1991) A study of the energetics and economics of microalgal mass culture with the marine

chlorophyte *Tetraselmis suecica*: implications for use of power plant stack gases. Biotechnol Bioeng 37:936–947
18. Morgan T, Grubb D, Santillan-Jimenez E, Crocker M (2010) Conversion of triglycerides to hydrocarbons over supported metal catalysts. Top Catal 53:820–829
19. Morgan T, Santillan-Jimenez E, Harman-Ware AE, Ji Y, Grubb D, Crocker M (2012) Catalytic deoxygenation of triglycerides to hydrocarbons over supported nickel catalysts. Chem Eng J 189–190:346–355
20. Nakamura T, Senior CL (2005) Recovery and sequestration of CO_2 from stationary combustion systems by photosynthesis of microalgae. DOE/PSI-1356/TR-2016; p 200
21. Nielsen SL, Enriquez S, Duarte CM, Sand-Jensen K (1996) Scaling maximum growth rates across photosynthetic organisms. Funct Ecol 10(2):167–175
22. Peng B, Yao Y, Zhao C, Lercher JA (2012) Towards quantitative conversion of microalgae oil to diesel-range alkanes with bifunctional catalysts. Angew Chem Int Ed 51:2072–2075
23. Peng B, Yuan X, Zhao C, Lercher JA (2012) Stabilizing catalytic pathways via redundancy: selective reduction of microalgae oil to alkanes. J Am Chem Soc 134:9400–9405
24. Santillan-Jimenez E, Crocker M (2012) Catalytic deoxygenation of fatty acids and their derivatives to hydrocarbon fuels via decarboxylation/decarbonylation. J Chem Technol Biotechnol 87:1041–1050
25. Santillan-Jimenez E, Morgan T, Shoup J, Harman-Ware AE, Crocker M (2013) Catalytic deoxygenation of triglycerides and fatty acids to hydrocarbons over Ni–Al layered double hydroxide. Catal Today, (in press)
26. Santillan-Jimenez E, Morgan T, Lacny J, Mohapatra S, Crocker M (2012) Catalytic deoxygenation of triglycerides and fatty acids over carbon-supported nickel. Fuel 103:1010–1017
27. Seambiotic http://www.seambiotic.com/. Accessed June 2013
28. Sheehan J, Dunahay T, Benemann J, Roessler P (1998) Look back at the US Department of Energy's Aquatic Species Program: biodiesel from algae; close-out report. NREL/TP-580-24190; p 325
29. Spolaore P, Joannis-Cassan C, Durn E, Isambert A (2006) Commercial applications of microalgae. J Biosci Bioeng 101(2):87–96
30. Tredici MR (2004) In: Richmond A (ed) Handbook of microalgal culture: biotechnology and applied phycology. Blackwell Science Ltd, Oxford, pp 178–214
31. Vunjak-Novakovic G, Kim Y, Wu XX, Berzin I, Merchuk JC (2005) Air-lift bioreactors for algal growth on flue gas: mathematical modeling and pilot-plant studies. Ind Eng Chem Res 44(16):6154–6163
32. Weissman JC, Goebel RP, Benemann JR (1988) Photobioreactor design: mixing, carbon, utilization, and oxygen accumulation. Biotechnol Bioeng 31:336–344
33. Williams PJB, Laurens LML (2010) Microalgae as biodiesel and biomass feedstocks: review and analysis of the biochemistry, energetics and economics. Energy Environ Sci 3(5):554–590
34. Zhao C, Brück T, Lercher JA (2013) Catalytic deoxygenation of microalgae oil to green hydrocarbons. Green Chem 15(1720): 1739

Characterization of polyethylene synthesized by zirconium single site catalysts

Abdulhamid A. Alsaygh · Jehan Al-hamidi · Fares D. Alsewailem · Ibrahim M. Al-Najjar · Vladimir L. Kuznetsov

Abstract New non-metallocene, bridged-zirconium center catalysts were prepared by Schiff base condensations reaction of two equivalents of appropriate 2-hydroxybenzaldehyde with one equivalent of 4,4′-methylene bis(2-methylcyclohexylamine), and subsequent metathesis reaction with $ZrCl_4(THF)_2$, and tested for ethylene polymerization reaction at different conditions. Catalyst productivities were found to be high, in the range of 29.5×10^3 kg/mole·Zr·h to 2.3×10^3 kg/mole·Zr·h, dependents on the polymerization condition. The molecular weight of the product, i.e. polyethylene, varied from 397,000 to 988,000. ^{13}C NMR study of the polymers indicate that fraction of branches and end groups percentage range from 0.01 to 0.55. X-ray diffraction analysis showed that polyethylene samples had highly crystalline structure within the orthorhombic space group (*Pnam*) and carbon–carbon interatomic distance of 0.154061 nm and the C–C–C angle of 112.192°.

Keywords Polyethylene · Zirconium · Single site catalysts

A. A. Alsaygh · F. D. Alsewailem · I. M. Al-Najjar
Petroleum and Petrochemicals Research Institute, King Abdulaziz City for Science and Technology (KACST), P.O. Box 6086, Riyadh 11442, Saudi Arabia
e-mail: aalsaygh@kacst.edu.sa

J. Al-hamidi
Chemistry Department, College of Science, Princess Nora Bint Abdulrahman University, Riyadh, Saudi Arabia

V. L. Kuznetsov
Department of Chemistry, University of Oxford, South Parks Road, Oxford OX1 3QR, UK

Introduction

The area of olefin polymerization catalysis has endured a remarkable increase in research outcome over the past 20 years, with many academic and industrial research centers involving in the design of new organometallic precatalysts for the controlled synthesis of polyolefin products. During that period, studies on the Group 4 metallocenes afforded much needed insight into the nature of the activated species. During the first half of the 1990s, interest grew in developing new generation "non-metallocene" catalysts, partly to avoid the growing patent minefield in Group 4 cyclopentadienyl systems, but also because late transition metal systems also offered the commercially attractive possibility of being able to incorporate polar comonomers into polyolefin materials, to give modified surface properties at low levels of comonomer incorporation, to change the bulk properties of the polyolefin at higher levels of incorporation.

The potential for this approach was realized with the discovery in the late 1990s of highly active ethylene polymerization catalysts based on iron, a metal with no previous track record in olefin polymerization. Another significant development over the past years or so has been the introduction of systems capable of catalyzing the living polymerization of olefinic monomers. The absence of chain-transfer or chain-termination processes allows access to polyolefinic materials with very high molecular weight (>600,000) and narrow molecular weight distributions (typically <1.1), block copolymers, and polymers with novel topologies [1, 2]. Ultrahigh molecular weight polyethylene is an important material for production of high performance fibers where the polymer chains are extended and aligned as in the crystalline material. The performance of polymers is improved as the chain length and crystallinity are improved [3].

Polyethylene is the simplest possible polymer; however, the simplicity of the molecular formula ($–CH_2–)_n$ does not reflect the real complexity of the material and its properties. Polyethylene is generally synthesized as a mixture of crystalline and amorphous components. The physical properties of the polyethylene are usually described using at least a two-phase model consisting of crystalline blocks in an amorphous matrix. [4] The crystalline phase ranges from around 50 to >98 % and its concentration generally increases with increasing molecular weight and decreases with increasing chain branching. As the temperature of a polymer is raised, the percentage of crystallinity diminishes and above the melting point the structure is entirely amorphous, and consists of a viscous liquid.

Polycrystalline polyethylene forms chained molecules that are distributed on lattice points with a rigorous periodicity, thus forming a long-range order crystalline structure. For the first time the basic structural features of polyethylene were determined, by means of powder diffraction, in 1939 by Bunn [5] who found that polyethylene crystallizes in the orthorhombic structure (space group *Pnam*) with two polymeric chains per unit cell. The orthorhombic unit cell, which is identical to the crystal structure of straight-chain alkanes, contains four CH_2 groups. Paraffin's are identical with polyethylene in regard to their molecular structure except for their molecular length; they crystallize in the same crystal form as the polyethylene analogues.

^{13}C NMR and X-ray powder diffraction (XRD) techniques are widely used in polymer science. Solution and melt ^{13}C NMR has been established as a useful tool for a quantitative determination of branch content of short- and long-chain branched polyethylenes while XRD can reveal important information on the morphology, structure and the degree of crystallinity of polythene [6–8]. In the present paper we report the synthesis of high molecular weight polyethylene using Zr-based single site catalyst and results of the use of melt-state ^{13}C MAS NMR and X-ray diffraction techniques for the determination of branch content of long-chain polyethylene's and their crystal structure.

Experimental

All operations involving air sensitive reagents and materials during catalysts precursor synthesis were carried out under nitrogen atmosphere using standard Schlenk, vacuum or dry box techniques. Solvents were dried and distilled under nitrogen prior to use of by passing through a column of activated alumina and a supported copper (Q5) catalyst. Imino-ligands were synthesized by condensation of two equivalents of appropriate 2-hydroxybenzaldehyde with one equivalent of 4, 4′-methylene bis(2-methylcyclohexylamine), and subsequent metathesis with $ZrCl_4(THF)_2$ [9].

The polymerization reactions were performed in stainless steel 1 L reactor, Polyclave Buchi glass uster. Solvent (Toluene) and MAO were loaded in the reactor in a glove box and removed after sealed. The reactor then was perched with ethylene. After reaching the reaction temperatures and pressure, catalyst was injected into the reactor. The polymerization product was transferred to 1 L flask with 100 mL methanol (5 % HCl). Polymer was isolated by filtration and washed with acidic methanol and dried under vacuum [10].

^{13}C MAS NMR measurements of melted polymers were performed using Varian Unity Inova 300 NMR Spectrometer operated at 75 MHz frequency and 2.58 kHz spin rate. Powdered polyethylene samples were placed into 7.5 mm ceramic rotors and heated to 170 °C during measurements. ^{13}C spectra were recorded with acquisition time of 100.0 ms, 90° pulse angle, 5 s recycle delay and 2,070 scans per spectra. Branch content was determined by numerical peak integration of branching ^{13}C NMR signals.

X-ray powder diffraction patterns of polymers were recorded using a Philips PW3040/60 X'pert PRO Console diffractometer equipped with a focus graphite monochromator and two Soller slits, 1 mm divergence slit and a 0.2 receiving slit. The spectra were recorded using Cu$K_{\alpha 1}$ (1.5406 Å) incident radiation; the tube voltage was 45 mV and tube current was 40 mA. The scan diffraction angle varied from 5° to 100° 2θ angular range and the samples were scanned at a rate of 0.008°/min in 2θ; the collection time was 3 h. The spectra were recorded in air at 25 °C using sample holders with a 1 mm deep depression filled with polymer powders. The sample holder was rotated at a rate of 0.5 Hz to enhance scattering signal.

Results and discussions

The ethylene polymerization activity of the zirconium dihalide complexes was investigated using MAO as cocatalyst. The activity of imino- of bridged-zirconium catalysts was found to be 29.5×10^3 to 2.3×10^3 kg $mmol^{-1}$ h^{-1} Zr^{-1}, which is considerably higher than the activities seen from (alpha-dimines) $ZrCl_2$ mononuclear catalysts [11]. The molecular weight of the polyethylene product is also significantly larger ranging between 397,000 and 988,000 g/mol (Table 1).

Results of ^{13}C MAS NMR of a series of molten polymers performed at 170 °C are shown in the last column of the Table 1. The total fraction of branches and end groups of main chains was determined by calculating the ratio between the area of peak at around 14.2 ppm (corresponding to CH_3 group shown as **1** in the scheme below) to

Table 1 Polymerization conditions, molecular weight and fraction of branches and end groups in several polyethylene samples

Sample	Polymerization conditions	Molecular wt (g/mol)	Fraction of branches and end groups (%)
1	40 Bar–24 °C–10 MAO	610,000	~0.02
2	10 Bar–24 °C–20 MAO	397,000	0.22
3	10 Bar–60 °C–10 MAO	–	0.21
4	10 Bar–24 °C–10 MAO	397,000	0.47
5	30 Bar–24 °C–10 MAO	696,000	0.04
6	20 Bar–24 °C–10 MAO	959,000	~0.01
7	10 Bar–40 °C–10 MAO	–	0.08
8	10 Bar–24 °C–7 MAO	–	0.15
9	10 Bar–24 °C–15 MAO	517,000	0.35

the area of main carbon peak at 30 ppm, which corresponds to long-chain CH_2 groups (δ and further carbons in the scheme below). Peaks corresponding to carbons 2 and 3 (23 and 32.3 ppm, respectively) are visible in all spectra and their observed intensities are the same as those for terminal carbon labeled as 1.

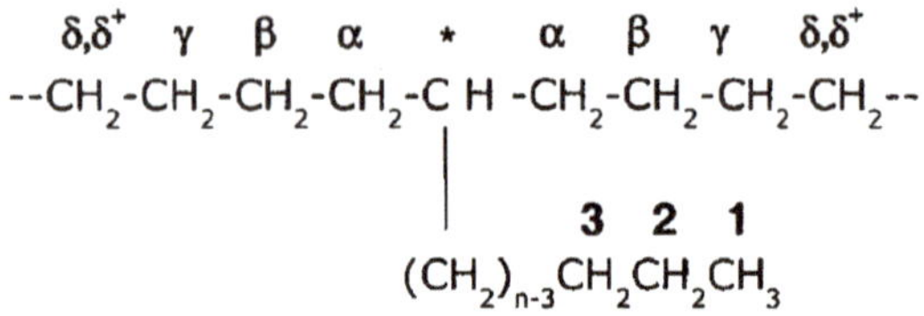

As seen from Table 1, the fraction of branches and end groups correlates well with molecular weight of polymers. Samples for tests 1, 5 and 6 have smallest fraction of branches and the largest molecular weight.

Typical results of X-ray diffraction measurements of polyethylene samples are presented in Fig. 1. All samples exhibit very similar diffraction patterns with sharp high-intensity peaks at 21.5° and 23.9° in 2θ and several lower intensity peaks in the range of 30°–55° and 74°–82° in 2θ. The spectra in the range of 90°–100° are not shown in Fig. 1 because they do not contain any well-defined diffraction peaks. For all the samples studied it was found that there are no appreciable changes in the position and the width of crystalline peaks.

The indexing of polyethylene XRD pattern is presented in Fig. 1 and agrees well with literature data [12]. All well defined peaks are indexed within the *Pnam* space group with the only exception of a weak shoulder at around 19° in 2θ which cannot be assigned to any *hkl* index within this space group. Some authors attribute the peaked feature of the maximum at about 19° in 2θ to the presence of a small amount of monoclinic polyethylene [13]. However, according to Fontana et al. [14] the monoclinic polyethylene phase can be formed only at high pressures (around 6 GPa) and the formation of this crystallographic modification during polyethylene synthesis (10–40 bar) is very unlikely.

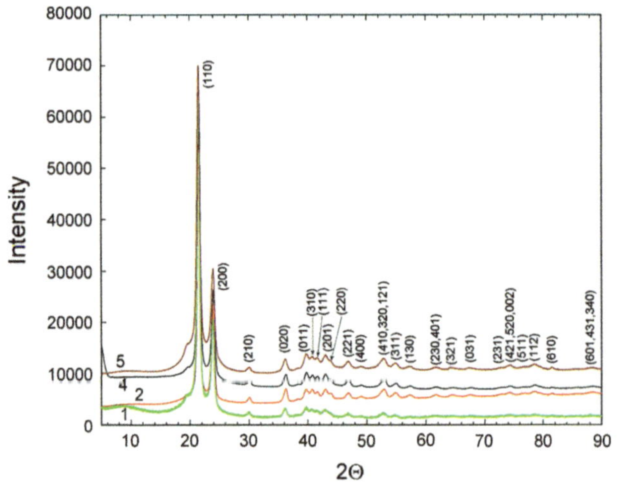

Fig. 1 XRD spectra of long-chain polyethylene samples 1, 2, 4 and 5 with all the diffraction peaks indexed within the *Pnam* space group

It should be noted that both crystalline orthorhombic polyethylene and amorphous polyethylene exhibit a diffraction peak at 19° in 2θ [15] and from our point of view, the weak shoulder at 19° corresponds to a broad peak of amorphous or semi-amorphous polyethylene phase.

Rietveld refinement of XRD spectra was performed using the structure refinement program GSAS [16], with the orthorhombic space group (*Pnam*) of crystalline polyethylene reported by Bunn [17] as the structural model. The peak shape was fitted with a convoluted pseudo-Voigt function to allow for broadening induced though finite crystallites. The preferred orientation of the films was not accounted for at this stage and it was assumed that the crystallite orientation in the polyethylene powders is completely random. The results of the Rietveld refinement of sample 5 are shown in Fig. 2 over the range of 5° to 100° in 2θ. Figure 3 displays refinement results over an enlarged area between 28° and 60° in 2θ. It is seen that there is a very good agreement between experimental and calculates intensities and positions of X-ray peaks with the exception of one peak at 43.5° in 2θ (Fig. 3).

Figures 4a and b depict the projections of packing of orthorhombic polyethylene chains along the *c* and *b* axes, respectively. The two planes containing the skeletal carbon chains within the unit cell are almost orthogonal to each other while all-*trans* chains are parallel, as seen in Fig. 4a. It has been shown that the setting angle of the molecular packing φ, i.e. the angle between the plane containing the ···C–C–C··· zigzag molecular chains and the *b* axis of the unit cell (see Fig. 3a), is in the range 41°–46° [5, 17]. Such differences may be attributed to the presence of an amorphous component that may significantly distort the diffraction profile in both peak position and intensity [18]. While the crystalline constituent gives sharp diffraction peaks, the amorphous phase produces three broad diffuse peaks at around 19°, 41° and 80° in 2Θ for CuK_α radiation.

The orthorhombic unit cell of branched polyethylene is expanded along axis *a* and, to a smaller degree, along axis *b*. At room temperature the unit cell parameters or linear polyethylene are $a = 0.74069$ nm, $b = 0.49491$ nm and $c = 0.25511$ nm [19]. The difference in *a* unit cell parameters might be due to a different degree of branching in polyethylene samples used by different authors.

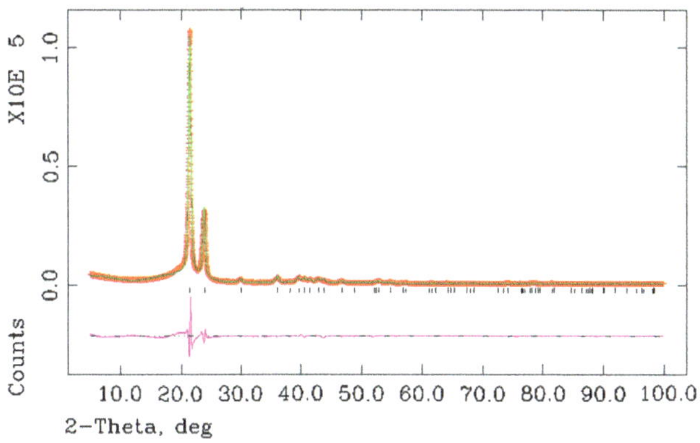

Fig. 2 Rietveld refinement of XRD spectrum of polyethylene sample 5. *Red crosses* experimental X-ray diffraction points, *green line* Rietveld refinement results, *pink line* difference line, *black ticks* calculated position of X-ray peaks

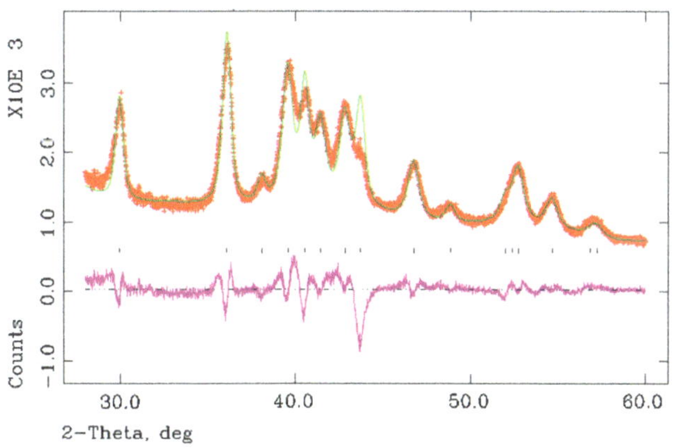

Fig. 3 Rietveld refinement of XRD spectrum of polyethylene sample 5. Enlarged area between 28° and 60° in 2θ is shown

With rising temperature the dimensions of the *a* and *b* axes change, while the *c* dimension does not change appreciably since it is determined by the strong intra-chain covalent C–C and C–H bonds unlike weak inter-chain van der Waals interactions which determine lattice parameters along the *a* and *b* axes. Due to a significant difference in the attractive bond forces between atoms parallel and perpendicular to the chain axis *c* the orthorhombic crystal phase of polyethylene, like most other crystalline polymers, has pronounced anisotropy of mechanical, optical and dielectric properties.

The simulated crystal structure of polyethylene sample 5 along and perpendicular to the *c* axis is presented in Figs. 5 and 6.

The calculated lattice parameters of sample 5 are as follows: $a = 0.74510 \pm 0.00003$ nm, $b = 0.49724 \pm 0.00002$ nm and $c = 0.25573 \pm 0.00004$ nm. The lattice parameters *a* and *b* are strongly affected by nonbonded interactions between the chains and, potentially, by the degree of branching. The *c* lattice parameter of polyethylene sample 5 which is mainly determined by the length of ···C–C–C··· zigzag chains is very close to the literature data for polyethylene samples [19] while *a* and *b* lattice parameters in sample 5 are significantly larger. On the basis of our results on ^{13}C MAS NMR investigation which shoved very small degree of branching ($\sim$0.01 %) in this sample it is unlikely that branching can explain larger values of both *a* and *b* lattice parameters in sample 5 comparing to those reported in the literature, for example $a = 0.74241(7)$ nm and $b = 0.49491(5)$ nm; $c = 0.25534(1)$ nm [13]. One of the possible reasons could be

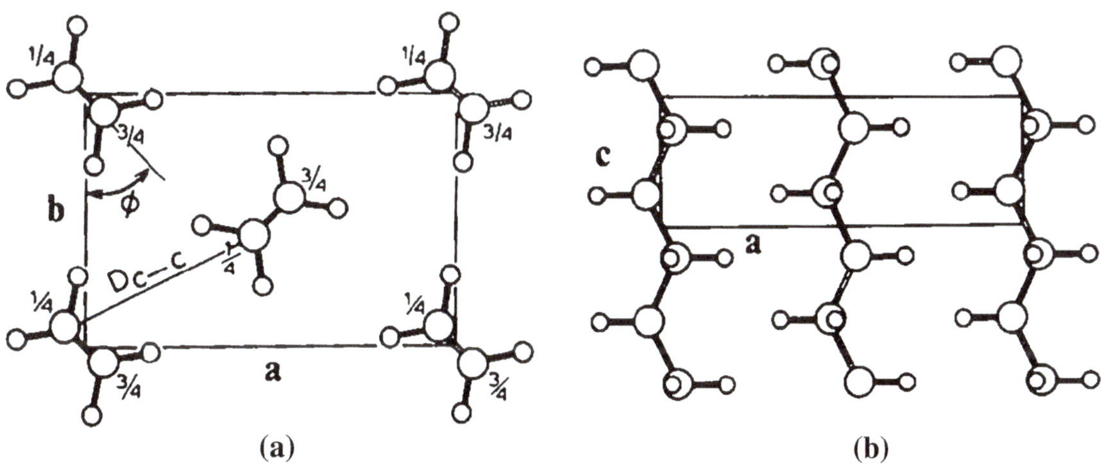

Fig. 4 Projection of polyethylene packing along the *c* axis **a** and along the *b* axis **b**. Polymer chain axes are parallel to axis *c*

Fig. 5 Simulated crystal structure of polyethylene sample 5 along the *c* axis

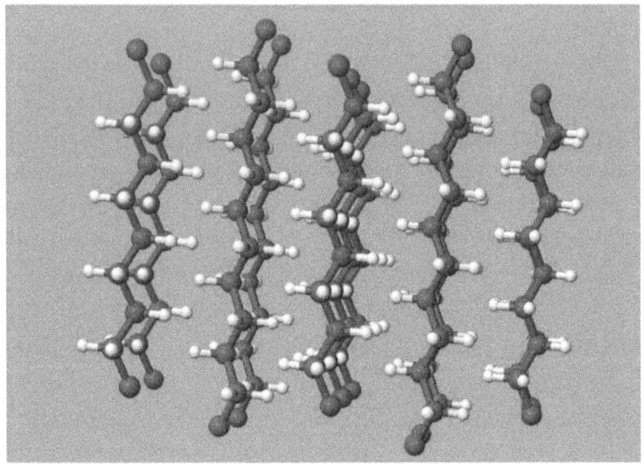

Fig. 6 Simulated crystal structure of polyethylene sample 5 perpendicular to axis *c*

the fact that literature X-ray diffraction data were obtained on re-melted polyethylene samples while we used as-synthesised polyethylene powder.

The calculated carbon–carbon interatomic distance in polyethylene sample 5 is 0.154061(1) nm and the C–C–C angle is 112.191(7)°. The calculated unit cell volume is 0.09474(7) nm^3 and calculated density of sample 5 is 0.9836 g/cm^3 which corresponds to a very high density polyethylene. Lattice parameters and densities of other samples in Table 1 are expected to be very similar to those of sample 5 as the positions of XRD diffraction peaks are almost identical.

Conclusion

In the polymerization of ethylene, the activity of the catalyst and the molecular weight of the polymer are highly dependent on the nature of the metal and the sizes of the group (R″) attached to the imino nitrogen atoms and the ortho constituent R. Developed Zr-based single site catalysts and optimized reaction condition enabled the synthesis of very high density polyethylene samples with very low branching content and high crystallinity.

Acknowledgments Authors would like to thank King Abdulaziz City for Science and Technology (KACST) for founding and supporting our collaborations with Oxford University. Solid-state NMR spectra were obtained at the EPSRC UK National Solid-state NMR Service at Durham.

References

1. Alsaygh A (2013) Distinct new catalysts for utilization of ethylene gas to polymer and oligomer. Appl Petrochem Res 3:101–105
2. Gibson VC, Spitzmesser SK (2003) Advances in non-metallocene olefin polymerization catalysis. Chem Rev 103:283–315
3. Crist B (1995) The ultimate strength and stiffness of polymers. Annu Rev Mater Sci 25:295–323
4. Young RJ, Lovell PA (1991) Introduction to polymers. Chapman, London
5. Bunn CW (1939) The crystal structure of long-chain normal paraffin hydrocarbons. Trans Faraday Soc 35:482–491
6. Hatfield GR, Killinge WE, Zeigler RC (1995) MeltlState ^{13}C MAS NMR determination of comonomer type and content in ethylene/α-Olefin copolymers. Anal Chem 67:3082–3085
7. Hay JN, Kemmish DJ, Langford JL, Rae AIM (1984) Polymer Commun 25:175
8. Ballirano P, Caminiti R, D'Ilario L, Martinelli A, Piozzi A, Maras A (1998) J Mater Sci 33:3519
9. Luo H-K, Schumann H (2005) New bi-nuclear and multi-nuclear α-diimine/nickel catalysts for ethylene polymerization. J Mol Catal A Chem 227(1–2):153–161 (And references therein)
10. Hsieh K-C, Chang C Jr, Lee M-T, Hu C-H, Hung C-H, Lee HM, Huang J-H, Wang M-H, Lee T-Y (2004) Hafnium chloride and hafnium methyl complexes bearing substituted pyrrolyl ligands: synthesis, characterization, and ethylene polymerization. Inorg Chim Acta 357:3517–3524
11. Britovsek GJP, Baugh SPD, Hoarau O, Gibson VC, Wass DF, White AJP, Williams DJ (2003) The role of bulky substituents in the polymerization of ethylene using late transition metal catalysts: a comparative study of nickel and iron catalyst systems. Inorg Chim Acta 345:279–291
12. Laridjani M, Leboucher P (2009) The structural dilemma of bulk polyethylene: an intermediary structure. PLoS ONE 4:e6228
13. Caminiti R, Pandolfi L, Ballirano P (2000) Structure of polyethylene from x-ray powder diffraction: influence of the amorphous fraction on data analysis. J Macromol Sci Phys B39:481–492
14. Fontana L, Vinh DQ, Santoro M, Scandolo S, Gorelli FA, Bini R, Hanfland M (2007) High-pressure crystalline polyethylene studied by x-ray diffraction and ab initio simulations. Phys Rev B 75:174112

15. VanderHart DL, Khoury F (1984) Polymer 25:1589
16. Larson AC, Von Dreele RB (2004) General structure analysis system (GSAS). Los Alamos National Laboratory Report LAUR, pp 86–748
17. Dorset DL, Moss B (1983) Polymer 24:291–294
18. Schwartz KB, Von Dreele RB (1997) Adv X-Ray Anal 39:515
19. Clark ES (2007) Chapter 38. Physical properties of polymers handbook. In: Mark JE (ed) Unit cell information on some important polymers, 2nd edn. Springer, New York

12

Characterising carbon deposited during pyrolysis gasoline hydrogenation: enhanced TPO methodologies

Javed Ali · S. David Jackson

Abstract In this paper, the use of temperature-programmed oxidation (TPO) is highlighted as a means of interrogating the surface of catalysts. Most studies only concern themselves with the active portion of the catalyst bed; in this study, we examine both the active part and the part that experiences equilibrated gas after reaction. The extent and nature of carbonaceous deposits on both sections of catalyst have been characterised by TPO and reveal that there is significant deposition of low H:C ratio carbonaceous species on catalyst that has only seen equilibrated reaction gas. This has potential implications for the life of the catalyst bed and the activity and selectivity observed.

Keywords PyGas · Hydrogenation · Temperature programmed oxidation · Carbon deposition

Introduction

Pyrolysis gasoline (PyGas) is a by-product of high temperature naphtha cracking during ethylene and propylene production [1, 2]. It is a mixture of highly unsaturated hydrocarbons ranging from C_5 to C_{12} and contains considerable amounts of aromatics, typically 40–80 % (benzene, toluene and xylene), together with paraffins, olefins and diolefins [3]. Composition of the PyGas produced depends on the feedstock and operating conditions and varies from plant to plant; however, a typical PyGas composition is given in Table 1 [8].

J. Ali · S. D. Jackson
Department of Chemistry, Centre for Catalysis Research, WestCHEM, University of Glasgow, G12 8QQ Glasgow, Scotland, UK

The main reason for the deactivation of catalysts during PyGas hydrogenation is the deposition of carbonaceous residue (coke) on the surface of the catalyst [10, 11]. The nature of the coke depends on the process conditions and length of reaction with the carbonaceous residues deposited, varying from hydrocarbons and poly-aromatics to graphitic coke. The deposition of coke decreases accessibility for the reactants to the catalyst surface and can block pores. Continued heavy build up of coke in pores may result in fracturing of the support material which can cause plugging of reactor voids [7, 13].

Carbonaceous residue formation is mainly due to the polymerisation of coke precursors present in PyGas such as styrene, di-olefins, mono-olefins, cyclo-olefins and aromatic species [14]. The deposited residue can convert to polyaromatic hydrocarbons, which then further convert to highly condensed polyaromatics and graphitic type coke. This is summarised in Fig. 1 [7, 9].

The nature of the carbonaceous residues also changes with reaction time. The change in the nature of the coke is due to polymerisation, multi-side reactions and loss of hydrogen content. The deposits convert to more condensed and hydrogen deficient coke with the passage of time and then finally form graphitic type coke.

The deposition of coke is highly dependent on the reaction conditions, especially on reaction temperature and hydrogen partial pressure. Increases in the reaction temperature of PyGas hydrogenation not only increase the amount of coke deposition but also result in a more condensed hydrogen deficient coke on the surface. Carbon laydown can be decreased by increasing hydrogen partial pressure [1].

Temperature-programmed oxidation (TPO) is a simple and important technique for the investigation into the quantity and nature of coke deposited [5, 6]. The amount of

Table 1 Typical composition of PyGas

PyGas components	Weight percent (wt%)
Benzene, toluene and xylenes	50
Olefins and dienes	25
Styrene and other aromatics	15
Paraffins and naphthenes	10

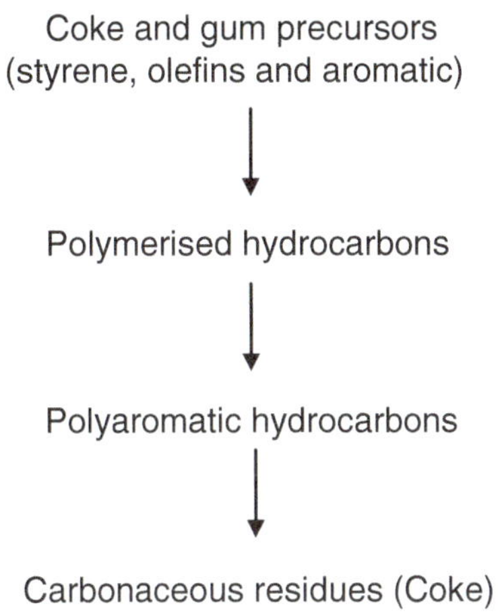

Fig. 1 Formation of carbonaceous residues (Coke)

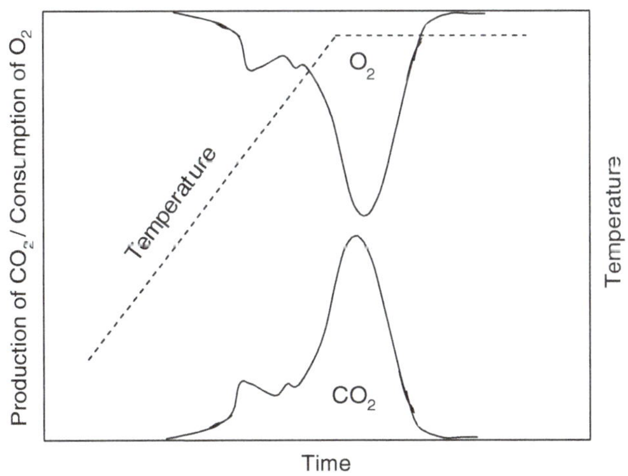

Fig. 2 Typical temperature programme oxidation profile [5, 6]

oxygen consumed and the amount of CO_2/CO produced indicate the amount of coke deposited on the surface of the catalyst. The oxygen consumption is commonly mirrored by the production of CO_2, as shown in Fig. 2.

The combustion of hydrogen rich coke, which is mostly present on the catalytic metal, occurs first. This is followed by the combustion of the hydrogen deficient coke present on the surface of the catalyst [4, 12]. The ratio of CO_2/H_2O and/or CO/H_2O can give insight into the carbon to hydrogen ratio present in the coke. A decrease in hydrogen to carbon ratio represents more condensed polyaromatic type of coke [6].

Table 2 Catalyst specifications

Catalyst	Metal loading (%wt)	Metal dispersion (%)	Surface area ($m^2\ g^{-1}$)	Pore volume ($cm^3\ g^{-1}$)
Ni/Al_2O_3	16.0	18.0	106	0.39

In this paper, we use TPO to probe the carbon deposit in both the reaction zone and the rest of the catalyst bed that is subjected to equilibrated gas in the typical fixed bed configuration found in the most chemical plants.

Experimental

The Ni/Al_2O_3 (HTC 400) catalyst used in this project was provided by Johnson Matthey. The characterisation data of catalyst provided by catalyst manufacturer are shown in Table 2.

Hydrogenation of the synthetic pyrolysis gasoline (PyGas) was investigated over the Ni/Al_2O_3 catalyst in a fixed-bed, high-pressure, continuous flow reactor. The catalyst (particle size between 250 and 425 microns) was added to the reactor tube and positioned in the middle of reactor tube. The catalyst was reduced in situ with a flow of hydrogen gas. The Ni/Al_2O_3 catalyst was reduced for 16 h at a temperature of 723 K. Following this, the reactor temperature was decreased to the required reaction temperature (413 K). The reactor was then pressurised to the desired pressure (20 barg). Once the reactor had obtained the desired conditions, the reaction was started with a particular flow of PyGas feed of 0.042 ml min^{-1} ($WHSV_{PyGas}$ 8 h^{-1}) and a flow of H_2/N_2 gas (10 ml min^{-1}). The PyGas was vaporised and mixed with hydrogen and then passed through the catalyst bed. The hydrogenated PyGas was condensed in a knockout pot, which was chilled to 273 K. Liquid samples were collected every 2 h during daytime and analysed using a Thermo Finnigan Focus G.C. The hydrogenated PyGas mainly condensed in the knockout pot; however, very small amounts of the reaction products remained in the gaseous phase and, therefore, the effluent gas was also analysed at regular intervals by online Agilent G.C. All reactions were run for 76 h. After finishing each reaction, nitrogen gas was passed through the catalyst over night so as to remove any species present in the reactor. The reactor was then cooled to room temperature in the flow of nitrogen gas and then an in situ TPO was performed on the post reaction catalyst. The gas was switched to a flow of 2 % O_2/Ar (40 ml min^{-1}) and the temperature increased to 723 K at a ramp of 5 K min^{-1} and then the reactor temperature was held constant at 723 K until all evolution of gases other

Table 3 Weight percent (%) composition of PyGas

Components	1-pentene	Cyclopentene	1-octene	Toluene	Styrene	Heptane	Decane
PyGas wt%	10.0	10.0	10.0	55.0	10.0	2.5	2.5

than O_2 and Ar ceased. The TPO process was monitored by an online mass spectrometer (QMS) in which the ion current of main species of mass fragments with m/z 28 (CO), 44 (CO_2), 32 (O_2), 18 (H_2O) was monitored. Other species with m/z 68 (cyclopentene), 70 (1-pentene), 72 (pentane), 78 (benzene), 92 (toluene), 98 (methylcyclohexane), 104 (styrene), 106 (ethylcyclo-hexane), 112 (1-octene), 114 (octane), 15 (CH_3), 29 (C_2H_5) and 2 (H_2) were also followed for detailed investigation of deposited coke.

A synthetic PyGas was used in this study and the composition is shown in Table 3. No benzene was used in this mix due to health and safety concerns: toluene was used as a replacement. Heptane and decane were used as internal standards for GC analysis.

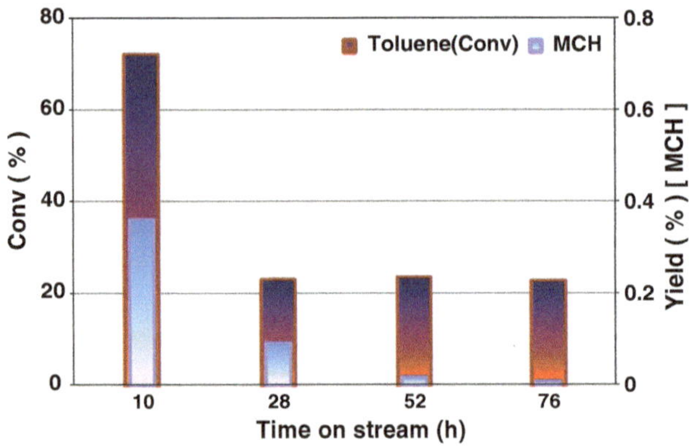

Fig. 3 Deactivation profile for toluene hydrogenation (component part of PyGas). Conditions: 413 K, WHSV 8 h^{-1}, H_2 pressure 5 barg, H_2:PyGas molar ratio = 5, overall pressure 20 barg

Results and discussion

During the course of the time on stream (TOS), the catalyst deactivated. A typical deactivation profile is shown in Fig. 3 with the hydrogenation of toluene to methylcyclohexane. There is a rapid deactivation over the first 24 h followed by a period of much slower deactivation. This behaviour was observed with all reactants. At the end of the 76-h TOS, the catalyst was subjected to an in situ TPO. The evolution of carbon dioxide and water is shown in Fig. 4, while the evolution of more complex species is shown in Fig. 5. In general, aliphatic species were not observed.

These graphs show that there has been considerable carbon deposition and that species such as benzene, which was not present in the feed, can be formed on the surface of the catalyst. As expected, the main species detected are aromatics, and very little of the fully hydrogenated species such as methylcyclohexane are seen.

In a typical fixed bed reactor on a chemical plant, the product flow will pass over a significant amount of catalyst that has been added to the reactor but only comes into action as the front end of the bed is deactivated. It is generally assumed that this part of the bed is unaffected by the product gases and is in a similar

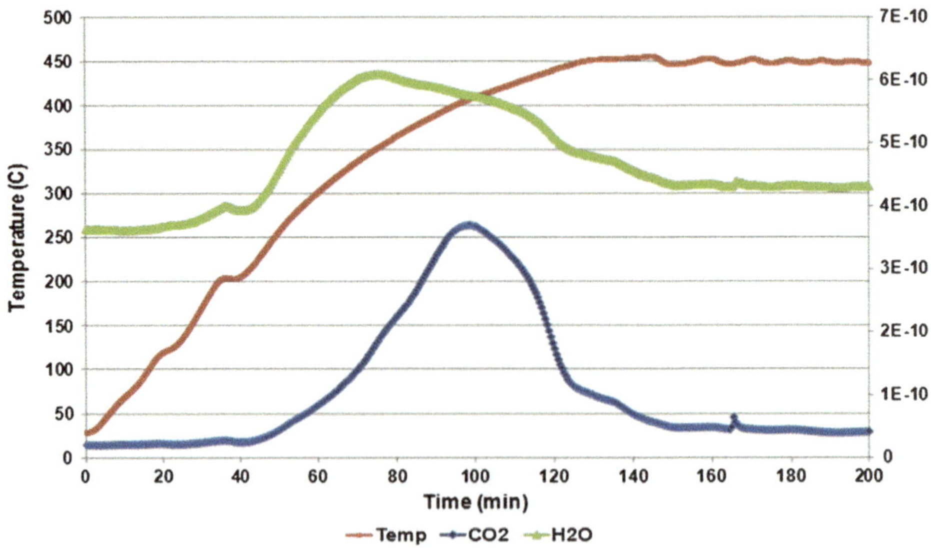

Fig. 4 Evolution of water and carbon dioxide as a function of temperature during TPO

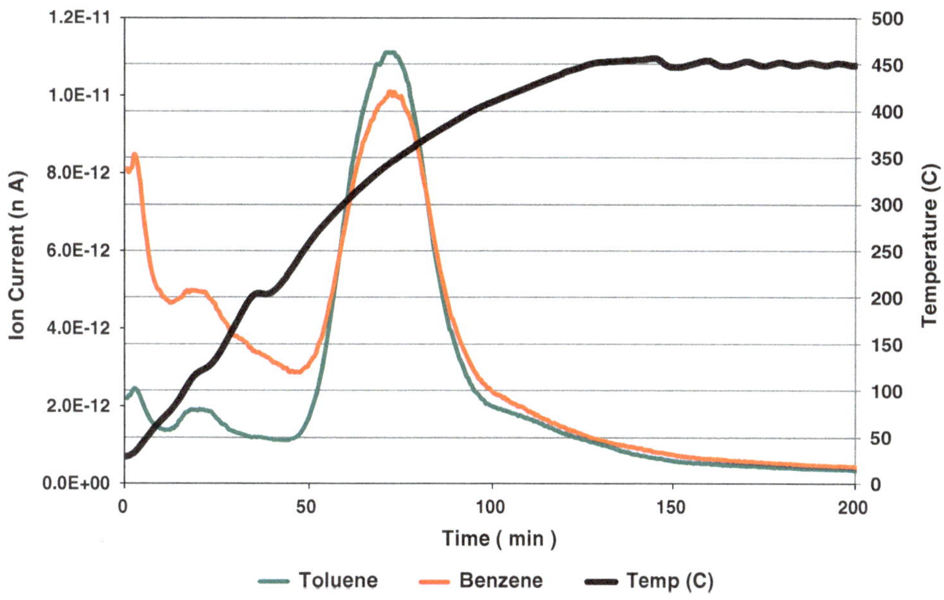

Fig. 5 Evolution of complex species during TPO

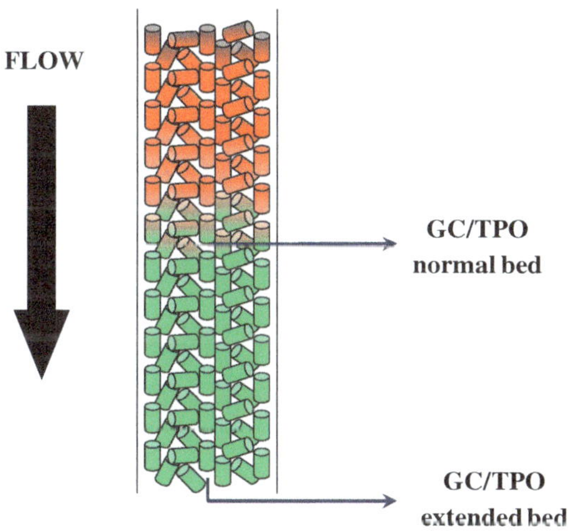

Fig. 6 Schematic of full-bed and half-bed reactor tests

Table 4 Conversion of PyGas components at normal-bed and extended-bed positions

Conversion (%)	Extended bed	Normal bed
1-pentene	90.6	87.3
1-octene	98.3	99.3
Cyclopentene	93.8	93.1
Toluene	24.1	23.1
Styrene	99.9	99.9

Conditions: 413 K, 20 barg pressure, 5 barg hydrogen, H_2:PyGas molar ratio = 5

state to fresh catalyst. To test this, the bed of catalyst was extended to double its length and used under the same conditions of flow, pressure and temperature. In this way, it would be possible to generate both GC and TPO data for both the normal bed and the extended bed (Fig. 6).

GC analysis was taken at the bottom of the normal bed and the extended bed to confirm whether the conversion was the same. The results are shown in Table 4 and confirm that the extent of hydrogenation changes very little between the normal-bed position and the extended-bed position as would be expected. In both cases, the catalyst was subjected to the same conditions and was on stream for the same length of time (76 h). TPOs were then run on both the normal and the extended bed. The normal-bed profiles are shown in Figs. 4 and 5. The extended-bed profiles are shown in Figs. 7 and 8. It is immediately clear that the profiles for the carbon dioxide and water have significantly altered. In Fig. 9, both the carbon dioxide evolution profiles for the normal and extended bed are compared. The carbon dioxide profile for the extended bed reveals that there has been considerable carbon deposition on the "unreactive" part of the bed and this is especially noticeable at the higher temperatures. By 450 °C, carbon dioxide evolution has almost ceased with the normal bed, whereas there is still significant amount of carbon combusting with the extended bed. The water profile is also different. The water evolution from the normal bed covers the same temperature range as the carbon dioxide, although the specific shape is different indicating a variety of carbonaceous species on the surface with different H:C ratios. However, the water profile from the extended bed does not mirror the temperature range of the carbon dioxide. In this case, the water evolution peaks at 350 °C and is almost back to

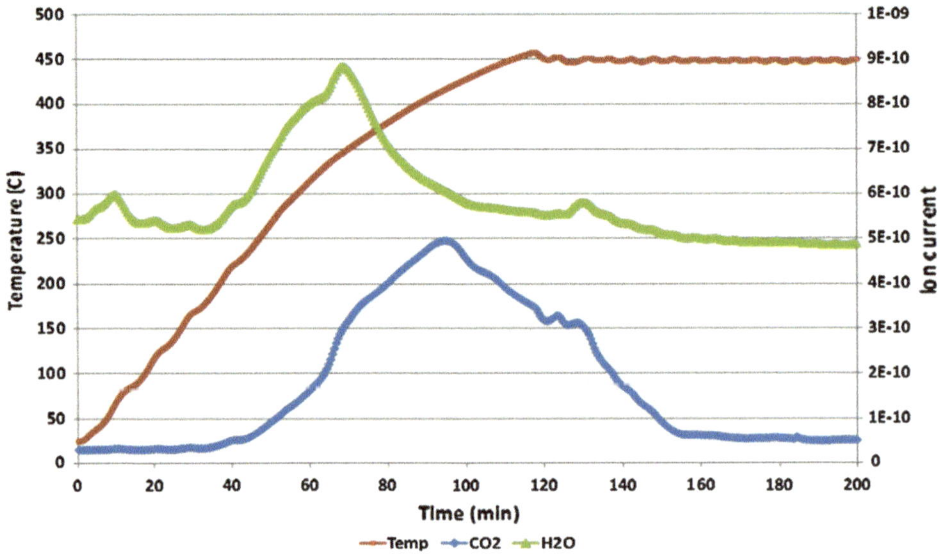

Fig. 7 Evolution of water and carbon dioxide as a function of temperature during TPO with an extended bed

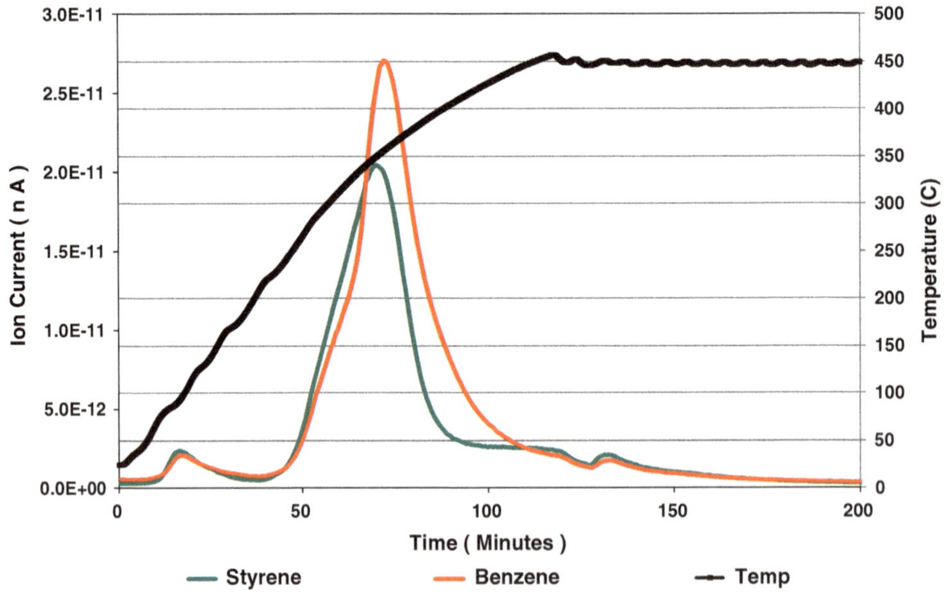

Fig. 8 Evolution of complex species during TPO with an extended bed

baseline by 450 °C. This change can give information related to the C:H ratio of the carbonaceous deposit. The carbonaceous deposit that is combusted at 450 °C on the extended bed produces carbon dioxide but very little water indicating that the deposit has a low H:C ratio, <1 suggesting the formation of polyaromatic or graphitic-like species. Note that this material is deposited in an atmosphere that has much less free hydrogen than that at the start of the bed. Even though much of the organic feed is now saturated, the lower gas phase hydrogen concentration does appear to affect the H:C ratio in a negative way. From the comparison of the water and carbon dioxide profiles, it is clear that on the normal bed the carbonaceous deposit has a much higher H:C ratio over the full profile and hence is easier to combust at lower temperatures.

In Fig. 10, the profiles obtained from benzene and toluene are compared and it can be seen that the extended bed has significantly more aromatic species than the normal bed but there is no evidence for desorption at higher temperatures.

Conclusions

Therefore using this enhanced TPO methodology, we have been able to show that in a standard fixed bed reactor, passing reaction equilibrated gas over fresh catalyst does not necessarily inhibit carbonaceous deposition and that the species deposited may be more deleterious to the catalyst activity than that deposited under reaction chemistry. This has potential implications for catalyst life in the plant and

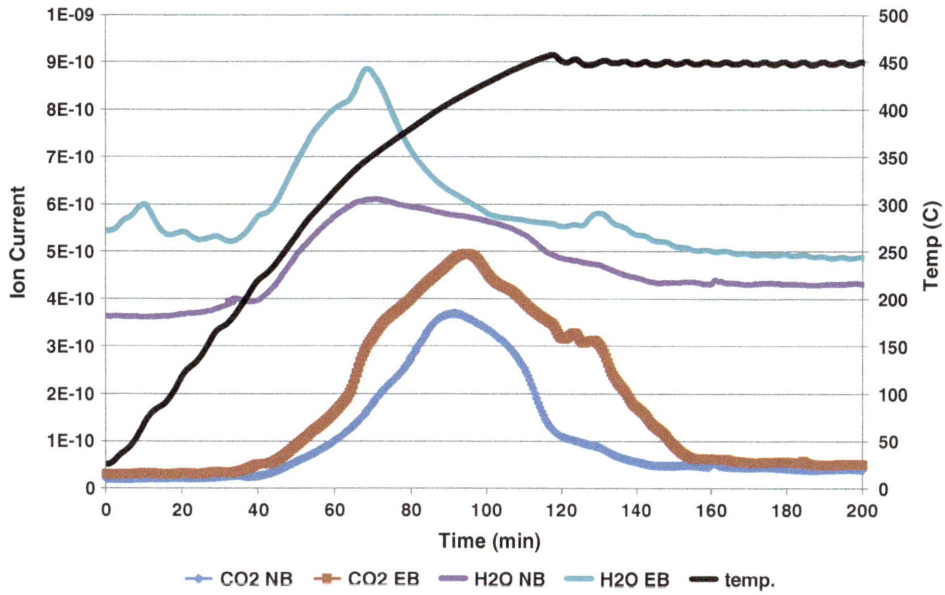

Fig. 9 Comparison of carbon dioxide and water evolution from the normal (NB) and extended (EB) beds

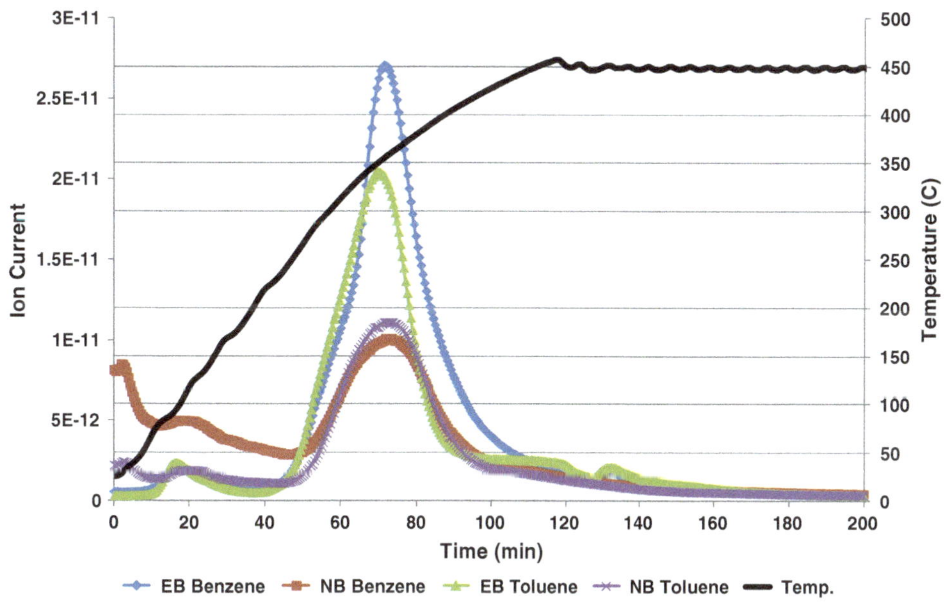

Fig. 10 Comparison of benzene and toluene evolution from the normal (NB) and extended (EB) beds

on the expected activity and selectivity. Using these techniques, a better understanding of the catalyst operating in the plant can be obtained.

References

1. Cheng YM, Chang JR, Wu JC (1986) Appl Catal 24:273–285
2. Reddy KM, Pokhriyal SK, Ratnasamy P, Sivasanker S (1992) Appl Catal A 83:1–13
3. (1979) US Patent 4150061, Process for converting pyrolysis gasoline to benzene and ethylbenzene-lean xylenes; assigned to Standard Oil Company (Indiana)
4. Anderson JA, Garcia MF (2005) Supported metals in catalysis. Imperial College Press, London
5. Barbier J (1986) Appl Catal 23:225–243
6. Barbier J, Churin E, Parera JM, Riviere J (1985) React Kinet Catal Lett 29:323–330
7. Bartholomew CH (2001) Appl Catal A 212:17–60
8. de Medeiros JL, Araujao OQF, Gaspar AB, Silva MAP, Britto JM (2007) Braz J Chem Eng 24:119–133
9. Hagen J (2006) Industrial catalysis: a practical approach, 2nd edn. Wiley-VCH, Mannheim
10. Le Page JF, Cosyns J, Courty EFP, Marcilly C, Martino JMG, Montarnal R, Sugier A, Landeghem HV (1987) Applied heterogeneous catalysis: design, manufacture, use of solid catalyst. Technip, Paris

11. Lloyd L (2008) Handbook of industrial catalysts. Springer, London
12. Porta F, Rossi M (2003) J Mol Catal A Chem 204–205:553–559
13. Richardson JT (1989) Principles of catalyst development. Plenum Press, New York
14. Weitkamp J, Raichle A, Traa Y (2001) Appl Catal A 222:277–297

13

A review of semi-rigid, stilbene-containing alternating copolymers

Alice M. Savage · Xu Zhou · Jing Huang · S. R. Turner

Abstract The synthesis and properties of sterically congested, stilbene-containing alternating copolymers are reviewed. Persistence lengths (2–6 nm) determined by size exclusion chromatography and small angle X-ray scattering techniques show that these are semi-rigid copolymers. Fully characterized polyanions and polyampholytes, prepared from organic-soluble precursors were studied, along with their respective salt and pH response behavior resulting from the semi-rigid polymer backbone. The solid-state characterization from these studies reveals a high degree of hindered rotation along the polymer backbone. The contorted structure and the hindered rotation of the polymer backbone generate inefficient chain packing, which leads to an increase of nanoporosity and higher surface areas. The semi-rigid stilbene-containing copolymers are a new class of copolymers where the increased polymer backbone rigidity leads to a wide range of polymer properties not attained with flexible polymers.

Keywords Stilbene · Alternating copolymers · Copolymerization · Maleic anhydride · Maleimides

Introduction

Substituted stilbenes (1,2-diphenylethylenes) are widely studied and used organic compounds for the optical properties arising from their conjugated ethylene structure, e.g. optical brighteners, dyes, etc. [1, 40]. Many applications of stilbene derivatives have resulted in a large library of substituted stilbene small molecules and facile synthetic techniques to enable synthesis of a variety of derivatives many of which can serve as copolymerizable monomers. However, the addition polymerization and copolymerization of this family of 1,2-disubstituted olefins, like all 1,2-substituted ethylene structures, is difficult due to the steric constraints of the 1,2-structure. Homopolymerization of stilbene and substituted stilbenes to the corresponding poly (benzyl) backbone has not been achieved; however, alternating copolymerization of stilbene both anionically [48] and radically [44] is well documented. In alternating copolymerizations, the cross-propagation step is greatly favored since the steric interactions in the cross propagation step are depressed compared to those required for a successful self-propagation of the stilbene monomers. In addition the electronic effects resulting from electron rich and electron poor comonomers cross-propagating with an electron rich or electron poor terminal radical favor fast cross-propagation. Stilbene copolymers have been claimed in the patent literature to be useful in lubricants [37], lubricating oil additives [2, 31], thermal stability additives for photoresists [42], and improved stain resistant polyamide textiles [11], although successful commercialization of these potential applications is not known.

The radical alternating copolymerization of unsubstituted stilbene is the most studied stilbene polymerization [6, 16, 35]. It was found in 1930 that stilbene copolymerized with maleic anhydride [44]. Since neither monomer can undergo radical homopolymerization, these are strictly alternating copolymers with no stilbene–stilbene or maleic anhydride-maleic anhydride dyads present [21]. This study attracted many groups to study the comonomer reactivity ratios [21], monomer sequences [16], thermal stability [32], and copolymerization mechanism [9, 17, 34]. However,

A. M. Savage · X. Zhou · J. Huang · S. R. Turner
Department of Chemistry MC0212 and Macromolecules and Interfaces Institute MC0344, Virginia Tech, Blacksburg, VA 24061, USA

surprisingly limited solubilities and uninteresting brittle materials have slowed detailed investigations of the copolymerizations and the properties of the resulting copolymers [41].

N-substituted maleimides readily radically polymerize with stilbene into predominately alternating copolymer structures—maleimides can radically homopolymerize, so some maleimide–maleimide dyads can be present in these copolymers [19, 35, 46]. Many of these copolymers have limited solubilities in common organic solvents so solution studies are limited.

Our recent work involves the design and synthesis of new stilbene comonomers with specifically chosen functional groups for effecting the copolymerization with maleic anhydride or N-substituted maleimides. Radical polymerization processes are compatible with many functional groups [5] and the alternating nature of the copolymerization enables precise placement of functionality into the copolymer backbones for study of the structure/property relationships of these novel sterically congested backbone polymers in both solution and the solid state.

Monomer substituent effect on copolymerization

The type and position of substituents on the aromatic groups of stilbene were found to have a significant effect on both the overall copolymerization rates as well as the solubility of the copolymers [24]. Several methyl-substituted stilbene monomers were prepared by using the Wittig-Horner reaction and the resulting monomers were copolymerized with maleic anhydride. Feed ratio/composition studies and ^{13}C NMR confirmed the strictly alternating sequences in the backbone [24]. The methyl-substitution increased the solubility of the copolymers in organic solvents by disrupting copolymer aggregation. This allowed for thorough characterization of these copolymers via SEC, NMR and DLS (dynamic light scattering). We note that with unsubstituted stilbene and maleic anhydride copolymers, strong aggregation in DLS was observed which likely accounts for the "insolubility" of these copolymers as reported in the literature. The location of the methyl groups on the phenyl rings was found to have a strong influence on the overall rate of copolymerization. Methyl groups in the para position led to increase copolymerization rates due to the inductively electron donating characteristics which lead to a more reactive stilbene radical to cross propagate with the electron deficient double bond of maleic anhydride. The ortho-methyl stilbene monomers were found to copolymerize at significantly slower rates due to the loss of conjugation of the aromatic groups with the radical due to steric interactions of the methyl groups forcing the phenyl groups out-of-plane (Fig. 1). DLS measurements showed that the para-substituted stilbene copolymers formed higher molecular weight aggregates in solution, where little aggregation was observed in the ortho substituted samples [24].

Strong electron donating groups, such as dialkyl amino substituents in the para position of stilbene showed extremely high copolymerization rates [29]. Unsymmetrical alkyl substituents on the stilbene comonomers were employed to obtain soluble copolymers. For all of the stilbene copolymers thermal analyses by DSC showed no observable T_g at temperatures <250 °C [29].

With the goal of synthesizing a family of precisely functionalized polyanions, both styrene and stilbene

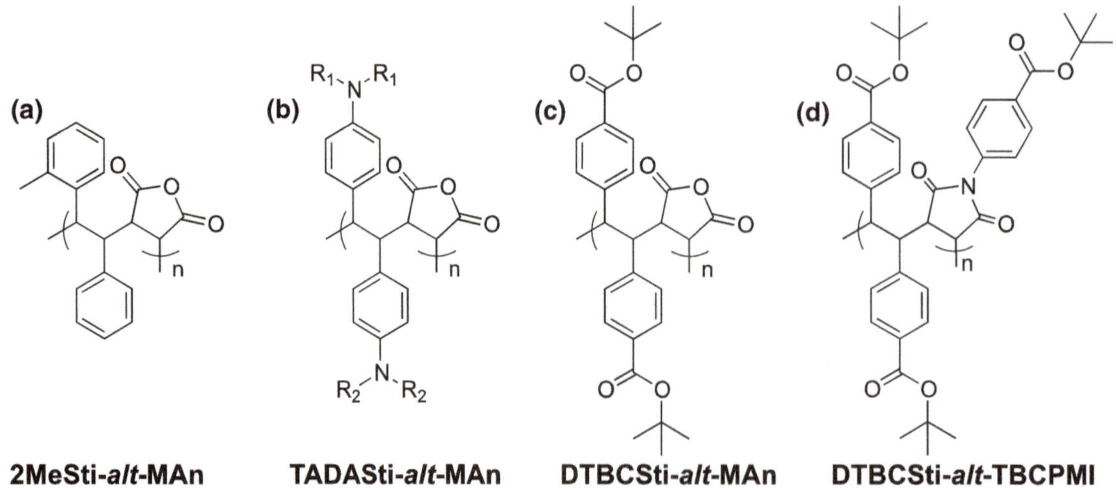

Fig. 1 Structures of organic-soluble, main-chain stilbene-containing copolymer precursors. **a** poly((*E*)-2-methylstilbene-*alt*-maleic anhydride) (2MeSti-*alt*-MAn) [24], **b** poly(*N*,*N*,*N*′,*N*′-tetraalkyl-4,4′-diaminostilbene-*alt*-maleic anhydride) (TADASti-*alt*-MAn) [29], **c** poly((*E*)-di-*tert*-butyl 4,4′-stilbenedicarboxylate-*alt*-maleic anhydride) (DTBCSti-*alt*-MAn) [22], and **d** poly((*E*)-di-*tert*-butyl 4,4′-stilbenedicarboxylate-*alt*-*tert*-butyl 4-maleimidobenzoate) (DTBCSti-*alt*-TBCPMI) [22]

monomers were synthesized with *tert*-butyl ester groups in the para positions [22]. Higher temperatures were required to obtain efficient copolymerizations because of the electron withdrawing nature of the para ester groups. The use of the *tert*-butyl ester copolymers, as precursors to the polyanions, enabled the accurate determination of the molar mass and molar mass distributions by SEC avoiding the known SEC analysis problems of polyelectrolytes. The *tert*-butyl ester groups were found to cleanly thermally decompose as determined by the expected weight loss for extrusion of isobutylene in thermal gravimetric analyses [22]. For study of the polyanions, deprotection with trifluoroacetic acid followed by neutralization with base and dialysis yielded the pure, well-characterized polyanions. The solution properties of the precursor *tert*-butyl ester copolymers and the polyanions were studied in detail [23].

Polymer rigidity

The early investigation of main-chain stilbene copolymers suggested that they possessed characteristics consistent with rigid rod-like structures. This led to further investigation into the chain stiffness of the copolymer backbone [28–30]. When examining the ^{1}H NMR spectra of any of these copolymers, the backbone and aromatic peaks appeared very broad [24]. When associated with rigid polymers and high molecular weight materials, broad peaks develop due to strong homonuclear dipolar couplings and slow ^{1}H relaxation rates [45]. With no observed T_g below 250 °C, evidence of hindered rotation and limited movement of the polymer backbone pointed to a more rigid polymer backbone with a sterically crowded configuration. In the case of poly (*N,N,N′,N′*-tetraethyl-4,4′-diaminostilbene-*alt*-maleic anhydride) (TEDASti-*alt*-MAn), the copolymer backbone likely conformed into a contorted, sterically crowded conformation with the maleic anhydride units existing predominately in the *cis* configuration [28] (Fig. 1b).

To confirm the rigidity or chain stiffness of the stilbene alternating copolymers, SEC and SAXS were used to directly measure the chain stiffness (persistence length, l_p) [25]. Defined as a measure of chain stiffness or rigidity, persistence lengths provide a direct comparison of intrinsic polymer chain stiffness with other flexible and rigid polymers. (*E*)-di-*tert*-butyl 4,4′-stilbenedicarboxylate and the styrene analog (*tert*-butyl 4-vinyl benzoate) were copolymerized with MAn and *tert*-butyl 4-maleimidobenzoate, respectively. This copolymer series was studied to investigate the effects of the added extra phenyl ring of stilbene monomer and cyclic monomer structure (MAn versus maleimide) on the persistence lengths [22, 25]. Based on

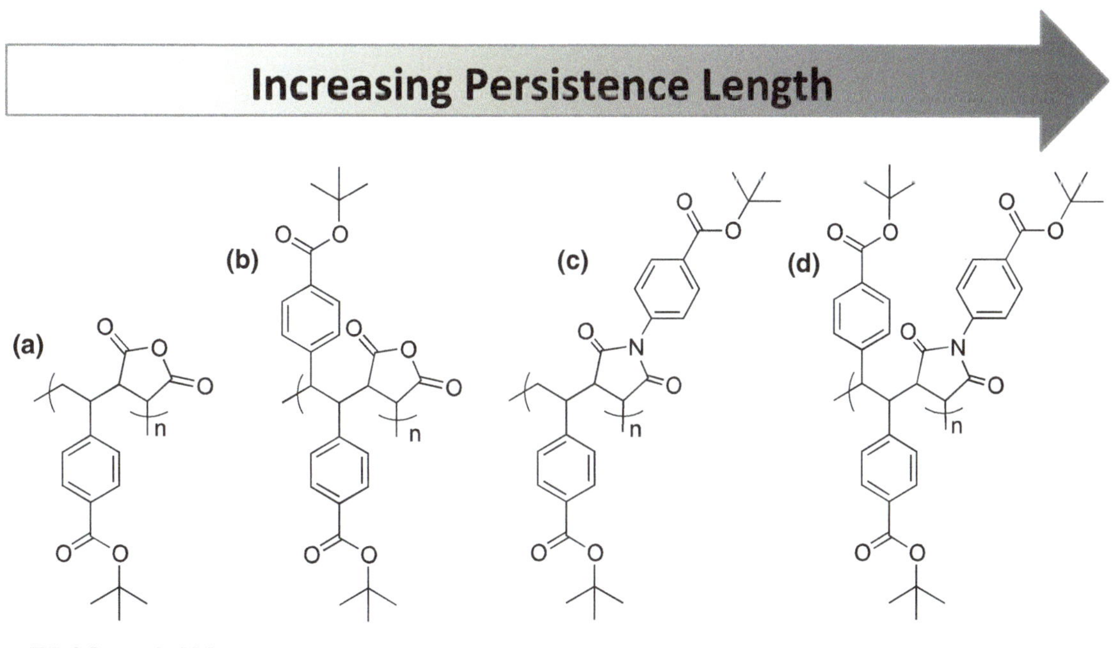

Fig. 2 A series of *tert*-butyl protected stilbene and styrene containing alternating copolymer precursors in order of increasing rigidity (increases *left to right*). **a** poly((*tert*-butyl 4-vinyl benzoate-*alt*-maleic anhydride) (CSty-*alt*-MAn), **b** poly((*E*)-di-*tert*-butyl 4,4′-stilbenedicarboxylate-*alt*-maleic anhydride) (DTBCSti-*alt*-MAn), **c** poly(*tert*-butyl 4-vinyl benzoate-*alt*-*tert*-butyl 4-maleimidobenzoate) (CSty-*alt*-TBCPMI) and **d** poly((*E*)-di-*tert*-butyl 4,4′-stilbenedicarboxylate-*alt*-*tert*-butyl 4-maleimidobenzoate) (DTBCSti-*alt*-TBCPMI) [25]

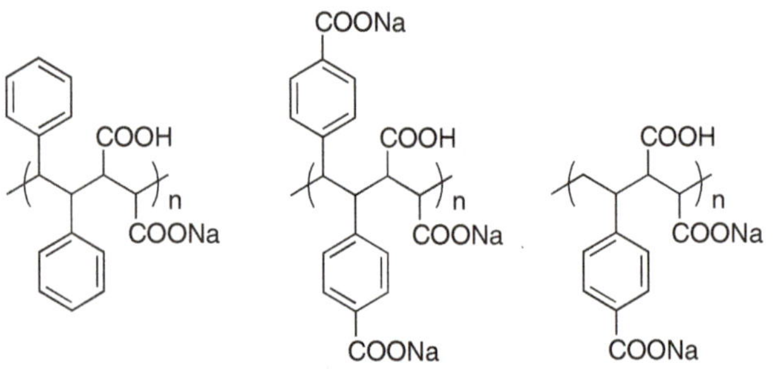

Sti-*alt*-MA **DCSti-*alt*-MA** **CSty-*alt*-MA**

Fig. 3 Structures of carboxylated stilbene-containing polyanions: poly(stilbene-*alt*-maleic acid) (Sti-*alt*-MA), poly((4,4′ stilbenedicarb oxylate)-*alt*-maleic acid) (DCSti-*alt*-MA), and poly((4-vinyl benzoic acid)-*alt*-maleic acid) (CSty-*alt*-MA) [23]

SEC and SAXS methods [3, 4, 38, 39], this copolymers series exhibited the persistence lengths between 2 and 6 nm. From this direct comparison and measurement, stilbene alternating copolymers are classified as semi-rigid polymers (Fig. 2). For example, a typical flexible copolymer, polystyrene, exhibits a persistence length of 0.9 nm [10] and a rigid copolymer, polyphenylene with sulfonate ester side groups exhibits a persistence length of 13 nm [14, 43]. Accordingly, the stilbene-containing copolymers appear 20 % more rigid than the styrene analogs and the maleimide-containing copolymers appear 50 % more rigid than MAn-containing copolymers of the same comonomer. The polymer physical properties that arise from these semi-rigid and sterically crowded copolymers involve unique solution and solid-state properties.

Solution properties

Using the same deprotection strategy above, a family of polyanion and polyampholyte copolymers were prepared with stilbene comonomers and the solution properties arising from the semi-rigid copolymer backbone were studied [22, 23, 30].

Polyanions

As mentioned above, carboxylated polyanions [22, 25] were synthesized from the deprotection of the precursor copolymers, and fully characterized with respect to molar mass, composition, and intrinsic polymer rigidity. Further physical aqueous solution properties evaluated using steady-state solution shear rheology, dynamic light scattering (DLS), and titration studies proved critical to understanding effects of polymer rigidity and conformation in aqueous solutions [23].

Using steady-state solution shear rheology, reduced viscosity plotted against polymer concentration revealed a

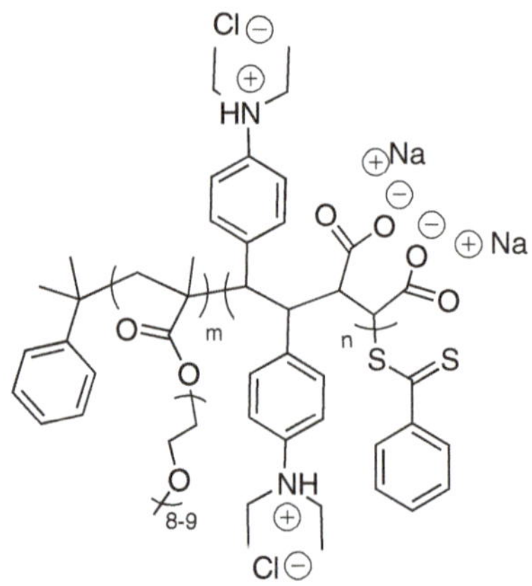

OEGMA-b-(TEDASti-*alt*-MA)

Fig. 4 Polyampholyte structure of poly(oligo-(ethylene glycol)methacrylate)-*b*-((*N*,*N*,*N*′,*N*′-tetraethyl-4,4′-diaminostilbene)-*alt*-maleic acid)) ((OEGMA-*b*-(TADASti-*alt*-MA))) [30]

prominent polyelectrolyte effect in deionized water for DCSti-*alt*-MA, CSty-*alt*-MA, and Sti-*alt*-MA shown in Fig. 3 [23]. Increasing the sodium chloride (NaCl) concentration suppressed the polyelectrolyte effect by decreasing the Debye Length and electrostatic repulsions as the polymer concentration decreased. These results were consistent with semi-rigid polymers.

The statistical segment length (L_k) for each copolymer with or without NaCl was measured by adding the intrinsic statistical segment length (L_0) to the electrostatic statistical segment length (L_e) [7, 8, 47]. In deionized water, the carboxylated copolymers behaved like rigid-rods due to strong electrostatic repulsions and the statistical segment lengths decreased as NaCl concentration increased. Thus, the polyanions behaved like semi-rigid copolymers in NaCl solutions and followed the same rigidity trends established for the intrinsic statistical segment lengths above [25].

To further characterize the polymer properties in aqueous solutions, pH titration was used to evaluate the dissociation behavior of the polyanions. Titration results revealed multiple inflection points arising from the two vicinal carboxylic acid groups. Due to the close proximity of the diacids, each carboxylic acid has a distinct pKa and dissociates at different pH. This is consistent with poly(styrene-*alt*-maleic acid) (Sty-*alt*-MA) titration results [12].

Polyampholytes

The first stilbene-containing polyampholyte diblock copolymers were prepared using RAFT (reversible addition fragmentation chain transfer) polymerization techniques [30]. The diblock copolymer, poly(OEGMA-*b*-(TADASti-*alt*-MAn)), consisted of a polymerized methoxy-capped

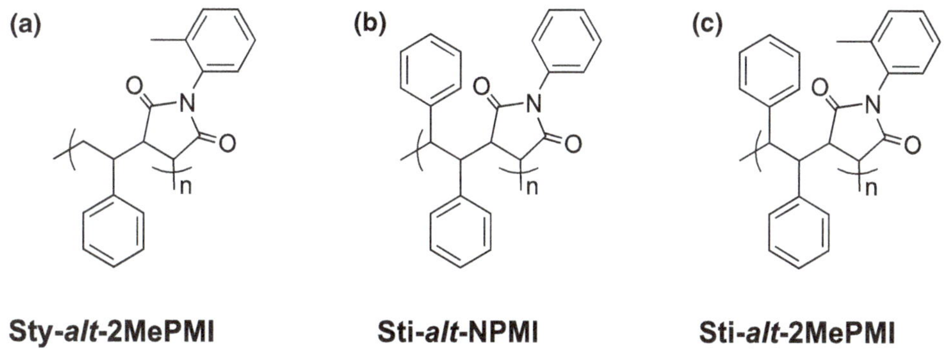

Fig. 5 Structures of maleimide copolymers with stilbene and styrene analogs. **a** poly(styrene-*alt*-*N*-(2-methylphenyl) maleimide) (Sty-*alt*-2MePMI); **b** poly(stilbene-*alt*-*N*-phenyl maleimide) (Sti–*alt*-NPMI); **c** poly(stilbene-*alt*-*N*-(2-methylphenyl) maleimide) (Sti-*alt*-2MePMI) [27]

oligo-(ethylene glycol)methacrylate) (OEGMA) block chain extended with *N,N,N′,N′*-tetraethyl-4,4′-diaminostilbene and MAn [30]. After molar mass determination in organic solvent, the block copolymer converted to its subsequent polyampholyte (Fig. 4) via simple acid hydrolysis. The resulting water-soluble polyampholyte ((OEGMA-*b*-(TADASti-*alt*-MA))) formed polyion complexes (PICs) and exhibited a pH and salt responsiveness atypical of polyampholytes.

To evaluate the PIC aggregate response to pH and salt concentration, DLS was used to monitor the aggregate size. The aggregate size increased as the pH approached the isoelectric point (IEP) of the polyampholyte. For more flexible and non-flexible polyampholytes, as the pH neared the IEP, the copolymer chains expand as the electrostatic attractive forces decrease [13, 26]. Atypical for polyampholytes, adding NaCl to the [(OEGMA-*b*-(TADASti-*alt*-MA))] copolymer solution increased the PIC size even as the electrostatic repulsions decrease. The "like-charge" attraction of the polymer chains causes polymer chains to aggregate and increase PIC size [15, 18, 20]. The semi-rigid polymer backbone of the TADASti-*alt*-MA block facilitated dipole–dipole interactions of condensed counter-ions and induced the "like-charge" attraction [18, 20]. Ongoing work involves investigating the effects of charge density, molar mass, and linear, water-soluble homopolymer blocks [36].

Solid-state properties

Forming brittle films when cast from solution, the semi-rigid copolymers offer unique solid-state properties like birefringence and nanoporosity as a result of the sterically crowded copolymer backbones. Hindered rotations and inefficient chain packing of the copolymer backbone significantly affects and enhances the solid-state properties.

A series of methyl-substituted maleimide copolymers with stilbene and styrene analog films exhibited large negative birefringence up to 0.02 for Sti-*alt*-2MePMI (Fig. 5) [27]. During casting of the film, the ortho-methyl forced the phenyl groups to align normal to the imide plane. Due to the anisotropy of the benzene ring, the alignment of the phenyl rings, and the restricted rotation induced a negative birefringence [33]. The same effect did not appear in Sti-*alt*-NPMI (Fig. 5b) without the ortho-methyl substituent; thus, the phenyl rings freely rotated. The less sterically crowded Sty-*alt*-2MePMI (Fig. 5a) showed a low negative birefringence of 0.002 which indicated that the sterically crowded polymer backbone influences the free rotation of the phenyl rings.

The nanoporous characteristics of the same series of *tert*-butyl carboxylate functionalized alternating copolymers using nitrogen adsorption/desorption isotherms and atomistic molecular simulations were studied (Fig. 2) [49]. These alternating copolymers showed a nanoporosity with Brunauer-Emmett-Teller (BET) surface area of 20 40 m^2/g. This is likely due to the insufficient chain packing of these semi-rigid polymer chains and the increase in the internal free volume leads to an increased surface area. Also, as the chain rigidity increased, then the BET surface areas increased for this copolymer series [49].

Conclusions

This review highlights the synthesis and properties of sterically crowded stilbene-containing copolymers. The alternating copolymerization gives precise control over monomer sequence, which affords precise placement of functional groups along the polymer backbone. With precise control over monomer sequence, stilbene-containing alternating copolymers afford semi-rigid copolymers. Both the solution and solid-state properties confirm the semi-rigid nature of the backbone. Continued research on stilbene-containing copolymers and block copolymers is focused on new polymeric derivatives to explore the wide range of solid state and solution properties that result from the semi-rigid backbone characteristics.

Acknowledgments National Science Foundation Grant Numbers DMR 9905231 and 1206409 supported fundamental studies on the synthesis, characterization, solution and solid properties of the alternating copolymers in this manuscript.

References

1. Henkel-Helios A (1959) Detergent compositions. United Kingdom Patent 810,151
2. Anderson WS (1967) Copolymers of ethylene and stilbene for use as lubricant additives. U.S. Patent 3,338,877
3. Beaucage G, Rane S, Sukumaran S, Satkowski MM, Schechtman LA, Doi Y (1997) Persistence length of isotactic poly(hydroxy butyrate). Macromolecules 30(14):4158–4162. doi:
4. Bohdanecky M (1983) New method for estimating the parameters of the wormlike chain model from the intrinsic viscosity of stiff-chain polymers. Macromolecules 16(9):1483–1492. doi:
5. Braunecker WA, Matyjaszewski K (2007) Controlled/living radical polymerization: features, developments, and perspectives. Prog Polym Sci 32(1):93–146. doi:
6. Chu Q, Lu H, He T, Ding M (2000) Synthesis and thermal polymerization of a novel stilbene-maleimide AB-monomer. Polym Bull 43(6):471–475. doi:
7. Davis RM, Russel WB (1986) On the theory of dilute polyelectrolyte solutions: extensions, refinements, and experimental tests. J Polym Sci B Polym Phys 24(3):511–533. doi:
8. Davis RM, Russel WB (1987) Intrinsic viscosity and Huggins coefficients for potassium poly(styrenesulfonate) solutions. Macromolecules 20(3):518–525. doi:
9. Ebdon JR, Hunt BJ, Hussein S (1987) Radical copolymerization of maleic-anhydride with *trans*-stilbene. Brit Polym J 19(3–4):333–337
10. Fetters L, Lohse D, Colby R (2007) Physical Properties of Polymers Handbook. In: Mark JE (ed). Springer, New York, p 447
11. Fitzgerald PH, Rao NS, Vinod YV, Henry GK, Prowse KS (1989) Stain-resistant aromatic/maleic anhydride polymers. EP Patent 0,329,899
12. Garrett ER, Guile RL (1951) Potentiometric titrations of a polydicarboxylic acid: maleic acid-styrene copolymer. J Am Chem Soc 73:4533–4535
13. Gotzamanis G, Tsitsilianis C (2006) Stimuli-responsive A-b-(B-co-C) diblock terpolymers bearing polyampholyte sequences. Macromol Rapid Commun 27(20):1757–1763. doi:
14. Guilleaume B, Blaul J, Ballauff M, Wittemann M, Rehahn M, Goerigk G (2002) The distribution of counterions around synthetic rod-like polyelectrolytes in solution, a study by small-angle X-ray scattering and by anomalous small-angle X-ray scattering. Eur Phys J E 8(3):299–309. doi:
15. Ha BY, Liu AJ (1997) Counterion-mediated attraction between two like-charged rods. Phys Rev Lett 79(7):1289–1292. doi:
16. Ha NTH (1998) Determination of triad sequence distribution of copolymers of maleic anhydride and its derivatives with donor monomers by 13C NMR spectroscopy. Polymer 40(4):1081–1086. doi:
17. Hallensleben ML (1973) Electron donor-acceptor complexes and polymerization. V. Copolymerization of maleicanhydride and *trans*-stilbene under ultraviolet irradiation. Eur Polymer J 9(3):227–231. doi:
18. Hampton KW Jr, Ford WT (2000) Styrylmethyl(trimethyl)ammonium methacrylate polyampholyte latexes. Macromolecules 33(20):7292–7299. doi:
19. Hayakawa K, Kawase K, Yamakita H (1985) On the reaction mechanism of gamma-ray solid-state copolymerization. J Polym Sci Polym Chem Ed 23(11):2739–2746. doi:
20. Lee K-C, Borukhov I, Gelbart William M, Liu Andrea J, Stevens Mark J (2004) Effect of mono- and multivalent salts on angle-dependent attractions between charged rods. Phys Rev Lett 93(12):128101
21. Lewis FM, Mayo FR (1948) Copolymerization. IX. Comparison of some *cis* and *trans* isomers. J Am Chem Soc 70:1533–1536
22. Li Y, Mao M, Matolyak LE, Turner SR (2012) Sterically crowded anionic polyelectrolytes with tunable charge densities based on stilbene-containing copolymers. ACS Macro Lett 1(2):257–260. doi:
23. Li Y, Savage AM, Zhou X, Richard Turner S, Davis RM (2013) Solution properties of stilbene-containing sterically crowded alternating polyanions. J Polym Sci B Polym Phys 51(21):1565–1570. doi:
24. Li Y, Turner SR (2010) Free radical copolymerization of methyl substituted stilbenes with maleic anhydride. Eur Polym J 46(4):821–828. doi:
25. Li Y, Zhang M, Mao M, Turner SR, Moore RB, Mourey TH, Slater LA, Hauenstein JR (2012) Chain stiffness of stilbene containing alternating copolymers by SAXS and SEC. Macromolecules 45(3):1595–1601. doi:
26. Lowe AB, McCormick CL (2002) Synthesis and solution properties of zwitterionic polymers. Chem Rev 102:4177–4189
27. Mao M, England J, Turner SR (2011) Alternating stilbene copolymers with negative birefringence. Polymer 52(20):4498–4502. doi:
28. Mao M, Kim C, Wi S, Turner SR (2008) Chain structure of substituted stilbene-maleic anhydride alternating copolymer probed by solid-state NMR. Macromolecules 41(2):387–389. doi:
29. Mao M, Turner SR (2006) Synthesis and characterization of highly functionalized polymers based on N, *N,N′,N′*-tetraalkyl-4,4′-diaminostilbene and maleic anhydride. Polymer 47(24):8101–8105. doi:
30. Mao M, Turner SR (2007) Aggregation of rod-coil block copolymers containing rigid polyampholyte blocks in aqueous solution. J Am Chem Soc 129(13):3832–3833. doi:
31. Marvel CS, Anderson WS (1954) Copolymerization of 1,3-butadiene with some polynuclear aromatic hydrocarbons. J Am Chem Soc 76:5434–5435. doi:
32. McNeill IC, Polishchuk AY, Zaikov GE (1995) Thermal degradation studies of alternating copolymers. III. The copolymer of maleic anhydride with *trans*-stilbene. Polym Degrad Stab 47(2):319–329. doi:
33. Riande E, Sáiz E (1992) Dipole moments and birefringence of polymers. Prentice Hall, Michigan
34. Rzaev ZM, Zeinalov IP, Medyakova LV, Babaev AI, Agaev MM (1981) Mechanism of radical copolymerization of maleic anhydride with *trans*-stilbene. Vysokomol Soedin Ser A 23(3):614–618
35. Rzaev ZMO, Milli H, Akovali G (1996) Complex-radical alternating copolymerization of *trans*-stilbene with N-substituted maleimides. Polym Int 41(3):259–265. doi:
36. Savage AM, Turner SR (2014). Unpublished work
37. Schaschel ET (1975) Reaction products of silicon monoxide and olefins. U.S. Patent 3,907,750

38. Schmid CW, Rinehart FP, Hearst JE (1971) Statistical length of DNA from light scattering. Biopolymers 10(5):883–893. doi:
39. Sharp P, Bloomfield VA (1968) Light scattering from wormlike chains with excluded volume effects. Biopolymers 6(8):1201–1211. doi:
40. Smtih RE (2000) Stilbene dyes. Kirk-Othmer encyclopedia of chemical technology, vol 27, 4th edn. John Wiley & Sons, Inc, New Jersey
41. Tanaka T, Vogl O (1982) Head-to-head poly(methyl cinnamate) from alternating stilbene-maleic anhydride copolymer, {poly[1,2-bis(methoxycarbonyl)-3,4-diphenyltetramethylene]}. Macromol Synth 8:21–24
42. Turner SR, Arcus R, Houle C, Schleigh W (1986) High-Tg base-soluble copolymers as novolac replacements for positive photoresists. Polym Sci Eng 26(16):1096–1100
43. Vanhee S, Rulkens R, Lehmann U, Rosenauer C, Schulze M, Koehler W, Wegner G (1996) Synthesis and characterization of rigid rod poly(p-phenylenes). Macromolecules 29(15):5136–5142. doi:
44. Wagner-Jauregg T (1930) Additive heteropolymerization. Berichte der Deutschen Chemischen Gesellschaft [Abteilung] B: Abhandlungen 63B:3213–3224
45. Weigand F, Fulber C, Blumich B, Spiess HW (1994) Spatially resolved NMR of rigid polymers and elastomers. Magn Reson Imaging 12(2):301–304. doi:
46. Yamakita H, Hayakawa K (1972) Copolymerization of *N*-ethylmaleimide with stilbenes. J Polym Sci A-1 Polym Chem 10(8):2223–2235. doi:
47. Yamazaki S, Muroga Y, Noda I (1999) Persistence lengths of ionenes in methanol. Langmuir 15(12):4147–4149. doi:
48. Yuki H, Okamoto Y, Tsubota K, Kosai K (1970) Anionic copolymerizations of *trans*-stilbene with butadiene, isoprene, and 2,3-dimethylbutadiene. Polym J 1(2):147–154. doi:
49. Zhou X, Li Y, Hart KE, Abbott LJ, Lin ZX, Svec F, Colina CM, Turner SR (2013) Nanoporous structure of semirigid alternating copolymers via nitrogen sorption and molecular simulation. Macromolecules 46(15):5968–5973. doi:

14

Catalysing sustainable fuel and chemical synthesis

Adam F. Lee

Abstract Concerns over the economics of proven fossil fuel reserves, in concert with government and public acceptance of the anthropogenic origin of rising CO_2 emissions and associated climate change from such combustible carbon, are driving academic and commercial research into new sustainable routes to fuel and chemicals. The quest for such sustainable resources to meet the demands of a rapidly rising global population represents one of this century's grand challenges. Here, we discuss catalytic solutions to the clean synthesis of biodiesel, the most readily implemented and low cost, alternative source of transportation fuels, and oxygenated organic molecules for the manufacture of fine and speciality chemicals to meet future societal demands.

Keywords Heterogeneous catalysis · Biofuels · Biodiesel · Selective oxidation · Alcohols

Introduction

Sustainability, in essence the development of methodologies to meet the needs of the present without compromising those of future generations has become a watchword for modern society, with developed and developing nations and multinational corporations promoting international research programmes into sustainable food, energy, materials and even city planning. In the context of energy and materials (specifically synthetic chemicals), despite significant growth in proven and predicted fossil fuel reserves over the next two decades, notably heavy crude oil, tar sands, deepwater wells, and shale oil and gas, there are great uncertainties in the economics of their exploitation via current extraction methodologies, and crucially, an increasing proportion of such carbon resources (estimates vary between 65 and 80 % [1–3]) cannot be burned without breaching the UNFCC targets for a 2 °C increase in mean global temperature relative to the pre-industrial level [4, 5]. There is clearly a tightrope to walk between meeting rising energy demands, predicted to rise 50 % globally by 2040 [6] and the requirement to mitigate current CO_2 emissions and hence climate change. The quest for sustainable resources to meet the demands of a rapidly rising global population represents one of this century's grand challenges [7, 8].

While many alternative sources of renewable energy have the potential to meet future energy demands for stationary power generation, biomass offers the most readily implemented, low cost solution to a drop-in transportation fuel for blending with/replacing conventional diesel [9] via carbohydrate hydrodeoxygenation (HDO) or lipid transesterification illustrated in Scheme 1. First generation bio-based fuels derived from edible plant materials received much criticism over the attendant competition between land usage for fuel crops versus traditional agricultural cultivation [10]. Deforestation practices, notably in Indonesia, wherein vast tracts of rainforest and peat land are being cleared to support palm oil plantations have also provoked controversy [11]. To be considered sustainable, second generation bio-based fuels and chemicals are sought that use biomass sourced from non-edible components of crops, such as stems, leaves and husks or cellulose from agricultural or forestry waste. Alternative non-food crops such as switchgrass or *Jatropha curcas* [12], which require minimal cultivation and do not compete with traditional arable land or drive deforestation, are other potential candidate biofuel

A. F. Lee
European Bioenergy Research Institute, Aston University, Aston Triangle, Birmingham B4 7ET, UK
e-mail: a.f.lee@warwick.ac.uk; a.f.lee@aston.ac.uk

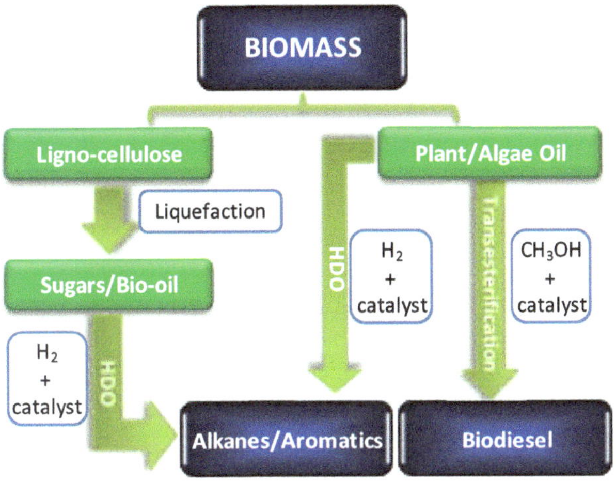

Scheme 1 Chemical conversion routes for the co-production of chemicals and transportation fuels from biomass

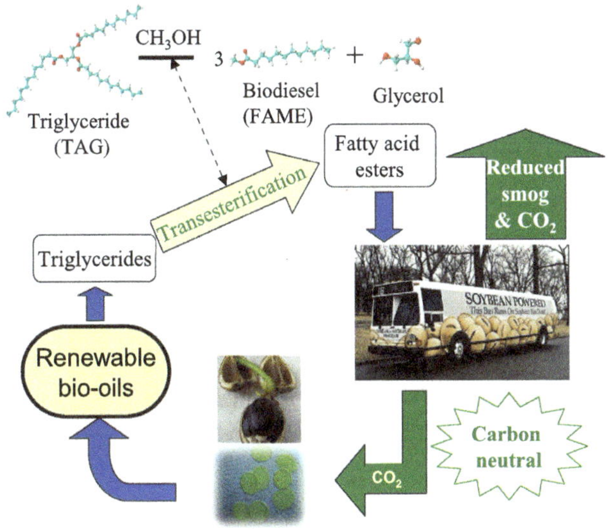

Scheme 2 Carbon cycle for biodiesel production from renewable bio-oils via catalytic transesterification

feedstocks. There is also growing interest in extracting bio-oils from aquatic biomass, which can yield 80–180 times the annual volume of oil per hectare than that obtained from plants [13]. Approximately 9 % of transportation energy needs are predicted to be met via liquid bio-fuels by 2030 [14]. While the abundance of land and aquatic biomass, and particularly of agricultural, forestry and industrial waste, is driving the search for technologies to transform lignocellulose into fuels and chemical, energy and atom-efficient processes to isolate lignin and hemicellulose from the more tractable cellulose component, remain to be identified [15]. Thermal pyrolysis offers one avenue by which to obtain transportation fuels, and wherein catalysis will undoubtedly play a significant role in both pyrolysis of raw biomass and subsequent upgrading of bio-oils via deoxygenation and carbon chain growth. Catalytic depolymerisation of lignin may also unlock opportunities for the production of phenolics and related aromatic compounds for fine chemical and pharmaceutical applications [16].

Biodiesel is a clean burning and biodegradable fuel which, when derived from non-food plant or algal oils or animal fats, is viewed as a viable alternative (or additive) to current petroleum-derived diesel [17]. Commercial biodiesel is currently synthesised via liquid base-catalysed transesterification of C_{14}–C_{20} triacylglyceride (TAG) components of lipids with C_1–C_2 alcohols [18–21] into fatty acid methyl esters (FAMEs) which constitute biodiesel as shown in Scheme 2, alongside glycerol as a potentially valuable by-product [22]. While the use of higher (e.g. C_4) alcohols is also possible [23], and advantageous in respect of producing a less polar and corrosive FAME [24] with reduced cloud and pour points [25], the current high cost of longer chain alcohols, and difficulties associated with separating the heavier FAME product from unreacted alcohol and glycerol, remain problematic. Unfortunately, homogeneous acid and base catalysts can corrode reactors and engine manifolds, and their removal from the resulting biofuel is particularly problematic and energy intensive, requiring aqueous quench and neutralisation steps which result in the formation of stable emulsions and soaps [9, 26, 27]. Such homogeneous approaches also yield the glycerine by-product, of significant potential value to the pharmaceutical and cosmetic industries, in a dilute aqueous phase contaminated by inorganic salts. Heterogeneous catalysis has a rich history of facilitating energy efficient selective molecular transformations and contributes to 90 % of chemical manufacturing processes and to more than 20 % of all industrial products [28, 29]. While catalysis has long played a pivotal role in petroleum refining and petrochemistry, in a post-petroleum era, it will face new challenges as an enabling technology to overcoming the engineering and scientific barriers to economically feasible routes to bio-fuels. The utility of solid base and acid catalysts for biodiesel production has been extensively reviewed [20, 30–33], wherein they offer improved process efficiency by eliminating the need for quenching steps, allowing continuous operation [34], and enhancing the purity of the glycerol by-product. Technical advances in catalyst and reactor design remain essential to utilise non-food based feedstocks and thereby ensure that biodiesel remains a key player in the renewable energy sector for the 21st century. Select pertinent developments in tailoring the nanostructure of solid acid and base catalysts for TAG transesterification to FAMEs and the related esterification of free fatty acid (FFAs) impurities common in bio-oil feedstocks are therefore discussed herein.

Biomass also offers the only non-fossil fuel route to organic molecules for the manufacture of bulk, fine and

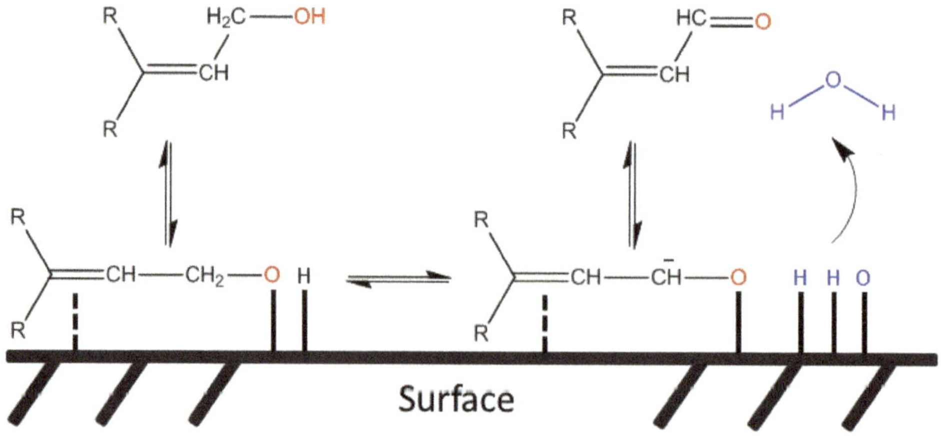

Scheme 3 Cartoon depicting the atom-efficient, chemoselective aerobic selective oxidation of allylic alcohols to aldehydes over a heterogeneous catalyst

speciality chemicals and polymers [35] required to meet societal demands for advanced materials [8, 36]. The production of such highly functional molecules, whether derived from petroleum feedstocks, requires chemoselective transformations in which e.g. specific heteroatoms or functional groups are incorporated or removed without compromising the underpinning molecular properties. The selective oxidation (selox) of alcohols, carbohydrates and related α,β-unsaturated substrates represent an important class of reactions that underpin the synthesis of valuable chemical intermediates [37, 38]. The scientific, technological and commercial importance of green chemistry presents a significant challenge to traditional selox methods, which previously employed hazardous and toxic stoichiometric oxidants including permanganates, chromates and peroxides, with concomitant poor atom efficiencies and requiring energy-intensive separation steps to obtain the desired carbonyl or acid product. Alternative heterogeneous catalysts utilising oxygen or air as the oxidant offer vastly improved activity, selectivity and overall atom efficiency in alcohol selox (Scheme 3), but are particularly demanding due to the requirement to activate molecular oxygen and C–O bonds in close proximity at a surface in a solid–liquid–gas environment [39–41], and must also be scalable in terms of both catalyst synthesis and implementation. For example, continuous flow microreactors have been implemented in both homogeneous and heterogeneous aerobic selox, providing facile catalyst recovery from feedstreams for the latter [42, 43], but their scale-up/out requires complex manifolding to ensure adequate oxygen dissolution uniform reactant mixing and delivery [44, 45]. Efforts to overcome mass transport and solubility issues inherent to 3-phase catalysed oxidations have centred around the use of supercritical carbon dioxide to facilitate rapid diffusion of substrates to and products from the active catalyst site at modest temperatures [46] affording enhanced turnover frequencies (TOFs), selectivity and on-stream performance versus conventional batch operation in liquid organic solvents [47–51].

The past decade has seen significant progress in understanding the fundamental mode of action of Platinum Group Metal heterogeneous catalysts for aerobic selox and the associated reaction pathways and deactivation processes [41]. This insight has been aided by advances in analytical methodologies, notably the development of in situ or *operando* (under working conditions) spectroscopic [52–54] /microscopic [55–58] tools able to provide quantitative, spatio-temporal information on structure–function relations of solid catalysts in the liquid and vapour phase. Parallel improvements in inorganic synthetic protocols offer finer control over preparative methods to direct the nanostructure (composition, morphology, size, valence and support architecture) of palladium catalysts [59–61] and thereby enhance activity, selectivity and lifetime in an informed manner. Ultimately, heterogeneous catalysts may offer significant advantages over homogeneous analogues in respect of initial catalyst cost, product separation, and metal recovery and recyclability [62]. Catalyst development can thus no longer be considered simply a matter of reaction kinetics, but as a clean technology wherein all aspects of process design, such as solvent selection, batch/flow operation, catalyst recovery and waste production and disposal are balanced [63]. The efficacy of Platinum Group Metals (PGMs) surfaces towards the liquid phase oxidation of alcohols has been known for over 50 years [64], and the development of heterogeneous platinum selox catalysts (and more recently coinage metals such as gold [65, 66]) the subject of recent reviews [39, 67–69] hence only palladium selox catalysis is described herein.

Heterogeneously catalysed routes to biodiesel

Solid acid catalysed biodiesel synthesis

A wide range of inorganic and polymeric solid acids are commercially available, however, their application for the transesterification of oils into biodiesel has only been recently explored, in part reflecting their lower activity compared with base-catalysed routes [27], in turn necessitating higher reaction temperatures to deliver suitable conversions. While their activities are generally low, solid acids have the advantage that they are less sensitive to FFA contaminants than their solid base analogues, and hence can operate with unrefined feedstocks containing 3–6 wt% FFAs [27]. In contrast to solid bases which require feedstock pretreatment to remove fatty acid impurities, solid acids are able to esterify FFAs through to FAME in parallel with transesterification major TAG components without soap formation and thus reduce the number of processing steps to biodiesel [70–72].

Mesoporous silicas from the SBA family [73] have been examined for biodiesel synthesis, and include materials grafted with sulfonic acid groups [74, 75] or SO_4/ZrO_2 surface coatings [76]. Phenyl and propyl sulfonic acid SBA-15 catalysts are particularly attractive materials with activities comparable to Nafion and Amberlyst resins in palmitic acid esterification [77]. Phenylsulfonic acid functionalised silica is reportedly more active than their corresponding propyl analogues, in line with their respective acid strengths but is more difficult to prepare. Unfortunately, conventionally synthesised sulfonic acid functionalised SBA-15 silicas with pore sizes below ~6 nm possess long, isolated parallel channels and suffer correspondingly slow in-pore diffusion and catalytic turnover in FFA esterification. However, poragens such as trimethylbenzene [78] triethylbenzene or triisopropylbenzene [79] can induce swelling of the Pluronic P123 micelles used to produce SBA-15, enabling ordered mesoporous silicas with diameters spanning 5–30 nm, and indeed ultra-large-pores with a BJH pore diameter as much as 34 nm [79]. This methodology was recently applied to prepare a range of large pore SBA-15 materials employing trimethylbenzene as the poragen, resulting in the formation of highly ordered periodic mesostructures with pore diameters of ~6, 8 and 14 nm [80]. These silicas were subsequently functionalised by mercaptopropyl trimethoxysilane (MPTS) and oxidised with H_2O_2 to yield expanded $PrSO_3$-SBA-15 catalysts which were effective in both palmitic acid esterification with methanol and tricaprylin and triolein transesterification with methanol under mild conditions. For both reactions, turnover frequencies dramatically increased with pore diameter, and all sulfonic acid heterogeneous catalysts significantly outperformed a commercial Amberlyst resin (Fig. 1). These rate enhancements are attributed to superior mass transport of the bulky FFA and triglycerides within the expanded $PrSO_3$-SBA-15. Similar observations have been made over Poly(styrenesulfonic acid)-functionalised ultra-large pore SBA-15 in the esterification of oleic acid with butanol [81].

Improving pore interconnectivity, for example through swapping the *p6 mm* architecture of SBA-15 for the *Ia3d* of KIT-6 was subsequently explored as an alternative means to enhance in-pore active site accessibility (Scheme 1) for FFA esterification [82]. KIT-6 mesoporous materials exhibit improved characteristics for biomolecule immobilisation [83] reflecting superior diffusion within the interconnected cubic structure. A family of pore-expanded propylsulfonic acid KIT-6 analogues were prepared via MPTS grafting and oxidation and screened for FFA

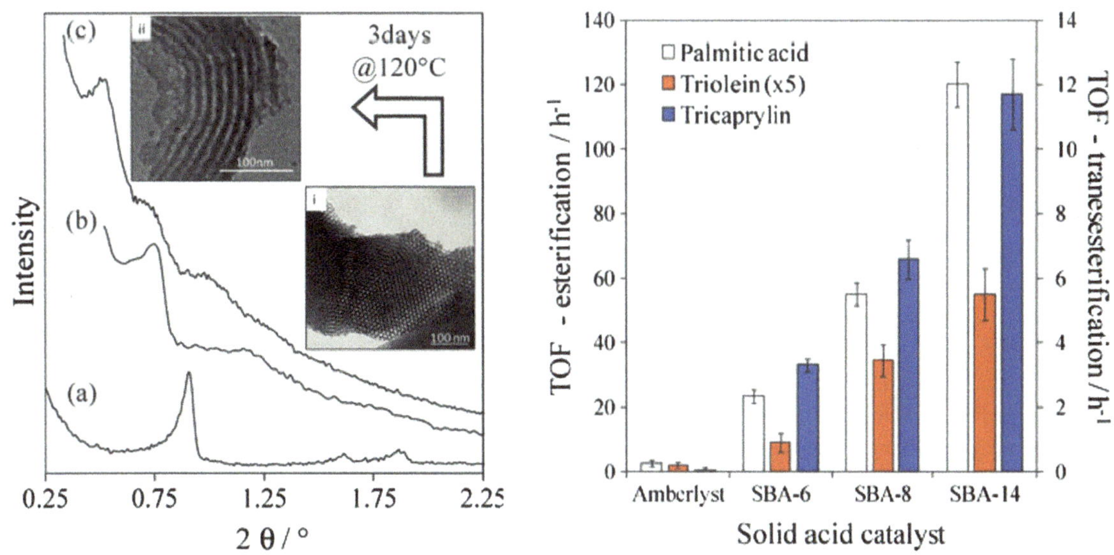

Fig. 1 (*Left*) Low angle powder X-ray diffraction patterns and transmission electron micrographs of propylsulfonic acid functionalised SBA-15 silicas as a function of pore diameter; and (*right*) corresponding catalytic activity in FFA esterification and TAG transesterification compared to a commercial solid acid resin. Adapted from reference [80] with permission from The Royal Society of Chemistry

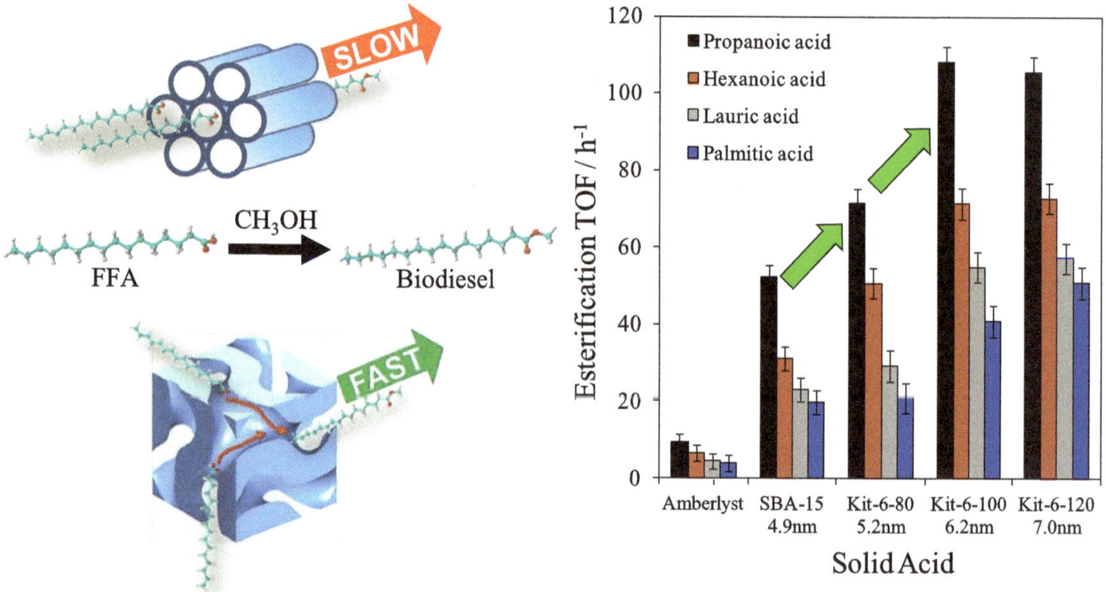

Fig. 2 Superior performance of interconnected, mesoporous propylsulfonic acid KIT-6 catalysts for biodiesel synthesis via FFA esterification with methanol versus non-interconnected mesoporous SBA-15 analogue. Adapted from reference [82]. Copyright 2012 American Chemical Society

esterification with methanol as a function of alkyl chain length under mild conditions. As-synthesised $PrSO_3H$-KIT-6 exhibited respective 40 and 70 % TOF enhancements toward propanoic and hexanoic acid esterification compared with a $PrSO_3H$-SBA-15 analogue of comparable (5 nm) pore diameter as a consequence of the improved mesopore interconnectivity. However, pore accessibility remained rate-limiting for esterification of the longer chain lauric and palmitic acids. Hydrothermal aging protocols facilitated expansion of the KIT-6 mesopore up to 7 nm, with consequent doubling of TOFs for lauric and palmitic acid esterification versus $PrSO_3H$-SBA-15 (Fig. 2).

While numerous solid acids have been applied for biodiesel synthesis [27, 32, 84], most materials exhibit micro- and/or mesoporosity which, as illustrated above, are not optimal for accommodating bulky C_{16}–C_{18} TAGs of FFAs. For example, incorporation of a secondary mesoporosity into a microporous H-β-zeolite to create a hierarchical solid acid significantly increased catalytic activity by lowering diffusion barriers [85]. Templated mesporous materials are widely used as catalyst supports [86, 87], with SBA-15 silicas popular candidates for reactions pertinent to biodiesel synthesis as previously discussed [75, 77, 88]. However, such surfactant-templated supports possessing long, isolated parallel and narrow channels are ill-suited to efficient in-pore diffusion of bio-oil feedstocks affording poor catalytic turnover. Further improvements in pore architecture are hence required to optimise mass transport of heavier bulky TAGs and FFAs commonly found in plant and algal oils. Simulations demonstrate that in the Knudsen diffusion regime [89], where reactants/products are able to diffuse enter/exit mesopores but experience moderate diffusion limitations, hierarchical pore structures may significantly improve catalyst activity. Materials with interpenetrating, bimodal meso–macropore networks have been prepared using microemulsion [90] or co-surfactant [91] templating routes and are particularly attractive for liquid phase, flow reactors wherein rapid pore diffusion is required. Liquid crystalline (soft) and colloidal polystyrene nanospheres (hard) templating methods have been combined to create highly organised, macro–mesoporous aluminas [92] and 'SBA-15 like' silicas [93] (Scheme 4), in which both macro- and mesopore diameters can be independently tuned over the range 200–500 and 5–20 nm, respectively. The resulting hierarchical pore network of a propylsulfonic acid functionalised macro–mesoporous SBA-15 is shown in Fig. 3, wherein macropore incorporation confers a striking enhancement in the rates of tricaprylin transesterification and palmitic acid esterification with methanol, attributed to the macropores acting as transport conduits for reactants to rapidly access $PrSO_3H$ active sites located within the mesopores.

The hydrophilic nature of polar silica surfaces hinders their application for reactions involving apolar organic molecules. This is problematic for TAG transesterification (or FFA esterification) due to preferential in-pore diffusion and adsorption of alcohol versus fatty acid components. Surface hydroxyl groups also favour H_2O adsorption, which if formed during FFA esterification can favour the reverse hydrolysis reaction and consequent low FAME

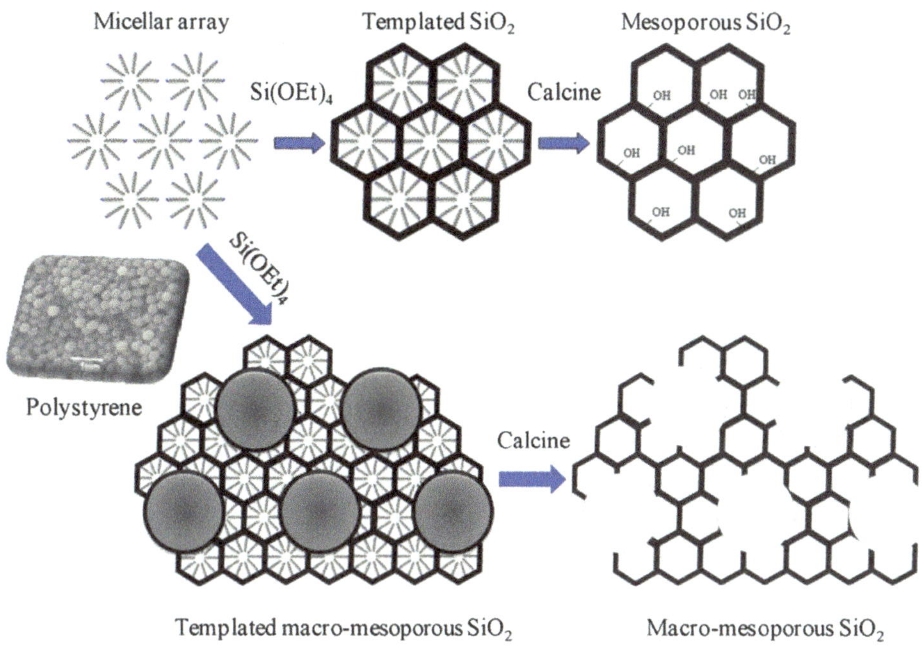

Scheme 4 Liquid crystal and polystyrene nanosphere dual surfactant/physical templating route to hierarchical macroporous–mesoporous silicas

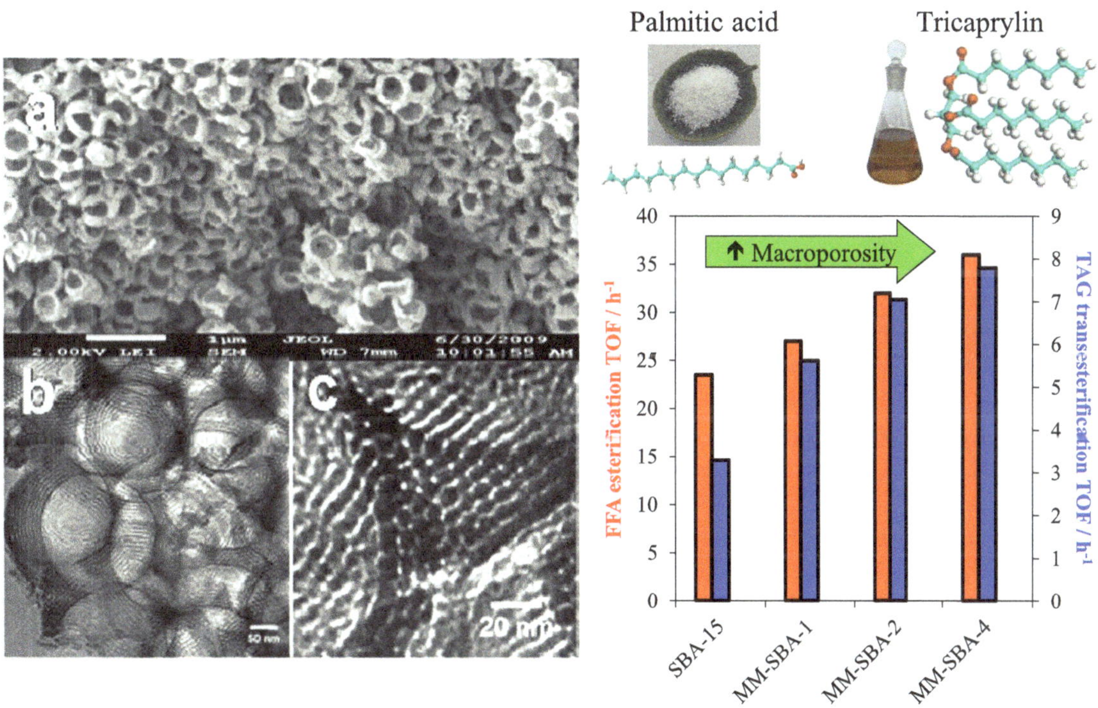

Fig. 3 (*Left*) SEM (**a**) and low and high magnification TEM (**b**, **c**) micrographs of a hierarchical macro–mesoporous Pr-SO_3H-SBA-15; (*right*) corresponding catalytic performance in palmitic acid esterification and tricaprylin transesterification with methanol as a function of macropore density versus a purely mesoporous Pr-SO_3H-SBA-15. Adapted from reference [93] with permission from The Royal Society of Chemistry

yields. Surface modification via the incorporation of organic functionality into polar oxide surfaces, or dehydroxylation, can lower their polarity and thereby increase initial rates of acid catalysed transformations of liquid phase organic molecules [94]. Surface polarity can also be tuned by incorporating alkyl/aromatic groups directly into the silica framework, for example polysilsesquioxanes can be prepared via the co-condensation of 1,4-bis(triethoxysilyl)benzene (BTEB), or 1,2-bis(trimethoxysilyl)-ethane (BTME), with TEOS and MPTS in the sol–gel process [95, 96] which enhances small molecule esterification [97] and etherification [98]. The incorporation of organic spectator

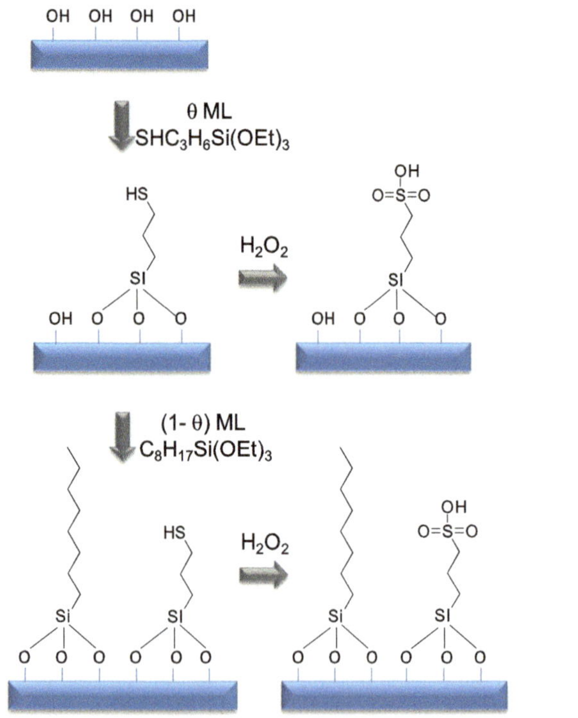

Scheme 5 Protocol for the synthesis of sulfonic acid and octyl co-functionalised sulfonic acid MCM-41catalysts. Adapted from reference [101] with permission from The Royal Society of Chemistry

groups (e.g. phenyl, methyl or propyl) during the sol–gel syntheses of SBA-15 [99] and MCM-41 [100] sulphonic acid silicas is achievable via co-grafting or simple addition of the respective alkyl or aryltrimethoxysilane during co-condensation protocols. An experimental and computational study of sulphonic acid functionalised MCM-41 materials was undertaken to evaluate the effect of acid site density and surface hydrophobicity on catalyst acidity and associated performance [101]. MCM-41 was an excellent candidate due to the availability of accurate models for the pore structure from kinetic Monte Carlo simulations [102], and was modified with surface groups to enable dynamic simulation of sulphonic acid and octyl groups co-attached within the MCM-41 pores. In parallel experiments, two catalyst series were investigated towards acetic acid esterification with butanol (Scheme 5). In one series, the propylsulphonic acid coverage was varied between θ (RSO_3H) = 0–100 % ML over the bare silica (MCM–SO_3H). For the second octyl co-grafted series, both sulfonic acid and octyl coverages were tuned (MCM–Oc–SO_3H). These materials allow the effect of lateral interactions between acid head groups and the role of hydrophobic octyl modifiers upon acid strength and activity to be separately probed.

To avoid diffusion limitations, butanol esterification with acetic acid was selected as a model reaction (Fig. 4). Ammonia calorimetry revealed that the acid strength of polar MCM–SO_3H materials increases from 87 to 118 kJ mol^{-1} with sulphonic acid loading. Co-grafted octyl groups dramatically enhance the acid strength of MCM–Oc–SO_3H for submonolayer SO_3H coverages, with $\_\Delta H_{ads}(NH_3)$ rising to 103 kJ mol^{-1}. The per site activity of the MCM–SO_3H series in butanol esterification with acetic acid mirrors their acidity, increasing with SO_3H content. Octyl surface functionalisation promotes esterification for all MCM–Oc–SO_3H catalysts, doubling the turnover frequency of the lowest loading SO_3H material. Molecular dynamic simulations indicate that the interaction of isolated sulphonic acid moieties with surface silanol groups is the primary cause of the lower acidity and activity of submonolayer samples within the MCM–SO_3H series. Lateral interactions with octyl groups help to re-orient sulphonic acid headgroups into the pore interior, thereby enhancing acid strength and associated esterification activity.

In summary, recent developments in tailoring the structure and surface functionality of sulfonic acid silicas have led to a new generation of tunable solid acid catalysts well-suited to the esterification of short and long chain FFAs, and transesterification of diverse TAGs, with methanol under mild reaction conditions. A remaining challenge is to extend the dimensions and types of pore-interconnectivities present within the host silica frameworks, and to find alternative low cost soft and hard templates to facilitate synthetic scale-up of these catalysts for multi-kg production. Surfactant template extraction is typically achieved via energy-intensive solvent reflux, which results in significant volumes of contaminated waste and long processing times, while colloidal templates often require high temperature calcination which prevents template recovery/re-use and releases carbon dioxide. Preliminary steps towards the former have been recently taken, employing room temperature ultrasonication in a small solvent volume to deliver effective extraction of the P123 Pluronic surfactant used in the preparation of SBA-15 in only 5 min, with a 99.9 % energy saving and 90 % solvent reduction over reflux methods, and without compromising textural, acidic or catalytic properties of the resultant Pr-SO_3H-SBA-15 in hexanoic acid esterification (Fig. 5) [103].

Solid base-catalysed biodiesel synthesis

Base catalysts are generally more active than acids in transesterification, and hence are particularly suitable for high purity oils with low FFA content. Biodiesel synthesis using a solid base catalyst in continuous flow, packed bed arrangement would facilitate both catalyst separation and co-production of high purity glycerol, thereby reducing production costs and enabling catalyst re-use. Diverse solid

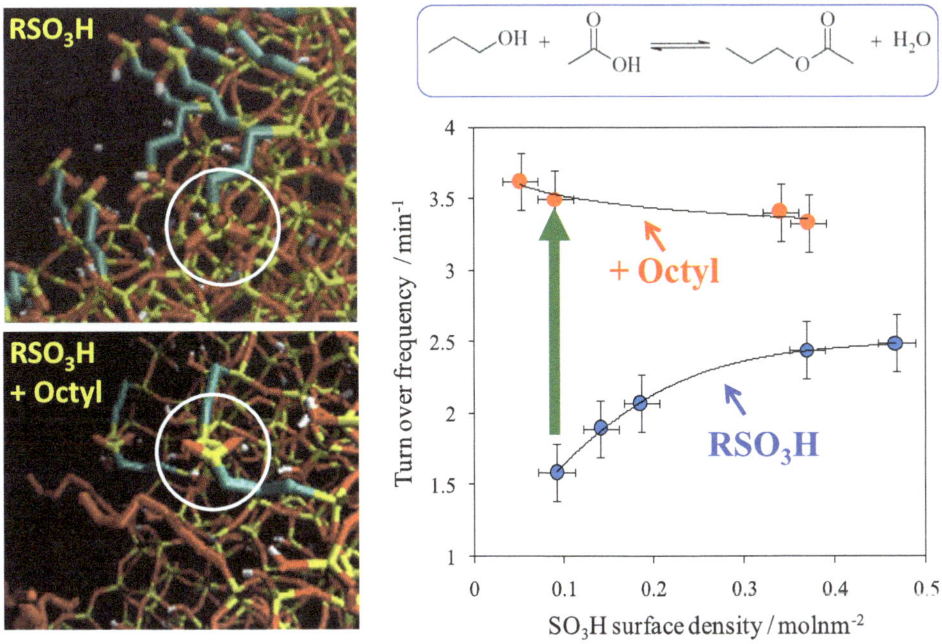

Fig. 4 (*Left*) Molecular dynamics simulations of MCM–SO_3H and MCM–Oc–SO_3H pore models highlighting the interaction between surface sulfonic acid and hydroxyl groups in the absence of co-grafted octyl chains; (*right*) influence of $PrSO_3H$ surface density and co-grafted octyl groups on catalytic performance in acetic acid esterification with butanol. Adapted from reference [101] with permission from The Royal Society of Chemistry

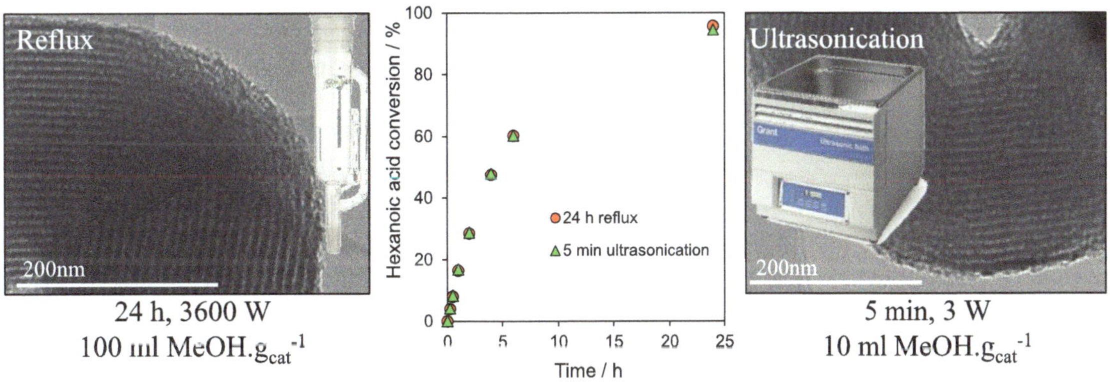

Fig. 5 Surfactant template extraction via energy/atom-efficient ultrasonication delivers a one-pot $PrSO_3H$-SBA-15 solid acid catalyst with identical structure and reactivity to that obtained by conventional, inefficient reflux. Adapted from reference [103] with permission from The Royal Society of Chemistry

base catalysts are known, notably alkali or alkaline earth oxides, supported alkali metals, basic zeolites and clays such as hydrotalcites and immobilised organic bases [104]. Basicity in alkaline earth oxides is believed to arise from M^{2+}–O^{2-} ion pairs present in different coordination environments [105]. The strongest base sites occur at low coordination defect, corner and edge sites, or on high Miller index surfaces. Such classic heterogeneous base catalysts have been extensively tested for TAG transesterification [106] and there are numerous reports on commercial and microcrystalline CaO applied to rapeseed, sunflower or vegetable oil transesterification with methanol [107, 108]. Promising results have been obtained, with 97 % oil conversion achieved at 75 °C [108], however, concern remains over Ca^{2+} leaching under reaction conditions and associated homogeneous catalytic contributions [109], a common problem encountered in metal catalysed biodiesel production which hampers commercialisation [110].

Alkali-doped CaO and MgO have also been investigated for TAG transesterification [111–113], with their enhanced basicity attributed to the genesis of O^- centres following the replacement of M^+ for M^{2+} and associated charge imbalance and concomitant defect generation. Optimum activity for Li-doped CaO occurs when a saturated Li^+ monolayer is formed (Fig. 6) [113], although leaching of the alkali promoter remains problematic [114].

It is widely accepted that the catalytic activity of alkaline earth oxide catalysts is very sensitive to their preparation, and corresponding surface morphology and/or defect density. For example, Parvulescu and Richards demonstrated the impact of the different MgO crystal facets upon the transesterification of sunflower oil by

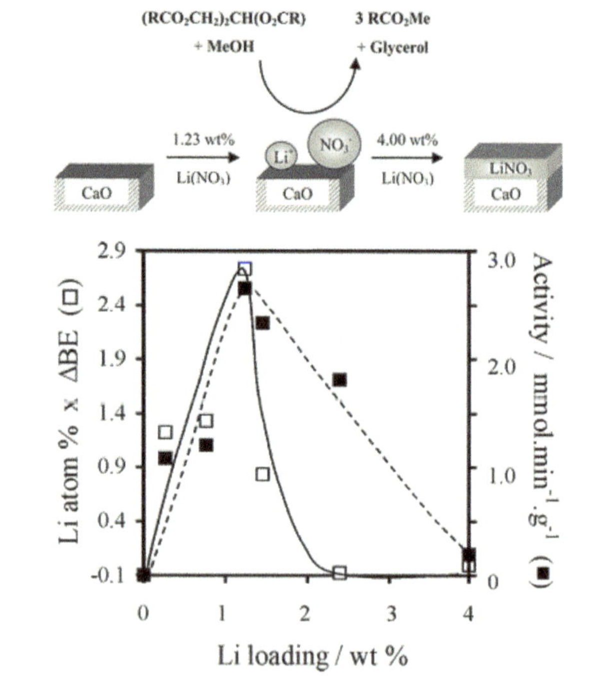

Fig. 6 Correlation between evolving surface composition, density of electronically perturbed Li^+ sites, and corresponding activity in tributyrin transesterification with methanol over Li-doped CaO as a function of Li loading. Adapted from reference [113] with permission from The Royal Society of Chemistry

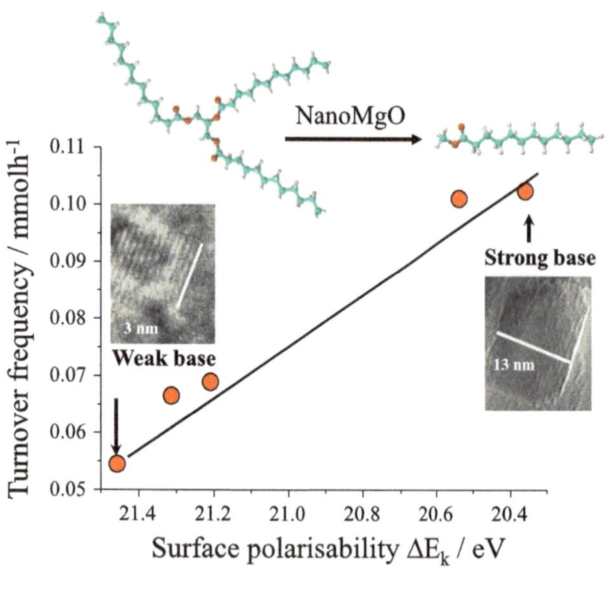

Fig. 7 Relationship between surface polarisability of MgO nanocrystals and their turnover frequency towards tributyrin transesterification. Adapted from reference [117] with permission from The Royal Society of Chemistry

comparing nanoparticles [115] versus (111) terminated nanosheets [116]. Chemical titration reveals that both morphologies possess two types of base sites, with the nanosheets exhibiting well-defined, medium-strong basicity consistent with their uniform exposed facets and

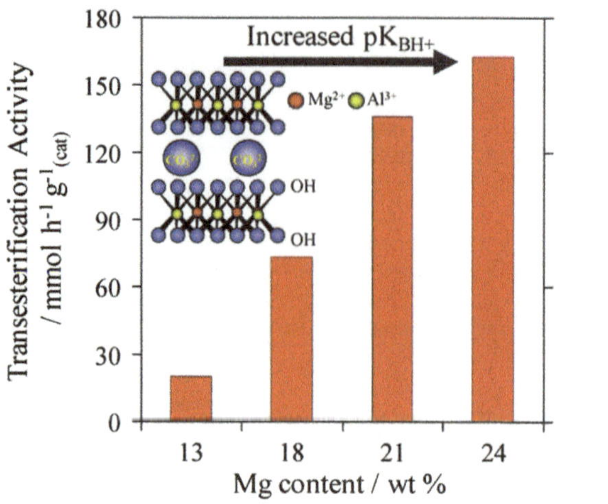

Fig. 8 Impact of Mg:Al hydrotalcite surface basicity on their activity towards tributyrin transesterification. Adapted from reference [118] with permission from Elsevier

which confer higher FAMe yields during sunflower oil transesterification. Subsequent synthesis, screening and spectroscopic characterisation of a family of size-/shape-controlled MgO nanoparticles prepared via a hydrothermal synthesis revealed small (<8 nm) particles terminate in high coordination (100) facets, and exhibit both weak polarisability and poor activity in tributyrin transesterification with methanol [117]. Calcination drives restructuring and sintering to expose lower coordination stepped (111) and (110) surface planes, which are more polarisable and exhibit much higher transesterification activities under mild conditions. A direct correlation was therefore observed between the surface electronic structure and associated catalytic activity, revealing a pronounced structural preference for (110) and (111) facets (Fig. 7).

Hydrotalcites are another class of solid base catalysts that have attracted recent attention because of their high activity and robustness in the presence of water and FFA [118, 119]. Hydrotalcites ($[M(II)_{1-x}M(III)_x(OH)_2]^{x+}$ $(A^{n-}_{x/n})$ mH_2O) adopt a layered double hydroxide structure with brucite-like $Mg(OH)_2$) hydroxide sheets containing octahedrally coordinated M^{2+} and M^{3+} cations and A^{n-} anions between layers to balance the overall charge [120], and are conventionally synthesised via co-precipitation from their nitrates using alkalis as both pH regulators and a carbonate source. Mg–Al hydrotalcites have been applied for TAG transesterification of poor and high quality oil feeds [121] such as refined and acidic cottonseed oil (9.5 wt% FFA), and animal fat feed (45 wt% water), delivering 99 % conversion within 3 h at 200 °C. It is important to note that many catalytic studies employing hydrotalcites for transesterification are suspect due to their use of Na or K hydroxide/carbonate solutions to precipitate

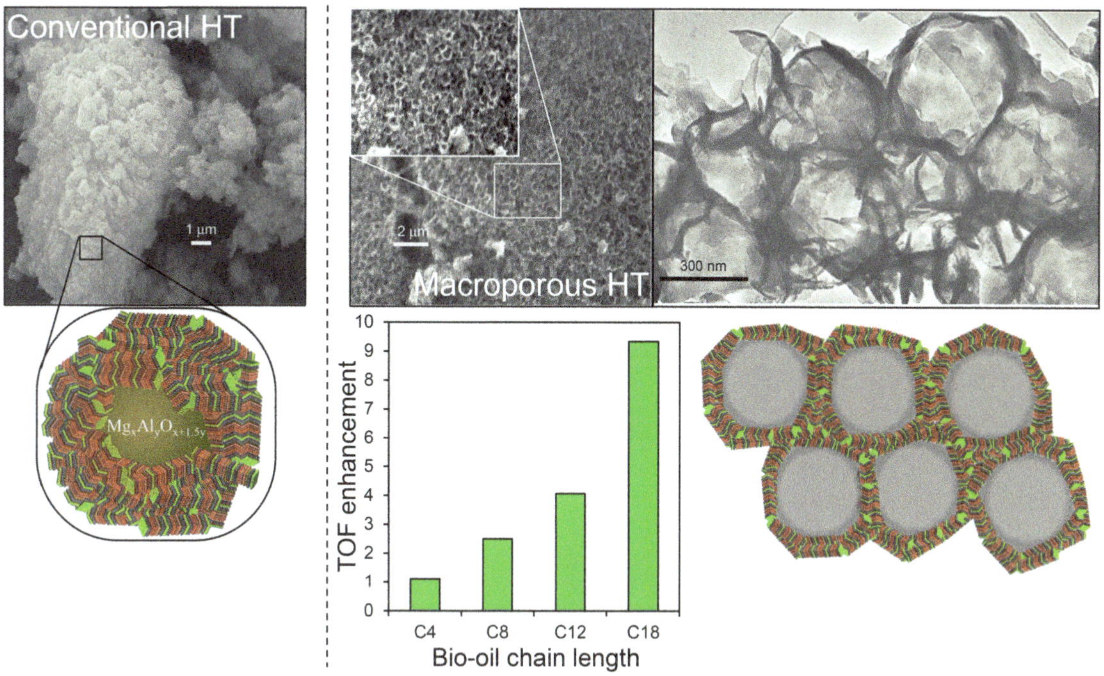

Fig. 9 Superior catalytic performance of a hierarchical macroporous–microporous Mg–Al hydrotalcite solid base catalyst for TAG transesterification to biodiesel versus a conventional microporous analogue. Adapted from reference [124] with permission from The Royal Society of Chemistry

the hydrotalcite phase. Complete removal of alkali residues from the resulting hydrotalcites is inherently difficult, resulting in parallel ill-defined homogeneous contributions to catalysis arising from leached Na or K [122, 123]. This problem has been overcome by the development of alkali-free precipitation routes using NH_3OH and NH_3CO_3, offering well-defined thermally activated and rehydrated Mg–Al hydrotalcites with compositions spanning $x = 0.25 - 0.55$ [118]. Spectroscopic measurements reveal that increasing the Mg:Al ratio enables the surface charge and accompanying base strength to be systematically enhanced, with a concomitant increase in the rate of tributyrin transesterification under mild reaction conditions (Fig. 8).

In spite of their promise for biodiesel production, conventionally prepared hydrotalcites are microporous, and hence poorly suited to application in the transesterification of bulky C_{16}–C_{18} TAG components of bio-oils. This problem was recently tackled by adopting the same hard templating method utilising polystyrene nanospheres described in Scheme 4 to incorporate macroporosity, and thus create a hierarchical macroporous–microporous hydrotalcite solid base catalyst [124]. The introduction of macropores as 'superhighways' to rapidly transport heavy TAG oil components to active base sites present at (high aspect ratio) hydrotalcite nanocrystallites, dramatically enhanced turnover frequencies for triolein transesterification compared with that achievable over an analogous Mg–Al microporous hydrotalcite (Fig. 9), reflecting superior mass transport through the hierarchical catalyst.

In terms of sustainability, it is important to find low cost routes to the synthesis of solid base catalysts that employ earth abundant elements. Dolomitic rock, comprising alternating $Mg(CO_3)$–$Ca(CO_3)$ layers, is structurally very similar to calcite ($CaCO_3$), with a high natural abundance and low toxicity, and in the UK is sourced from quarries working Permian dolomites in Durham, South Yorkshire and Derbyshire [125]. In addition to uses in agriculture and construction, dolomite finds industrial applications in iron and steel production, glass manufacturing and as fillers in plastics, paints, rubbers, adhesives and sealants. Catalytic applications for powdered, dolomitic rock offer the potential to further valorise this readily available waste mineral, and indeed dolomite has shown promise in biomass gasification [126] as a cheap, disposable and naturally occurring material that significantly reduces the tar content of gaseous products from gasifiers. Dolomite has also been investigated as a solid base catalyst in biodiesel synthesis [127], wherein fresh dolomitic rock comprised approximately 77 % dolomite and 23 % magnesian calcite. High temperature calcination induced Mg surface segregation, resulting in MgO nanocrystals dispersed over $CaO/(OH)_2$ particles, while the attendant loss of CO_2 increases both the surface area and basicity. The resulting calcined dolomite proved an effective catalyst for the transesterification of C_4, C_8 and TAGs with methanol and longer chain C_{16-18}

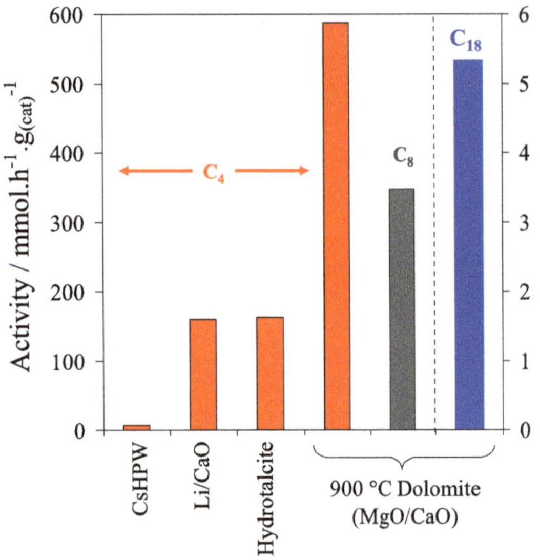

Fig. 10 Catalytic activity of calcined Dolomite for the transesterification of short and long chain TAGs with methanol benchmarked against literature solid acid and base catalysts. Reproduced from reference [127] with permission from The Royal Society of Chemistry

components present within olive oil, with TOFs for tributyrin conversion to methyl butanoate the highest reported for any solid base (Fig. 10). The slower transesterification rates for bulkier TAGs were attributed to diffusion limitations in their access to base sites. Calcined dolomite has also shown promise in the transesterification of canola oil with methanol, achieving 92 % FAME after 3 h reaction with 3 wt% catalyst [128].

In summary, a host of inorganic solid base catalysts have been developed for the low temperature transesterification of triglyceride components of bio-oil feedstocks, offering activities far superior to those achieved via alternative solid acid catalysts to date. However, leaching of alkali and alkaline earth elements and associated catalyst recycling remains a challenge, while improved resilience to water and fatty acid impurities in plant, algal and waste oils feedstocks is required to eliminate additional esterification pre-treatments. To date, only a handful of biodiesel production processes employing heterogeneous catalysts have been commercialised, notably the Esterfip-H process developed by Axens and IFP which utilises a mixture of ZnO and alumina and is operated on a 200 kton per annum scale with parallel production of high quality glycerine [129].

Palladium catalysed aerobic alcohol selox

Particle size effects

Within nanocatalysis, the particle size is a well-documented key parameter influencing both activity and selectivity. This reflects the combination of quantum and geometric effects associated with the respective evolution of electronic properties from atomic like to delocalised bands, and shifting population of low to high coordination surface atoms, with increasing nanoparticle size and dimensionality. Kaneda et al. [130] hypothesised that the unique reactivity of 2060 atom Pd clusters supported on titania towards aromatic alcohol selox arose from a distribution of Pd^0, Pd^+ and Pd^{2+} surfaces sites, with π-bonding interactions between the phenyl group and Pd^{2+} species facilitating subsequent oxidative addition of the O–H bond by neighbouring Pd^0 and eventual β–hydride elimination. Surface hydride was hypothesised to react with oxygen from a neighbouring Pd_2O centre forming H_2O and regenerating the metal site. Optimal activity for cinnamyl alcohol selox to cinnamaldehyde coincided with clusters possessing the maximum fraction of Pd^+ character.

Particle size dependency was also reported for the catalytic transformation of benzyl alcohol over Pd nanoparticles dispersed on alumina, SiO_2 and NaX zeolite supports [131, 132]. For Pd/NaX and Pd/SiO_2-Al_2O_3, benzyl alcohol selox was fastest over particles between 3 and 5 nm, whereas geraniol and 2-octanol were structure-insensitive. Systematic studies of particle size effects in cinnamyl and crotyl alcohol selox over amorphous and mesostructured alumina and silica supports have likewise uncovered pronounced size effects in both initial selox rates and TOFs [133–136], which increase monotonically with shrinking nanoparticle diameters (even down to single atoms) [137]. HAADF–STEM analysis reveals atomically dispersed palladium exhibits maximal rates towards benzyl, cinnamyl and crotyl alcohols, with selectivities to their corresponding aldehydes >70 %. The origin of such size effects is revisited below. The use of colloidal Pd nanoclusters for aqueous phase alcohol selox is limited [138–140], wherein Pd aggregation and Pd black formation hinders catalytic performance. However, the successful stabilisation of 3.6 nm Pd nanoclusters is reported using an amphiphilic nonionic triblock copolymer, Pluronic P123; in the selective oxidation of benzyl alcohol, 100 % aldehyde selectivity and high selox rates are achievable, with high catalytic activity maintained with negligible sintering after 13 recycling reactions [141].

Surface reaction mechanism

The rational design and optimisation of palladium selox catalysts require a microscopic understanding of the active catalytic species responsible for alcohol and oxygen activation, and the associated reaction pathway to the aldehyde/ketone products and any competing processes. A key characteristic of palladium is its ability to perform selox chemistry at temperatures between 60 and 160 °C and with

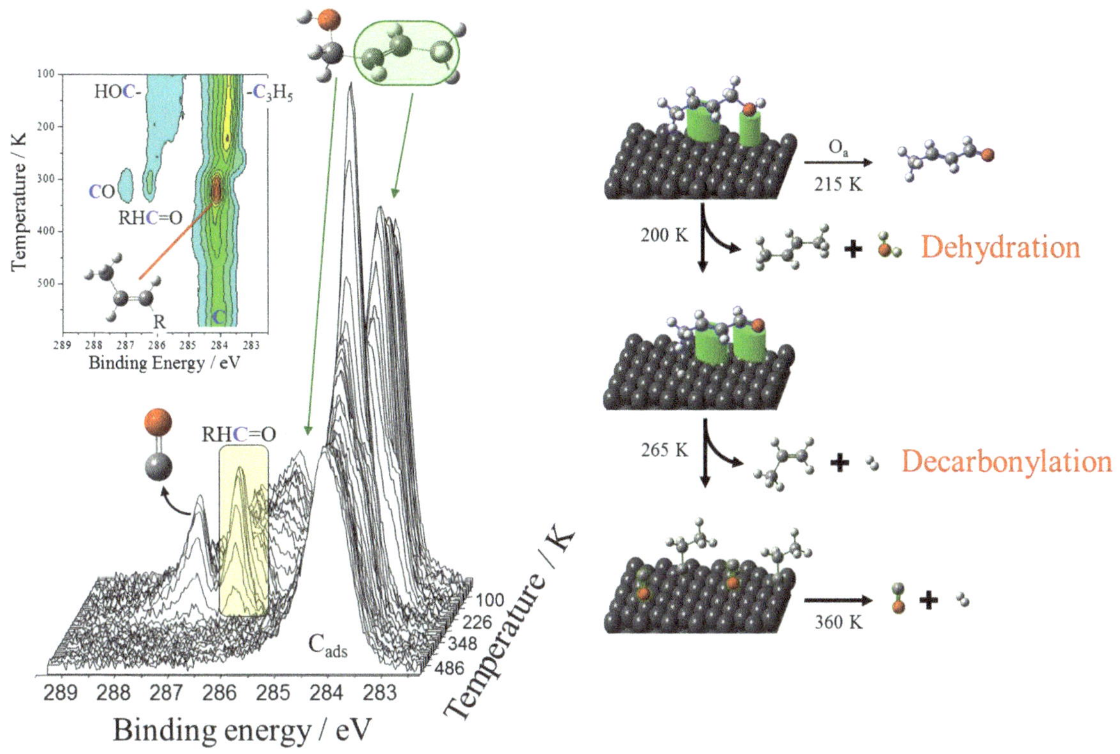

Fig. 11 Temperature-programmed C 1s XP spectra of a reacting crotyl alcohol adlayer over Pd(111) highlighting the primary dehydrogenation pathway and competing decarbonylation pathways. Adapted from reference [143]. Copyright 2007 American Chemical Society

ambient oxygen pressure [39, 142] via the wi dely accepted oxidative dehydrogenation route illustrated in Scheme 3 [39, 67]. Whether O–H or C–H scission of the α-carbon is the first chemical step remains a matter of debate, since the only fundamental studies over well-defined Pd(111) surfaces to date employed temperature-programmed XPS [143] and metastable de-excitation spectroscopy (MDS) [144] with temporal resolutions on the second → minute timescale, over which loss of both hydrogens appears coincident. However, temperature-programmed mass spectrometric [145] and vibrational [146] studies of unsaturated C_1–C_3 alcohols implicate O–H cleavage and attendant alkoxy formation over Pd single crystal surfaces as the first reaction step [142, 147]. It is generally held that the resultant hydrogen adatoms react with dissociatively absorbed oxygen to form water, which immediately desorbs at ambient temperature thereby shifting the equilibrium to carbonyl formation [39, 67]. Temperature-programmed XPS studies of crotyl alcohol adsorbed over clean Pd(111) [143] prove that oxidative dehydrogenation to crotonaldehyde occurs at temperatures as low as −60 °C (Fig. 11), with alcohol dehydration to butane only a minor pathway. These ultra-high vacuum measurements also revealed that reactively formed crotonaldehyde undergoes a competing decarbonylation reaction over metallic palladium above 0 °C yielding strongly bound CO and propylidene which may act as site-blockers poisoning subsequent catalytic selox cycles, coincident with evolution of propene into the gas phase. Unexpectedly, pre-adsorbed atomic oxygen switched-off undesired decarbonylation chemistry, promoting facile crotonaldehyde desorption.

Nature of the active site

The preceding observation that surface oxygen is not only critical for the removal of hydrogen adatoms but also to suppress decarbonylation of selox products over metallic palladium is in excellent agreement with an in situ ATR-IR study of cinnamyl alcohol selox over Pd/Al_2O_3 [148]. In related earlier investigations employing aqueous electrochemical protocols, the same researchers postulated that oxidative dehydrogenation of alcohols requires PGM catalysts in a reduced state, hypothesising that 'over-oxidation' was responsible for deactivation of palladium selox catalysts [69]. A subsequent operando X-ray absorption spectroscopy (XAS) study by Grunwaldt et al. [150], bearing remarkable similarity to an earlier study to the author of this review [149], evidenced in situ reduction of oxidised palladium in an as-prepared Pd/Al_2O_3 catalyst during cinnamyl alcohol oxidation within a continuous flow fixed-bed reactor. Unfortunately the reaction kinetics were not measured in parallel to explore the impact of palladium reduction, however, a follow-up study of 1-phenylethanol selox employing the same reactor configuration (and oxygen-deficient conditions) evidenced a strong interplay between selox conversion/selectivity and palladium oxidation state [151]. It was concluded that metallic Pd was the catalytically active species, an assertion re-affirmed in subsequent in situ ATR-IR/XAS measurements of benzyl [152–154] and cinnamyl alcohol [155] selox in toluene and under supercritical CO_2, respectively, wherein the C=O stretching intensity was assumed to track alcohol conversion. It is interesting to note that the introduction of oxygen to the reactant feed in these infrared studies dramatically improved alcohol conversion/aldehyde production (Fig. 12), which was attributed to hydrogen abstraction from the catalyst surface [156, 157] rather than to a change in palladium oxidation state. In contrast to their liquid phase experiments, high pressure XANES and EXAFS measurements of Pd/Al_2O_3 catalysed benzyl alcohol selox under supercritical CO_2 led Grunwaldt and Baiker to conclude that maximum activity arose from particles mainly oxidised in the surface/shelfedge [48].

In a parallel research programme, the author's group systematically characterised the physicochemical properties of palladium nanoparticles as a function of size over non-reducible supports to quantify structure–function relations in allylic alcohol selox [133–137, 158, 159]. The combination of XPS and XAS measurements revealed that freshly prepared alumina [134, 137] and silica [135, 158] supported nanoparticles are prone to oxidation as their diameter falls below ~4 nm, with the fraction of PdO proportional to the support surface area and interconnectivity. Complementary kinetic analyses uncovered a direct correlation between the surface PdO content and activity/TOFs towards cinnamyl and crotyl alcohol selox [134, 137]. Operando liquid phase XAS of Pd/C and Pd/Al_2O_3-SBA-15 catalysts during cinnamyl alcohol selox evidenced in situ reduction of PdO (Fig. 13), however, by virtue of simultaneously measuring the rate of alcohol selox, Lee et al. were able to prove that this oxide → metal structural transition was accompanied by coincident deactivation. Together these findings strongly implicate a (surface) PdO active phase, consistent with surface science predictions that metallic palladium favours aldehyde decarbonylation and consequent self-poisoning by CO and organic residues [143, 160], akin to that reported during fatty acid decarboxylation over Pd/MCF [161].

To conclusively establish whether oxide or metal is responsible for alcohol selox catalysed by dispersed palladium nanoparticles, a multi-dimensional spectroscopic investigation of vapour phase crotyl alcohol selox was undertaken (since XAS is an averaging technique a complete understanding of catalyst operation requires multiple analytical techniques [162–164]). Synchronous, time-resolved DRIFTS/MS/XAS measurements of supported and colloidal palladium were performed in a bespoke environmental cell [165] to simultaneously interrogate adsorbates on the catalyst surface, Pd oxidation state and reactivity under transient conditions in the absence of competitive solvent effects [166, 167]. Under mild reaction

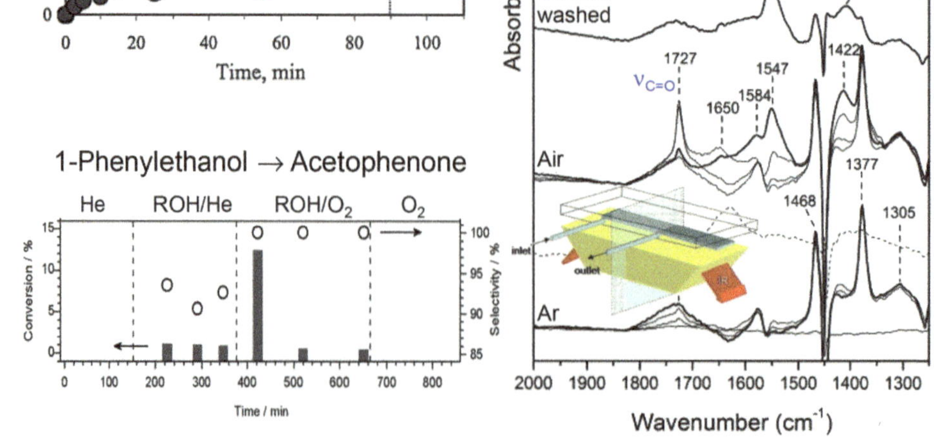

Fig. 12 Impact of oxygen on the selective oxidation of (*top left*) cinnamyl alcohol; (*bottom left*) 1-phenylethanol; and (*right*) 2-octanol. Adapted from references [148, 151, 154] with permission from Elsevier

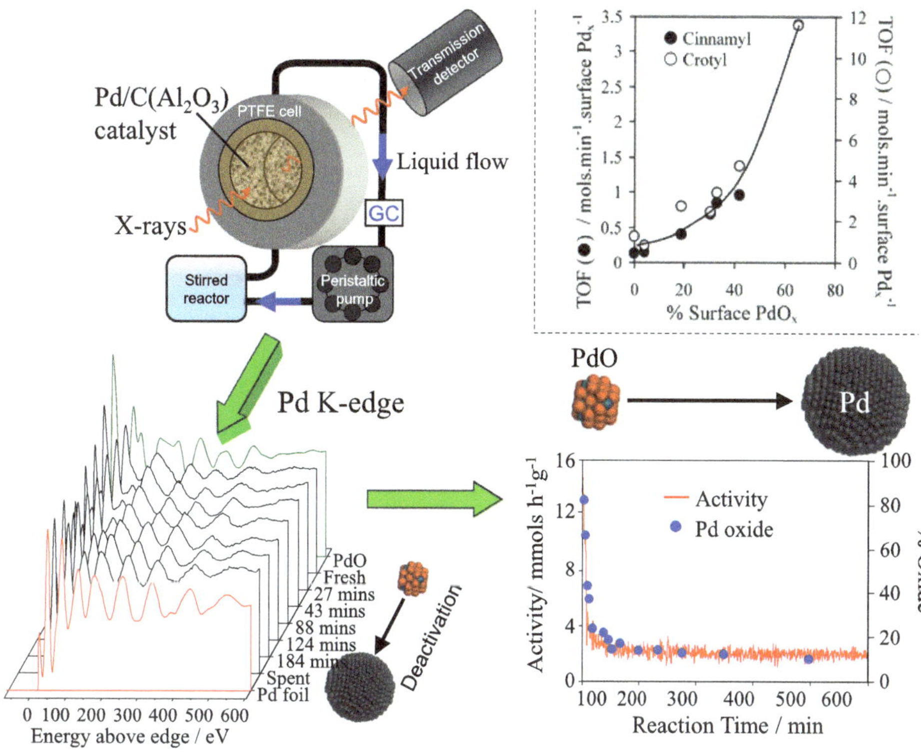

Fig. 13 (*Top right*) Dependence of allylic alcohol selox rate upon surface PdO; (*top left*) schematic of operando liquid phase reactor; (*bottom left*) evolution of Pd K-edge XAS of Pd/Al_2O_3 catalyst during cinnamyl alcohol aerobic selox; (*bottom right*) temporal correspondence between Pd oxidation state and selox activity in cinnamyl alcohol selox. Adapted from references [133, 134] with permission from The Royal Society of Chemistry

temperatures, palladium nanoparticles were partially oxidised, and unperturbed by exposure to sequential alcohol or oxygen pulses (Fig. 14). Crotonaldehyde formed immediately upon contact of crotyl alcohol with the oxide surface, but only desorbed upon oxygen co-adsorption. Higher reaction temperatures induced PdO reduction in response to crotyl alcohol exposure, mirroring that observed during liquid phase selox, however, this reduction could be fully reversed by subsequent oxygen exposure. Such reactant-induced restructuring was exhibited by all palladium nanoparticles, but the magnitude was inversely proportional to particle size [168]. These dynamic measurements decoupled the relative reactivity of palladium oxide from metal revealing that PdO favoured crotyl alcohol selox to crotonaldehyde and crotonic acid, whereas metallic palladium drove secondary decarbonylation to propene and CO in accordance with surface science predictions [143].

Recent ambient pressure XPS investigations of crotyl alcohol/O_2 gas mixtures over metallic and oxidised Pd(111) single crystal surfaces confirmed that only two-dimensional Pd_5O_4 and three-dimensional PdOx surfaces were capable of crotonaldehyde production (Fig. 15) [169]. However, even under oxygen-rich conditions, on-stream reduction of the Pd_5O_4 monolayer oxide occurred >70 °C accompanied by surface poisoning by hydrocarbon residues. In contrast, PdOx multilayers were capable of sustained catalytic turnover of crotyl alcohol to crotonaldehyde, conclusively proving surface palladium oxide as the active phase in allylic alcohol selox.

Establishing support effects

Anchoring Pd nanoparticles onto support structures offers an effective means to tune their physicochemical characteristics and prevent on-stream deactivation e.g. by sintering. Supports employing porous architectures, acid/base character and/or surface redox chemistry e.g. strong metal support interaction (SMSI), afford further opportunities to influence catalyst reactivity [170–173]. Mesoporous silicas are widely used to disperse metal nanoparticles [135, 136, 171, 174, 175]. The transition from low surface area, amorphous silica (200 m^2g^{-1}) to two-dimensional non-interconnected pore channels (SBA-15) [73] and three-dimensional interconnected porous frameworks (SBA-16, KIT-6) [73, 176, 177] improved the dispersion of Pd nanoparticles and hence degree of surface oxidation and thus activity in allylic alcohol selox (Fig. 16), but had little impact on the mass transport of small alcohols to/from the active site. [135, 136] The high thermal and chemical stability of such mesoporous silica [178, 179] makes such supports well-suited to commercialisation. Pd nanoparticles confined within such mesoporous silicas demonstrate good selectivity in crotyl and cinnamyl alcohol selox to their respective aldehydes (>70 %), and excellent TOFs of 7,000 and 5,000 h^{-1} for the respective alcohols. Similar activities are reported for secondary and tertiary allylic alcohols, highlighting the versatility of silica supported Pd nanoparticles [51, 135, 136, 180–182]. Incorporation of macropores into SBA-15 via dual hard/soft templating to

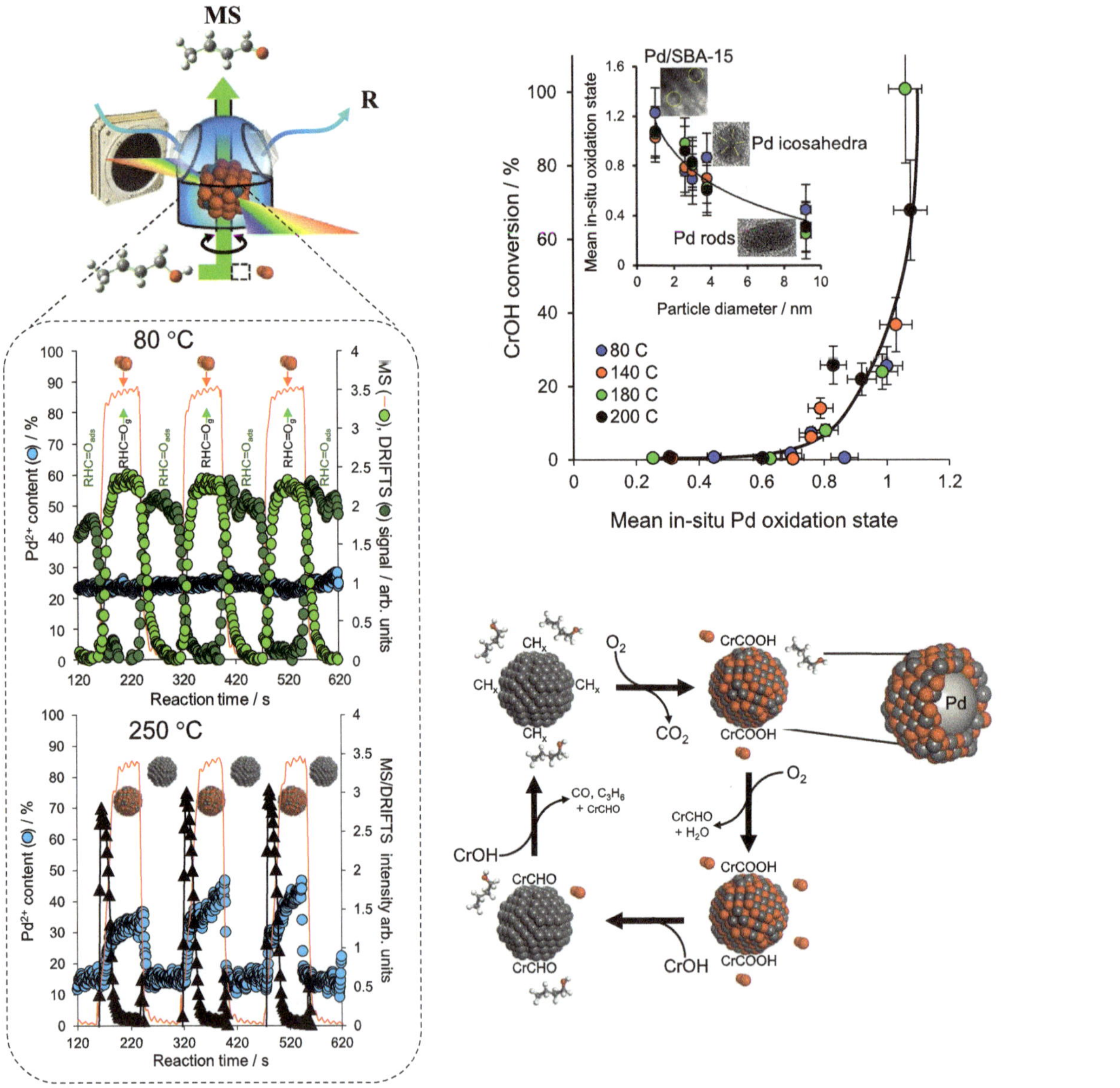

Fig. 14 (*Left*) Cartoon of operando DRIFTS/MS/XAS reaction cell and resulting temperature dependent behaviour of Pd oxidation state and associated reactivity towards crotyl alcohol oxidation over a Pd/meso-Al_2O_3 catalyst—only selective oxidation over surface PdO occurs at 80 °C, whereas crotonaldehyde decarbonylation and combustion dominate over Pd metal at 250 °C; (*top right*) relationship between Pd oxidation derived in situ and crotyl alcohol conversion; (*bottom right*) summary of reaction-induced redox processes in Pd-catalysed crotyl alcohol selox. Adapted with permission from references [166, 168]. Copyright 2011 and 2012 American Chemical Society

form a hierarchically ordered macroporous–mesoporous Pd/SBA-15 was recently shown to promote the catalytic selox of sterically challenging sesquiterpenoid substrates such as farnesol and phytol via (1) stabilising PdO nanoparticles and (2) dramatically improving in-pore diffusion and access to active sites [158].

The benefits of mesostructured supports are not limited to silica, with ultra-low loadings of palladium impregnated onto a surfactant-templated mesoporous alumina (350 $m^2\ g^{-1}$) generating atomically dispersed Pd^{2+} centres [137]. Such single-site catalysts were 10 times more active in crotonaldehyde and cinnamaldehyde production than comparable materials employing conventional (100 $m^2\ g^{-1}$) γ-alumina, owing to the preferential genesis of higher concentrations of electron-deficient palladium [134, 137], due to either pinning at cation vacancies or metal → support charge transfer [183]. These Pd/meso-Al_2O_3 catalysts exhibited similar TOFs to their silica counterparts (7,080 and

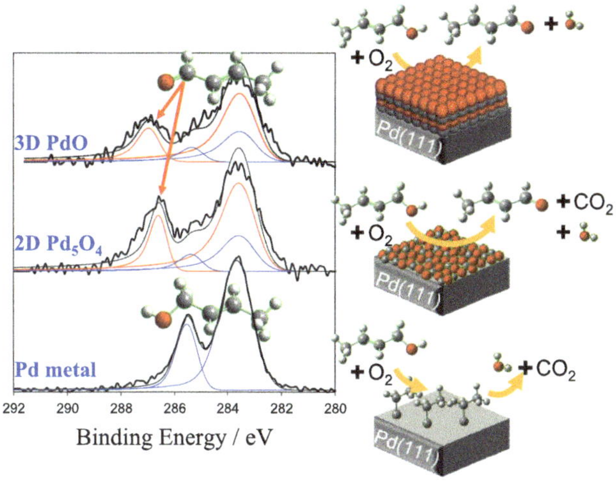

Fig. 15 (*Left*) C 1s XP spectra of crotyl alcohol/O_2 gas mixture over metallic and oxidised Pd(111) surfaces; (*right*) differing reactivity of palladium metal and oxide surfaces. Adapted from reference [169]. Copyright 2012 American Chemical Society

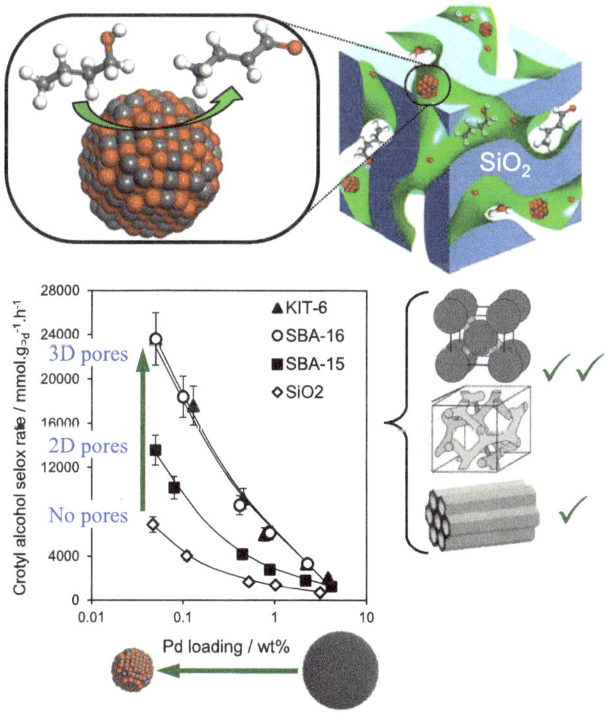

Fig. 16 Comparative activity of Pd nanoparticles dispersed over amorphous, 2D non-interconnected SBA-15 and 3D interconnected SBA-16 and KIT-6 mesoporous silicas in the selective aerobic oxidation of crotyl alcohol. Adapted from reference [135]. Copyright 2011 American Chemical Society

4,400 h^{-1} for crotyl and cinnamyl alcohol selox, respectively) [137], consistent with a common active site and reaction mechanism (Fig. 17).

Mesoporous titania and ceria have also attracted interest as novel catalyst supports. The oxygen storage capacity of ceria-derived materials is of particular interest due to their facile $Ce^{3+} \leftrightarrow Ce^{4+}$ redox chemistry [173, 184–188]. Sacrificial reduction of the ceria supports by reactively formed hydrogen liberated during the oxidative dehydrogenation of alcohols could mitigate in situ reduction of oxidised palladium, and hence maintain selox activity and catalyst lifetime, with Ce^{4+} sites regenerated by dissociatively adsorbed gas phase oxygen [187, 189, 190]. Due to its high density, conventional nanocrystalline cerias possess meagre surface areas (typically ~5 m^2g^{-1}), hence Pd/CeO_2 typically exhibit poor selox behaviour due to their resultant low nanoparticle dispersions which favour (self-poisoning) metallic Pd [189, 191, 192].

Bimetallic palladium selox catalysts

Incorporation of a second metal into palladium catalysts can improve both alcohol selox stability and selectivity. Typical promoters such as Ag, Bi, Pb and Sn [157, 193–196], enhance oxidation performance towards challenging substrates such as propylene glycol [197] as well as allylic and benzylic alcohols. Wenkin et al. [194] reported glucose oxidation to gluconates was increased by a factor of 20 over Pd–Bi/C catalysts (Bi/Pd_s = 0.1) versus Pd/C counterparts. In situ XAS and attenuated total reflection infrared spectroscopy (ATR-IR) suggested that Bi residing at the catalyst surface protects palladium from deactivation by either over-oxidation (a hypothesis since disproved [166, 167, 169]) or site-blocking by aromatic solvents [153]. Prati et al. [200] first reported significant rate enhancements and resistance to deactivation phenomena in the liquid phase selox of D-sorbitol to gluconic/gulonic acids upon addition of Au to Pd/C and Pt/C materials [198], subsequently extended to polyol and long chain aliphatic alcohols [199]. A strong synergy between Pd and Au centres was also demonstrated by Hutchings et al., wherein Au–Pd alloy nanoparticles supported on titania exhibited increased reactivity towards a diverse range of primary, allylic and benzylic alkyl alcohols compared to monometallic palladium analogues. The versatility of Au–Pd catalysts has also been shown in selox of saturated hydrocarbons [201], ethylene glycol [202], glycerol [203] and methanol [204], wherein high selectivity and resistance to on-stream deactivation is noted.

The effect of Au–Pd composition has been extensively studied for bimetallic nanoparticles stabilised by PVP surfactants [205]. An optimal Au:Pd composition of 1:3 was identified for 3 nm particles towards the aqueous phase aerobic selox of benzyl alcohol, 1-butanol, 2-butanol, 2-buten-1-ol and 1,4-butanediol; in each case the bimetallic catalysts were superior to palladium alone. Mertens et al. [206] examined similar systems utilising 1.9 nm nanoparticles, wherein an optimal Au content of around 80 % was

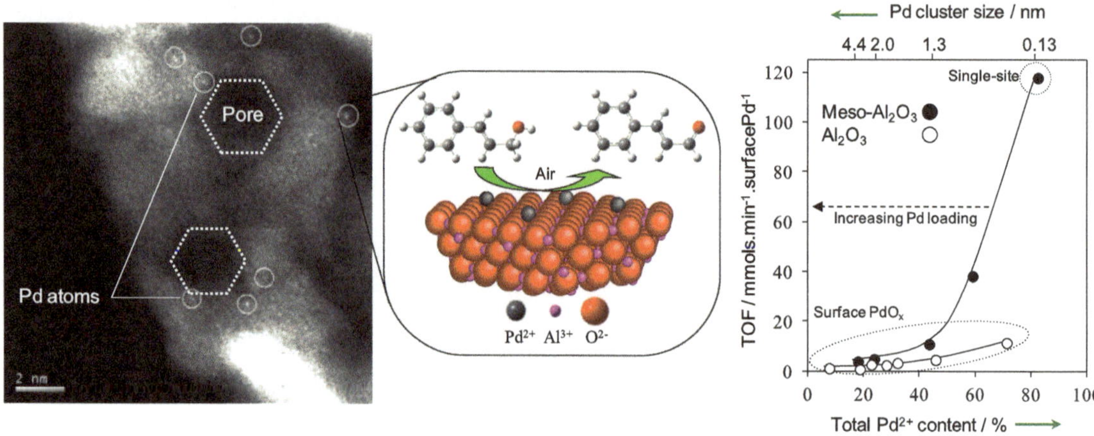

Fig. 17 (*Left*) HAADF–STEM image of atomically dispersed Pd atoms on a mesoporous Al_2O_3 support; and (*right*) associated relationship between Pd^{2+} content/dispersion and activity in crotyl alcohol selox over Pd/alumina catalysts. Adapted with permission from reference [137]. Copyright Wiley–VCH Verlag GmbH & Co. KGaA

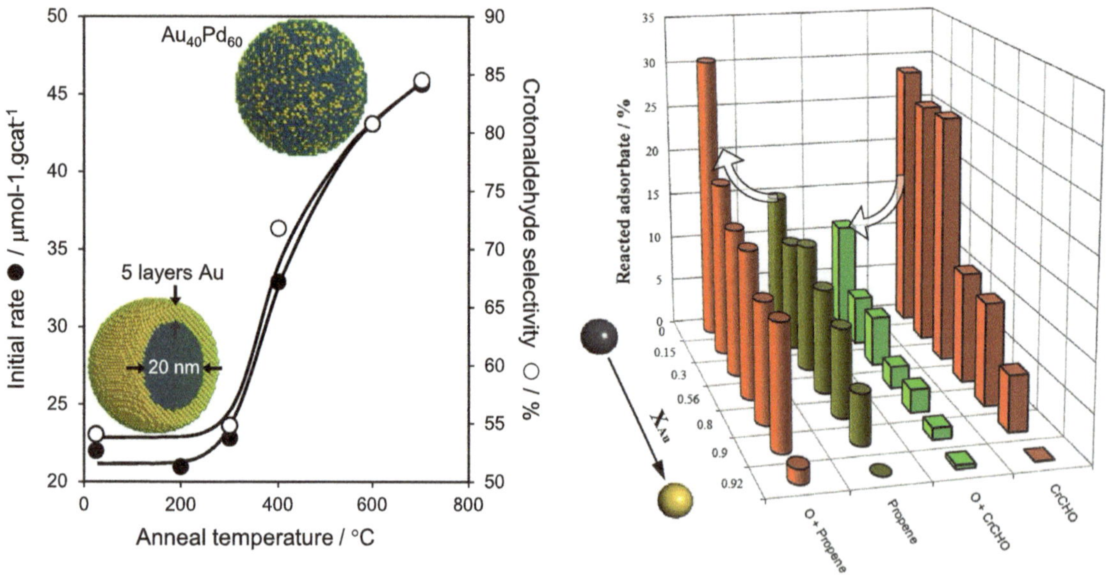

Fig. 18 Impact of thermally induced Au–Pd alloying of (*left*) titania-supported Au shell–Pd core nanoparticles on crotyl alcohol aerobic selox adapted from reference [208], with permission from Elsevier; and (*right*) ultrathin gold overlayers on Pd(111) on crotonaldehyde and propene decomposition with/without co-adsorbed oxygen, adapted from reference [160] with permission from the PCCP Owner Societies

determined for benzyl alcohol selox. The synergic interaction between Au and Pd therefore appears interdependent on nanoparticle size. It is well-known that the catalytic activity of Au nanoparticles increases dramatically <2 nm [207], hence it is interesting to systematically compare phase separated and alloyed catalysts. The author's group prepared titania-supported Au shell (5-layer)-Pd core (20 nm) bimetallic nanoparticles for the liquid phase selox of crotyl alcohol and systematically studied the evolution of their bulk and surface properties as a function of thermal processing by in situ XPS, DRIFTS, EXAFS, XRD and ex-situ HRTEM. Limited Au/Pd alloying occurred below 300 °C in the absence of particle sintering [208]. Higher temperatures induced bulk and surface alloying, with concomitant sintering and surface roughening. Migration of Pd atoms from the core to the surface dramatically enhanced activity and selectivity, with the most active and selective surface alloy containing 40 atom % Au (Fig. 18). This discovery was rationalised in terms of complementary temperature-programmed mass spectometric studies of crotyl alcohol and reactively formed intermediates over Au/Pd(111) model single crystal catalysts which reveal that gold–palladium alloys promote desorption of the desired crotonaldehyde selox product while co-adsorbed oxygen adatoms actually suppress aldehyde combustion. In contrast, the combustion of propene, the undesired secondary product of crotonaldehyde decarbonylation, is enhanced by co-adsorbed oxygen [160].

Scott et al. prepared the inverse Au core-Pd shell nanoparticles and explored the catalytic cycle for alcohol selox to assess their associated stability [205, 209–212]. In situ Pd–K and Pd–L_{III} edge XAS of a Au nanoparticle/Pd(II) salt solution were undertaken to discriminate two possible reaction mechanisms. No evidence was found that crotyl alcohol oxidation was accompanied by Pd^{2+} reduction onto Au nanoparticles, resulting in the formation of a metallic Pd shell (with oxygen subsequently regenerating electron-deficient palladium), and therefore proposed β-H elimination as the favoured pathway. Scott and co-workers proposed that the Au core prevents the re-oxidation of surface Pd^0 atoms; no Pd–O and Pd–Cl contributions were observed by EXAFS.

In summary, the selective oxidation of complex alcohol substrates can be accomplished through Pd-mediated heterogeneous catalysis with high turnover and product selectivity. Application of in situ and operando techniques, such as X-ray and IR spectroscopies, has elucidated the mechanism of alcohol oxidative dehydrogenation and competing aldehyde decarbonylation. Surface PdO has been identified as the active catalytic species, and deactivation the result of reduction to metallic palladium and concomitant self-poisoning by strongly bound CO and carbonaceous residues. Breakthroughs in analytical tools and synthetic approaches to engineering nanoporous supports and shape/size controlled nanoparticles have delivered significant progress towards improved atom and energy efficiency and catalyst stability, however, next generation palladium selox catalysts necessitate improved synthetic protocols to create higher densities of ultra-dispersed Pd^{2+} centres with superior resistance to on-stream reduction under atmospheric oxygen.

Acknowledgments A.F.L. thanks the EPSRC (EP/G007594/3) for financial support and a Leadership Fellowship, and acknowledges the invaluable contributions of Prof Karen Wilson (European Bioenergy Research Institute, Aston University).

References

1. S. Kretzmann, http://priceofoil.org/2013/11/26/new-analysis-shows-growing-fossil-reserves-shrinking-carbon-budget/
2. C.C. Authority, Reducing Australia's greenhouse gas emissions—targets and progress review draft report, commonwealth of Australia, 2013
3. C.C. Secretariat, The critical decade 2013 climate change science, risks and responses, commonwealth of Australia, 2013
4. I.E. Agency, prospect of limiting the global increase in temperature to 2 °C is getting bleaker, http://www.iea.org/newsroomandevents/news/2011/may/name,19839,en.html
5. I.E. Agency, Redrawing the energy climate map, 2013
6. U.S.E.I. Administration, international energy outlook 2013, 2013
7. Armaroli N, Balzani V (2007) Angew Chem Int Ed 46:52–66
8. Azadi P, Inderwildi OR, Farnood R, King DA (2013) Renew Sustain Energy Rev 21:506–523
9. Demirbas A (2007) Energy Policy 35:4661–4670
10. McLaughlin DW (2011) Conserv Biol 25:1117–1120
11. Danielsen F, Beukema H, Burgess ND, Parish F, BrÜHl CA, Donald PF, Murdiyarso D, Phalan BEN, Reijnders L, Struebig M, Fitzherbert EB (2009) Conserv Biol 23:348–358
12. Achten WMJ, Verchot L, Franken YJ, Mathijs E, Singh VP, Aerts R, Muys B (2008) Biomass Bioenergy 32:1063–1084
13. Mata TM, Martins AA, Caetano NS (2010) Renew Sustain Energy Rev 14:217–232
14. BP, BP energy outlook 2030, 2011
15. Sheldon RA (2014) Green Chem 16:950–963
16. Pandey MP, Kim CS (2011) Chem Eng Technol 34:29–41
17. Knothe G (2010) Top Catal 53:714–720
18. Climent MJ, Corma A, Iborra S, Velty A (2004) J Catal 221:474–482
19. Constantino U, Marmottini F, Nocchetti M, Vivani R (1998) Eur J Inorg Chem 10:1439–1446
20. Narasimharao K, Lee A, Wilson K (2007) J Biobased Mater Bioenergy 1:19–30
21. Othman MR, Helwani Z, Martunus, Fernando WJN (2009) Appl Organomet Chem 23:335–346
22. Liu Y, Lotero E, Goodwin JG, Mo X (2007) Appl Catal A 33:138–148
23. Geuens J, Kremsner JM, Nebel BA, Schober S, Dommisse RA, Mittelbach M, Tavernier S, Kappe CO, Maes BUW (2007) Energy Fuels 22:643–645
24. Hu, Du, Tang Z, Min (2004) Ind Eng Chem Res 43:7928–7931
25. Knothe G (2005) Fuel Process Technol 86:1059–1070
26. Ma F, Hanna MA (1999) Bioresour Technol 70:1–15
27. Lotero E, Liu Y, Lopez DE, Suwannakarn K, Bruce DA, Goodwin JG (2005) Ind Eng Chem Res 44:5353–5363
28. Thomas JM (2012) Proc R Soc A Math Phys Eng Sci 468:1884–1903
29. Somorjai GA, Frei H, Park JY (2009) J Am Chem Soc 131:16589–16605
30. Luque R, Herrero-Davila L, Campelo JM, Clark JH, Hidalgo JM, Luna D, Marinas JM, Romero AA (2008) Energy Environ Sci 1:542–564
31. Luque R, Lovett JC, Datta B, Clancy J, Campelo JM, Romero AA (2010) Energy Environ Sci 3:1706–1721
32. Dacquin J-P, Lee AF, Wilson K (2010) Thermochemical conversion of biomass to liquid fuels and chemicals. The Royal Society of Chemistry, UK, pp 416–434
33. Wilson K, Lee AF (2012) Catal Sci Technol 2:884–897
34. Eze VC, Phan AN, Pirez C, Harvey AP, Lee AF, Wilson K (2013) Catal Sci Technol 3:2373–2379
35. Chen G-Q, Patel MK (2011) Chem Rev 112:2082–2099
36. Bozell JJ, Petersen GR (2010) Green Chem 12:539–554
37. Sheldon RA (2005) Green Chem 7:267–278
38. Sheldon RA, Arends I, Hanefeld U (2007) Green chemistry and catalysis. Wiley-VCH Verlag GmbH & Co. KGaA, Weinheim
39. Mallat T, Baiker A (2004) Chem Rev 104:3037–3058
40. Lee AF (2014). In: Wilson K, Lee AF (eds) Heterogeneous catalysts for clean technology: spectroscopy, design, and

monitoring chapt. 2. Wiley-VCH Verlag GmbH & Co. KGaA, Weinheim, pp 11–33
41. Vinod CP, Wilson K, Lee AF (2011) J Chem Technol Biotechnol 86:161–171
42. Kobayashi S, Miyamura H, Akiyama R, Ishida T (2005) J Am Chem Soc 127:9251–9254
43. Kaizuka K, Lee KY, Miyamura H, Kobayashi S (2012) J Flow Chem 2:1
44. Ye XA, Johnson MD, Diao TN, Yates MH, Stahl SS (2010) Green Chem 12:1180–1186
45. Ayude A, Cechini J, Cassanello M, Martinez O, Haure P (2008) Chem Eng Sci 63:4969–4973
46. Tschan R, Wandeler R, Schneider MS, Schubert MM, Baiker A (2001) J Catal 204:219–229
47. Caravati M, Grunwaldt JD, Baiker A (2004) Catal Today 91–2:1–5
48. Grunwaldt JD, Caravati M, Baiker A (2006) J Phys Chem B 110:9916–9922
49. Burgener M, Tyszewski T, Ferri D, Mallat T, Baiker A (2006) Appl Catal A Gen 299:66–72
50. Hou ZS, Theyssen N, Leitner W (2007) Green Chem 9:127–132
51. Hou Z, Theyssen N, Brinkmann A, Klementiev KV, Grünert W, Bühl M, Schmidt W, Spliethoff B, Tesche B, Weidenthaler C (2008) J Catal 258:315–323
52. Beale AM, Jacques SDM, Weckhuysen BM (2010) Chem Soc Rev 39:4656–4672
53. Grunwaldt JD, Schroer CG (2010) Chem Soc Rev 39:4741–4753
54. Lee AF (2012) Aust J Chem 65:615–623
55. Gai PL, Sharma R, Ross FM (2008) MRS Bull 33:107–114
56. Jungjohann KL, Evans JE, Aguiar JA, Arslan I, Browning ND (2012) Microsc Microanal 18:621–627
57. Browning ND, Bonds MA, Campbell GH, Evans JE, Lagrange T, Jungjohann KL, Masiel DJ, Mckeown J, Mehraeen S, Reed BW, Santala M (2012) Curr Opin Solid State Mat Sci 16:23–30
58. Boyes ED, Ward MR, Lari L, Gai PL (2013) Ann Phys Berlin 525:423–429
59. Xiong Y, Cai H, Wiley BJ, Wang J, Kim MJ, Xia Y (2007) J Am Chem Soc 129:3665–3675
60. Zhang H, Jin M, Xiong Y, Lim B, Xia Y (2012) Acc Chem Res 46:1783–1794
61. Xia X, Choi S-I, Herron JA, Lu N, Scaranto J, Peng H-C, Wang J, Mavrikakis M, Kim MJ, Xia Y (2013) J Am Chem Soc 135:15706–15709
62. Astruc D, Lu F, Aranzaes JR (2005) Angew Chem Int Ed 44:7852–7872
63. Hermans I, Spier ES, Neuenschwander U, Turrà N, Baiker A (2009) Top Catal 52:1162–1174
64. Heyns K, Paulsen H (1957) Angew Chem 69:600–608
65. Liu XY, Madix RJ, Friend CM (2008) Chem Soc Rev 37:2243–2261
66. Dimitratos N, Lopez-Sanchez JA, Hutchings GJ (2012) Chem Sci 3:20–44
67. Besson M, Gallezot P (2000) Catal Today 57:127–141
68. Kluytmans J, Markusse A, Kuster B, Marin G, Schouten J (2000) Catal Today 57:143–155
69. Mallat T, Baiker A (1994) Catal Today 19:247–283
70. Narasimharao K, Brown DR, Lee AF, Newman AD, Siril PF, Tavener SJ, Wilson K (2007) J Catal 248:226–234
71. Suwannakarn K, Lotero E, Ngaosuwan K, Goodwin JG (2009) Ind Eng Chem Res 48:2810–2818
72. Kouzu M, Nakagaito A, Hidaka J-s (2011) Appl Catal A 405:36–44
73. Zhao D, Huo Q, Feng J, Chmelka BF, Stucky GD (1998) J Am Chem Soc 120:6024–6036
74. Mbaraka IK, Shanks BH (2005) J Catal 229:365–373
75. Melero JA, Bautista LF, Morales G, Iglesias J, Briones D (2008) Energy Fuels 23:539–547
76. Chen X-R, Ju Y-H, Mou C-Y (2007) J Phys Chem C 111:18731–18737
77. Mbaraka IK, Radu DR, Lin VSY, Shanks BH (2003) J Catal 219:329–336
78. Chen D, Li Z, Wan Y, Tu X, Shi Y, Chen Z, Shen W, Yu C, Tu B, Zhao D (2006) J Mater Chem 16:1511–1519
79. Cao L, Man T, Kruk M (2009) Chem Mater 21:1144–1153
80. Dacquin JP, Lee AF, Pirez C, Wilson K (2012) Chem Commun 48:212–214
81. Martin A, Morales G, Martinez F, van Grieken R, Cao L, Kruk M (2010) J Mater Chem 20:8026–8035
82. Pirez C, Caderon J-M, Dacquin J-P, Lee AF, Wilson K (2012) ACS Catal 2:1607–1614
83. Vinu A, Gokulakrishnan N, Balasubramanian VV, Alam S, Kapoor MP, Ariga K, Mori T (2008) Chem A Eur J 14:11529–11538
84. Melero JA, Iglesias J, Morales G (2009) Green Chem 11:1285–1308
85. Carrero A, Vicente G, Rodríguez R, Linares M, del Peso GL (2011) Catal Today 167:148–153
86. Ying JY, Mehnert CP, Wong MS (1999) Angew Chem Int Ed 38:56–77
87. Lu Y (2006) Angew Chem Int Ed 45:7664–7667
88. Garg S, Soni K, Kumaran GM, Bal R, Gora-Marek K, Gupta JK, Sharma LD, Dhar GM (2009) Catal Today 141:125–129
89. Gheorghiu S, Coppens M-O (2004) AIChE J 50:812–820
90. Zhang X, Zhang F, Chan K-Y (2004) Mater Lett 58:2872–2877
91. Sun J-H, Shan Z, Maschmeyer T, Coppens M-O (2003) Langmuir 19:8395–8402
92. Dacquin J-P, Dhainaut JRM, Duprez D, Royer SB, Lee AF, Wilson K (2009) J Am Chem Soc 131:12896–12897
93. Dhainaut J, Dacquin J-P, Lee AF, Wilson K (2010) Green Chem 12:296–303
94. Wilson K, Rénson A, Clark JH (1999) Catal Lett 61:51–55
95. Rác B, Hegyes P, Forgo P, Molnár Á (2006) Appl Catal A 299:193–201
96. Yang Q, Liu J, Yang J, Kapoor MP, Inagaki S, Li C (2004) J Catal 228:265–272
97. Yang Q, Kapoor MP, Shirokura N, Ohashi M, Inagaki S, Kondo JN, Domen K (2005) J Mater Chem 15:666–673
98. Morales G, Athens G, Chmelka BF, van Grieken R, Melero JA (2008) J Catal 254:205–217
99. Margolese D, Melero JA, Christiansen SC, Chmelka BF, Stucky GD (2000) Chem Mater 12:2448–2459
100. Díaz I, Márquez-Alvarez C, Mohino F, Pérez-Pariente JN, Sastre E (2000) J Catal 193:283–294
101. Dacquin J-P, Cross HE, Brown DR, Duren T, Williams JJ, Lee AF, Wilson K (2010) Green Chem 12:1383–1391
102. Schumacher C, Gonzalez J, Wright PA, Seaton NA (2005) J Phys Chem B 110:319–333
103. Pirez C, Wilson K, Lee AF (2014) Green Chem 16:197–202
104. Ono Y, Baba T (1997) Catal Today 38:321–337
105. Hattori H (1995) Chem Rev 95:537–558
106. Albuquerque MCG, Azevedo DCS, Cavalcante CL Jr, Santamaría-González J, Mérida-Robles JM, Moreno-Tost R, Rodríguez-Castellón E, Jiménez-López A, Maireles-Torres P (2009) J Mol Catal A: Chem 300:19–24
107. Peterson GR, Scarrah WP (1984) J Am Oil Chem Soc 61:1593–1597
108. Verziu M, Coman SM, Richards R, Parvulescu VI (2011) Catal Today 167:64–70
109. Granados ML, Alonso DM, Alba-Rubio AC, Mariscal R, Ojeda M, Brettes P (2009) Energy Fuels 23:2259–2263

110. Di Serio M, Tesser R, Casale L, D'Angelo A, Trifuoggi M, Santacesaria E (2010) Top Catal 53:811–819
111. MacLeod CS, Harvey AP, Lee AF, Wilson K (2008) Chem Eng J 135:63–70
112. Montero J, Wilson K, Lee A (2010) Top Catal 53:737–745
113. Watkins RS, Lee AF, Wilson K (2004) Green Chem 6:335–340
114. Alonso DM, Mariscal R, Granados ML, Maireles-Torres P (2009) Catal Today 143:167–171
115. Verziu M, Cojocaru B, Hu J, Richards R, Ciuculescu C, Filip P, Parvulescu VI (2008) Green Chem 10:373–381
116. Zhu K, Hu J, Kübel C, Richards R (2006) Angew Chem Int Ed 45:7277–7281
117. Montero JM, Gai P, Wilson K, Lee AF (2009) Green Chem 11:265–268
118. Cantrell DG, Gillie LJ, Lee AF, Wilson K (2005) Appl Catal A 287:183–190
119. Di Serio M, Ledda M, Cozzolino M, Minutillo G, Tesser R, Santacesaria E (2006) Ind Eng Chem Res 45:3009–3014
120. Cavani F, Trifirò F, Vaccari A (1991) Catal Today 11:173–301
121. Barakos N, Pasias S, Papayannakos N (2008) Bioresour Technol 99:5037–5042
122. Fraile JM, García N, Mayoral JA, Pires E, Roldán L (2009) Appl Catal A 364:87–94
123. Cross HE, Brown DR (2010) Catal Commun 12:243–245
124. Woodford JJ, Dacquin J-P, Wilson K, Lee AF (2012) Energy Environ Sci 5:6145–6150
125. D Highley, A Bloodworth, R Bate Dolomite-mineral planning factsheet, British Geological Survey, 2006
126. Sutton D, Kelleher B, Ross JRH (2001) Fuel Process Technol 73:155–173
127. Wilson K, Hardacre C, Lee AF, Montero JM, Shellard L (2008) Green Chem 10:654–659
128. Ilgen O (2011) Fuel Process Technol 92:452–455
129. Scharff Y, Asteris D, Fédou S (2013) OCL 20:D502
130. Choi K-M, Akita T, Mizugaki T, Ebitani K, Kaneda K (2003) New J Chem 27:324–328
131. Chen J, Zhang Q, Wang Y, Wan H (2008) Adv Synth Catal 350:453–464
132. Li F, Zhang Q, Wang Y (2008) Appl Catal A 334:217–226
133. Lee AF, Wilson K (2004) Green Chem 6:37
134. Lee AF, Hackett SF, Hargreaves JS, Wilson K (2006) Green Chem 8:549–555
135. Parlett CMA, Bruce DW, Hondow NS, Lee AF, Wilson K (2011) ACS Catal 1:636–640
136. Parlett CMA, Bruce DW, Hondow NS, Newton MA, Lee AF, Wilson K (2013) Chem Cat Chem 5:939–950
137. Hackett SF, Brydson RM, Gass MH, Harvey I, Newman AD, Wilson K, Lee AF (2007) Angew Chem 119:8747–8750
138. Baeza J, Calvo L, Gilarranz M, Mohedano A, Casas J, Rodriguez J (2012) J Catal 293:85–93
139. Wang X, Yang H, Feng B, Hou Z, Hu Y, Qiao Y, Li H, Zhao X (2009) Catal Lett 132:34–40
140. Yudha S, Dhital RN, Sakurai H (2011) Tetrahedron Lett 52:2633–2637
141. Dun R, Wang X, Tan M, Huang Z, Huang X, Ding W, Lu X (2013) ACS Catal 3:3063–3066
142. Weldon MK, Friend CM (1996) Chem Rev 96:1391–1412
143. Lee AF, Chang Z, Ellis P, Hackett SF, Wilson K (2007) J Phys Chem C 111:18844–18847
144. Naughton J, Pratt A, Woffinden CW, Eames C, Tear SP, Thompson SM, Lee AF, Wilson K (2011) J Phys Chem C 115:25290–25297
145. Davis JL, Barteau MA (1987) Surf Sci 187:387–406
146. Davis JL, Barteau MA (1990) Surf Sci 235:235–248
147. Zaera F (2003) Catal Lett 91:1–10
148. Keresszegi C, Burgi T, Mallat T, Baiker A (2002) J Catal 211:244–251
149. Lee AF (2001) Abstr Pap Am Chem Soc 221:U335–U336
150. Grunwaldt J-D, Keresszegi C, Mallat T, Baiker A (2003) J Catal 213:291–295
151. Keresszegi C, Grunwaldt J-D, Mallat T, Baiker A (2004) J Catal 222:268–280
152. Grunwaldt J-D, Caravati M, Baiker A (2006) J Phys Chem B 110:25586–25589
153. Mondelli C, Ferri D, Grunwaldt J-D, Krumeich F, Mangold S, Psaro R, Baiker A (2007) J Catal 252:77–87
154. Keresszegi C, Ferri D, Mallat T, Baiker A (2005) J Catal 234:64–75
155. Caravati M, Meier DM, Grunwaldt J-D, Baiker A (2006) J Catal 240:126–136
156. Mallat T, Baiker A (1995) Catal Today 24:143–150
157. Mallat T, Bodnar Z, Hug P, Baiker A (1995) J Catal 153:131–143
158. Parlett CMA, Keshwalla P, Wainwright SG, Bruce DW, Hondow NS, Wilson K, Lee AF (2013) ACS Catal 3:2122–2129
159. Parlett CMA, Durndell LJ, Wilson K, Bruce DW, Hondow NS, Lee AF (2014) Catal Commun 44:40–45
160. Naughton J, Lee AF, Thompson S, Vinod CP, Wilson K (2010) Phys Chem Chem Phys 12:2670–2678
161. Ping EW, Pierson J, Wallace R, Miller JT, Fuller TF, Jones CW (2011) Appl Catal A 396:85–90
162. Newton MA, Jyoti B, Dent AJ, Fiddy SG, Evans J (2004) Chem Commun 2004:2382–2383
163. Ferri D, Baiker A (2009) Top Catal 52:1323–1333
164. Newton MA (2009) Top Catal 52:1410–1424
165. Newton MA, Dent AJ, Fiddy SG, Jyoti B, Evans J (2007) Catal Today 126:64–72
166. Lee AF, Ellis CV, Naughton JN, Newton MA, Parlett CM, Wilson K (2011) J Am Chem Soc 133:5724–5727
167. Parlett CMA, Gaskell CV, Naughton JN, Newton MA, Wilson K, Lee AF (2013) Catal Today 205:76–85
168. Gaskell CV, Parlett CMA, Newton MA, Wilson K, Lee AF (2012) ACS Catal 2:2242–2246
169. Lee AF, Naughton JN, Liu Z, Wilson K (2012) ACS Catal 2:2235–2241
170. Tauster S (1987) Acc Chem Res 20:389–394
171. Ghedini E, Menegazzo F, Signoretto M, Manzoli M, Pinna F, Strukul G (2010) J Catal 273:266–273
172. Hicks RF, Qi H, Young ML, Lee RG (1990) J Catal 122:295–306
173. Beckers J, Rothenberg G (2010) Green Chem 12:939–948
174. Inumaru K, Nakamura K, Ooyachi K, Mizutani K, Akihara S, Yamanaka S (2010) J Porous Mater 18:455–463
175. Liu Y-M, Cao Y, Yi N, Feng W-L, Dai W-L, Yan S-R, He H-Y, Fan K-N (2004) J Catal 224:417–428
176. Kim T-W, Kleitz F, Paul B, Ryoo R (2005) J Am Chem Soc 127:7601–7610
177. Zhao D, Feng J, Huo Q, Melosh N, Fredrickson GH, Chmelka BF, Stucky GD (1998) Science 279:548–552
178. Sneh O, George SM (1995) J Phys Chem 99:4639–4647
179. Kim JM, Kwak JH, Jun S, Ryoo R (1995) J Phys Chem 99:16742–16747
180. Kaneda K, Fujii M, Morioka K (1996) J Org Chem 61:4502–4503
181. Polshettiwar V, Varma RS (2009) Org Biomol Chem 7:37–40
182. Peterson KP, Larock RC (1998) J Org Chem 63:3185–3189
183. Behafarid F, Ono LK, Mostafa S, Croy JR, Shafai G, Hong S, Rahman TS, Bare SR, Cuenya BR (2012) Phys Chem Chem Phys 14:11766–11779
184. Yuan Q, Duan HH, Li LL, Sun LD, Zhang YW, Yan CH (2009) J Colloid Interface Sci 335:151–167

185. Lykhach Y, Staudt T, Lorenz MPA, Streber R, Bayer A, Steinrück H-P, Libuda J (2010) Chem Phys Chem 11:1496–1504
186. Tana, Zhang M, Li J, Li H, Li Y, Shen W (2009) Catal Today 148:179–183
187. Bensalem A, Bozon-Verduraz F, Perrichon V (1995) J Chem Soc Faraday Trans 91:2185–2189
188. Harrison B, Diwell A, Hallett C (1988) Plat Met Rev 32:73–83
189. Zhu Y, Zhang S, Shan J-j, Nguyen L, Zhan S, Gu X, Tao F (2013) ACS Catal 3:2627–2639
190. Maillet T, Madier Y, Taha R, Barbier J Jr, Duprez D (1997) Stud Surf Sci Catal 112:267–275
191. Oh S-H, Hoflund GB (2006) J Phys Chem A 110:7609–7613
192. Badri A, Binet C, Lavalley J-C (1996) J Chem Soc Faraday Trans 92:1603–1608
193. Lee AF, Gee JJ, Theyers HJ (2000) Green Chem 2:279–282
194. Wenkin M, Ruiz P, Delmon B, Devillers M (2002) J Mol Catal A: Chem 180:141–159
195. Alardin F, Delmon B, Ruiz P, Devillers M (2000) Catal Today 61:255–262
196. Keresszegi C (2004) J Catal 225:138–146
197. Pinxt H, Kuster B, Marin G (2000) Appl Catal A 191:45–54
198. Dimitratos N, Porta F, Prati L, Villa A (2005) Catal Lett 99:181–185
199. Prati L, Villa A, Campione C, Spontoni P (2007) Top Catal 44:319–324
200. Enache DI, Edwards JK, Landon P, Solsona-Espriu B, Carley AF, Herzing AA, Watanabe M, Kiely CJ, Knight DW, Hutchings GJ (2006) Science 311:362–365
201. Long J, Liu H, Wu S, Liao S, Li Y (2013) ACS Catal 3:647–654
202. Griffin MB, Rodriguez AA, Montemore MM, Monnier JR, Williams CT, Medlin JW (2013) J Catal 307:111–120
203. Liu Z, Xu J, Zhang H, Zhao Y, Yu B, Chen S, Li Y, Hao L, Green Chem 2013
204. Wang R, Wu Z, Chen C, Qin Z, Zhu H, Wang G, Wang H, Wu C, Dong W, Fan W (2013) Chem Commun 49:8250–8252
205. Hou W, Dehm NA, Scott RW (2008) J Catal 253:22–27
206. Mertens P, Corthals S, Ye X, Poelman H, Jacobs P, Sels B, Vankelecom I, De Vos D (2009) J Mol Catal A: Chem 313:14–21
207. Hutchings GJ (2005) Catal Today 100:55–61
208. Lee AF, Ellis CV, Wilson K, Hondow NS (2010) Catal Today 157:243–249
209. Scott RW, Wilson OM, Oh S-K, Kenik EA, Crooks RM (2004) J Am Chem Soc 126:15583–15591
210. Scott RW, Sivadinarayana C, Wilson OM, Yan Z, Goodman DW, Crooks RM (2005) J Am Chem Soc 127:1380–1381
211. Dash P, Bond T, Fowler C, Hou W, Coombs N, Scott RWJ (2009) J Phys Chem C 113:12719–12730
212. Maclennan A, Banerjee A, Hu Y, Miller JT, Scott RWJ (2013) ACS Catal 3:1411–1419

15

The catalytic stability of some selected bifunctional nanoporous-based catalysts in the hydroisomerisation of n-C7 and the effect of post-synthesis modification techniques

Faisal M. Alotaibi · Raed H. Abudawood · Hamid A. Al-Megren · Mohammed C. Al-Kinany · Essam H. Jamea · Arthur A. Garforth

Abstract In this work, some commercial and in-house nanoporous-based catalysts, such as USY, beta and mordenite zeolites, and mesoporous aluminosilicate molecular sieves such as MCM-48 and SBA-15, loaded with metals and acting as mono- and bimetallic bifunctional catalysts, were used for hydroisomerisation experiments in a fixed-bed reactor at pressures between 1 and 15 bar and at feed space time ranging from 2.57 to 10.26 h^{-1} (35.14–140.6 kg s mol^{-1}) to hydroisomerise *n*-heptane over a temperature range of 210–270 °C. The effect of post-synthesis treatments of micro- and mesoporous catalysts was examined, regarding their activity, selectivity and stability, such as acid and steam dealumination techniques, acid leaching via a chelating agent, bimetal loading techniques, different platinum loading methods, and composite or hybrid catalyst generation. Results show that pore architecture is the most important factor affecting coke formation and deactivation in zeolite catalysts. It was found that those catalysts with high Si/Al ratios and those which had been acid-leached or steamed showed better activity, higher selectivity towards isomeric products and better time stability. Moreover, the balance between the number of metal sites and the number of acid sites played an important role in determining the activity, selectivity and stability of the bifunctional catalysts. Higher metal loading improves catalytic stability, due to a better balance and closeness of the catalytic functions. Moreover, the bimetallic catalyst improves the formation of smaller metal particles and better dispersion, which may affect selectivity and stability.

Keywords The time-on-stream stability · Alkanes' hydroisomerisation · Catalyst deactivation · Zeolites · Bifunctional catalysts · Nanoporous catalysts · Bimetallic catalysts

F. M. Alotaibi · H. A. Al-Megren · M. C. Al-Kinany
Petrochemicals Research Institute, King Abdulaziz City for Science and Technology (KACST), P.O.Box 6086, Riyadh 11442, Saudi Arabia

R. H. Abudawood
Research and Development Center, Saudi Aramco, Dhahran 31311, Saudi Arabia

E. H. Jamea
SABIC Technology Center, Saudi Basic Industries Corporation, Riyadh 11551, Saudi Arabia

A. A. Garforth
School of Chemical Engineering and Analytical Science, The University of Manchester, Manchester M13 9PL, UK

Introduction

Tailoring catalysts to optimise their chemical, physical and mechanical features is the most widely used method of limiting a reduction in catalytic activity. With relatively slow coke formation, the fine tuning of the strength and density of acid sites in zeolites is essential to achieve a sufficiently high activity for the desired reaction. Furthermore, limitations to the diffusion of reactant, product and coke precursor molecules can be reduced by the creation of mesoporosity within zeolite crystals to produce nano-sized crystals, which will slow down coke formation. Appropriate selection of operating conditions is a crucial factor in promoting desorption of the desired products and thus limiting deactivation of the catalyst [1].

Despite the inevitable loss of catalytic activity (catalyst deactivation) in most industrial processes, there are many ways to minimise this process and thus extend the lifetime of a catalyst. As the old saying goes, "an ounce of prevention is worth a pound of cure" [2]. The zeolite pore architecture has been found to play an important role for both the deactivation rate as a result of coking and also the composition of the carbonaceous deposits, such that zeolites with relatively large supercage pore systems, such as Y zeolites, are inherently favourable to coking reactions, which in turn leads to faster deactivation. In contrast, beta zeolite exhibits much better coking resistance, even though both zeolites have three-dimensional systems of channels [3].

The Pt/H-MOR catalyst is the most widely recognised example of a zeolitic isomerisation catalyst. It was first developed to isomerise smaller alkanes such as *n*-pentane and *n*-hexane for the Shell Hysomer process and can bring about a high isomer yield [4]. However, due to the strong tendency of middle paraffin isomerisation (C7–C9 carbon chain length) to undergo cracking, no hydroisomerisation process has yet been used industrially [5]. Strongly acidic zeolites such as H-mordenite have weaknesses as isomerisation catalysts due to the rapidity of their deactivation, particularly by olefinic carbonaceous species. There is therefore a great incentive for academic and industrial researchers to discover and develop new catalysts that are active, selective and able to resist deactivation [6, 7]. There have been detailed descriptions in the literature of how post-synthesis treatments by steam and acid of several zeolites can reduce the concentration of lattice aluminium and improve thermal/hydrothermal stability and catalytic performance whilst retaining the zeolite crystal structure. It must also be noted that mesopores form as a result of the dealumination of H-mordenite, significantly diminishing the coke's deactivating effect [8, 9].

The vast majority of microporous zeolites have pore diameters below 15 Å, leading to very high mass transfer limitations; hence, the intracrystalline diffusion of species inside the pores of zeolites is severely limited, which affects their catalytic activity and significantly reduces their product selectivity, because the increased residence time of the products in the zeolite pores favours undesired reactions such as over-cracking. Because of this limitation, there has been much work to find creative ways to eliminate the mass transfer problem. Moreover, the formation of mesopores can favour the formation of more branched or bulkier products. Evidently, generating a network of mesopores in zeolites crystals would offer numerous advantages for a great number of reactions [10]. Mesoporous aluminosilicate molecular sieves, such as SBA-15 and MCM-48, which have high surface area, large pore diameters and narrow pore size distribution and which possess uniform hexagonal and cubic one- and three-dimensional pore systems, respectively, are thus supposedly more resistant to pore blocking, which allows faster diffusion of reactants. However, they have been shown to be inferior to microporous zeolites in their acid strength and despite their larger pore diameters, which make them more able to deal with bulky molecules, they do not show higher activity than zeolites, e.g. in heavy oil cracking and the catalytic conversion of large molecules in vacuum gas oil and distillation residues. Thus, many significant improvements in both the hydrothermal stability and acidity of mesostructured aluminosilicates have been recently reported [11–13].

In this work, the time-on-stream (TOS) behaviour and deactivation rates of some commercial USY, beta and mordenite zeolites and mesoporous aluminosilicate molecular sieves such as Al-MCM-48 and Al-SBA-15 that have been loaded with platinum metal, with a wide range of Si/Al ratios, pore sizes and acidity characteristics, are evaluated in the hydroisomerisation of n-C7 when reaction conditions such as temperature, pressure and contact time are varied. n-C7 was used as a representative model compound for naphtha. A number of post-synthetic methods have been adopted to improve the catalytic properties of zeolites, such as acidity, porosity and thermal and hydrothermal stability, as these modifications are hard to achieve directly during zeolite synthesis. The main methods are ion exchange, dealumination by acid leaching or acid chelating, and stabilising through hydrothermal treatment (steaming), mono- and bimetallic effects.

Experimental

Catalyst preparation

Some commercial USY, beta, mordenite, and mesoporous aluminosilicates samples were provided by different suppliers [14–20], as shown in Table 1. These zeolite supports were designed and modified by various post-synthesis techniques so that they could be used for multiple purposes; for instance, high alkane hydroconversions like cracking, isomerisation and alkylation. They were supplied in powder form without binder materials. The in-house mordenite zeolite sample was synthesised with different properties from those of the commercial mordenite samples, then post-synthesis modified by acid and steam dealumination techniques under different treatment conditions.

USY-A, USY-C, 320HOA and 350HOA samples were subjected to post-synthetic modification by treating with AHFS to remove any EFAL. Three series of bifunctional catalysts supported on 350HOA USY were also prepared: monometallic types containing only platinum or nickel were prepared by the ion-exchange method, while the method used for bimetallic ones containing both platinum and nickel was simultaneous competitive ion exchange.

Table 1 List of catalyst supports investigated in the TOS study

	Zeolite trade name	Zeolite type	Post-synthesis modification	Supplier
1	USY-A	USY	Mildly steamed	Crosfield
2	USY-C	USY	Severely steamed	Crosfield
3	320HOA	USY	Moderately steamed	Tosoh corporation
4	350HOA	USY	Moderately steamed and acid leached	Tosoh corporation
5	CP814E	BEA	As synthesised	Zeolyst international
6	Al-MCM-48	Mesoporous aluminosilicates	Aluminated by grafting	Claytec
7	Al-SBA-15	Mesoporous aluminosilicates	Aluminated by grafting	Claytec

Finally, two series of dealuminated mordenite catalysts were prepared by acid leaching under different conditions and by applying the steaming dealumination technique at different temperatures to the in-house synthesised mordenite in its ammonium form.

Hydrothermal synthesis of mordenite zeolite

The recipe for mordenite zeolite was obtained from a verified synthesis handbook [21]. Gel calculations to determine the required weights of each reactant were performed using the synthesis molar gel composition of 6 Na_2O:Al_2O_3:30 SiO_2:780 H_2O. Therefore, calculated weights of deionised water and sodium hydroxide (99 wt% NaOH) from Merck and sodium aluminate (50.9 wt% Al_2O_3 + 31.2 wt% Na_2O + 17.9 wt% H_2O) from Sigma-Aldrich were mixed in a plastic bottle until they were completely dissolved. The required amount of Ludox AS-40 colloidal silica (40 wt% silica + 60 wt% water) from Sigma-Aldrich was then added to the solution and all were stirred using a magnetic stirrer for 12 h, after which the mixture was poured into a Teflon-lined stainless-steel autoclave and heated in an oven at 180 °C for crystallisation. After 24 h crystallisation time, the mixture was taken out of the oven, decanted and centrifuged several times while washing with deionised water after each centrifugation until the pH of the decanted solution dropped below 9. Finally, the samples were dried at 110 °C overnight until they were ready for characterisation. The optimum mordenite phase with the highest purity and crystallinity was optimised against some synthesis parameters while keeping the composition ratio of Na_2O–Al_2O_3–SiO_2–H_2O constant at the above molar gel composition. Three factors were optimised: the ageing time of the synthesis gel, the crystallisation time and the crystallisation temperature. It was found that the optimum synthesis conditions were an ageing time of 12 h, crystallisation time of 24 h and crystallisation temperature of 180 °C. This procedure was repeated more than 10 times and reproducible results were obtained, in terms both of the yields, crystallinity, bulk and framework Si/Al ratio and of the crystal morphology and size.

Post-synthesis modification techniques

Ion exchange

The procedure of ion exchange is carried out simply by immersing zeolite in a solution of the salt of the targeted cation at room temperature or a higher temperature if the exchange rate is to be improved [22], since the Brønsted acid site is generated after the modification of the parent zeolite in its sodium form by first exchanging the sodium ion with ammonium, then calcining the zeolite in the ammonium form at high temperatures to transform it into the hydrogen form. One gram of the parent commercial zeolites in their sodium form—320HOA and CBV10A—and in-house as-synthesised mordenite zeolite were dissolved in 100 ml of 0.5 M ammonium nitrate, (99 wt% purity, Sigma-Aldrich) and heated at 80 °C under reflux conditions and constant stirring for 1 h. This process was performed three times (multiple ion exchange) assuming a complete uptake of ammonium in the catalysts. The mixtures obtained were centrifuged, washed with deionised water until pH was 7, then dried at 100 °C overnight. After three ion exchanges, residual Na was determined by inductively coupled plasma (ICP) elemental analysis to be below 0.05 wt%.

Acid leaching

The treatment of zeolites by acid leaching leads to dealumination, which generally improves their efficiency in processes that result in rapid catalyst deactivation through coking, such as cracking. Acid leaching results in an increase in the Si/Al ratio through the removal of tetrahedral aluminium atoms from the framework and the subsequent introduction of mesopores into the structure, thus enhancing the thermal stability of the zeolite. It has the advantage of not forming EFAL inside the zeolite structure, as happens when Al is removed by steaming [23–26]. Acid attack was carried out by heating a suspension of in-house as-synthesised mordenite in its ammonium form (NH_4-MOR) in nitric acid or hydrochloric acid (BDH Chemicals) under reflux conditions and constant stirring under the acid leaching conditions listed in Table 2. The suspension was then filtered, washed with distilled water and dried at 110 °C overnight until it was ready for characterisation.

Table 2 Conditions of acid leaching by mineral acids

Condition	NH_4-MOR (wt)	Acid type	Acid conc. (molar)	Acid, vol. (ml)	Leaching time (h)	Temp (°C)
1	1	HCl	6	50	12	25
2	1	HCl	6	50	24	25
3	1	HCl	6	100	24	25
4	1	HCl	12.1 (conc)	100	24	120
5	1	HNO_3	10	100	16	25
6	1	HNO_3	6	50	24	25
7	1	HNO_3	15.8 (conc)	100	24	25
8	1	HNO_3	15.8 (conc)	50	24	25
9	1	HNO_3	15.8 (conc)	100	24	120
10	1	HNO_3	6	100	24	120
11	1	HNO_3	6	100	48	120
12	2	HNO_3	10	100	24	120
13	1	HNO_3	10	100	16.5	120
14	1	HNO_3	10	100	48	120
15	1	HNO_3	10	100	87	120
16	1	HNO_3	10	100	168	120

Table 3 Conditions of acid leaching by AHFS

Sample	AHFS, wt (g)	320USY, wt (g)	Mixing time (h)	Temp (°C)	Mixture volume (ml)
Optimising the mixture volume					
F1	0.6	2	1	25	25
F2	0.6	2	1	25	50
F3	0.6	2	1	25	75
F4	0.6	2	1	25	100
Optimising the mixing time					
F2	0.6	2	1	25	50
F5	0.6	2	3.5	25	50
F6	0.6	2	7	25	50
F7	0.6	2	14	25	50
Optimising the AHFS weight					
F8	0.3	2	1	25	50
F2	0.6	2	1	25	50
F9	1.2	2	1	25	50
Optimising the mixture temperature					
F2	0.6	2	1	25	50
F10	0.6	2	1	50	50
F11	0.6	2	1	75	50
F12	0.6	2	1	100	50

AHFS solution was used to remove EFAL from the USY samples: USY-A, USA-C, 320HOA and 350HOA. The 320HOA sample containing 2 wt% sodium with the ammonium form as the starting material was fine tuned using AHFS as a chelating compound to provide the best conditions to remove only the EFAL whilst maintaining the aluminium framework and original zeolite crystallinity. Thus, four main factors were considered when optimising the chemical treatments: the AHFS and USY mixture volume, the treatment duration time, the zeolite-to-AHFS weight ratio and the temperature of chemical treatment, as shown in Table 3. The zeolite-to-AHFS mass ratio was optimised in the range of 1.67–6.67. For AHFS treatments, 2 g of the steamed zeolite USY samples was placed in AHFS (99.999 % trace metals basis, Sigma-Aldrich) solution and the suspension was heated to the required temperature and stirred whilst being heated with refluxing. The suspension was filtered, washed by warm deionised water to remove any residual fluoride species until pH was 7, and then dried at 100 °C overnight. The solid obtained was subjected to three ion exchanges in ammonium nitrate solution in order to eliminate sodium and any residual fluoride species.

Steaming

Steaming or hydrothermal treatment dealumination involves calcining zeolites in the ammonium (or hydrogen) form, in the presence of steam at high temperatures. This results in the removal of aluminium atoms from the zeolite framework and their replacement by silicon atoms, thus increasing the Si/Al ratio of the framework. However, the aluminium atoms which have been removed remain in the zeolite as EFAL species, which can lead to enhanced acidity through the formation of Lewis acid sites [23]. The formation of EFAL species occurs through the migration of aluminium from the framework into the outer surface of crystals and is rate-controlled, which means that it can be accelerated or slowed down by changing the steaming temperature [27]. The in-house as-synthesised NH_4-MOR was dealuminated by steaming at 500, 600 and 700 °C for 12 h at each steaming temperature, using a fixed-bed laboratory microreactor. Deionised water was fed to the reactor using a high-performance liquid chromatography (HPLC) pump at a controlled flow rate of 0.125 ml min^{-1}. Nitrogen carrier gas (BOC, oxygen free) was dried by passage through a bed of molecular sieves and fed into the reactor at a controlled flow rate of 187.5 ml min^{-1} using a mass flow controller. The catalyst (1 g, bed particle size of 40 and 60 mesh) was located in the reactor between two beds of glass wool in the centre of the heated stable zone of the furnace. In a typical experiment, the NH_4-MOR was heated from ambient temperature to the required reactor temperature of 500 °C at a rate of 3 °C min^{-1} in a flow of air (50 ml min^{-1}) at ambient pressure for 2 h. When these conditions had stabilised, the water was introduced to the reactor. At the end of the required time period (12 h for steaming), the water feed was stopped and the catalyst was then cooled in a stream of dry nitrogen.

Metal loading

Metals from group VIII, such as platinum and nickel, are commonly loaded onto the zeolites used in commercial refining processes as bifunctional catalysts, having both a hydrogenation/dehydrogenation function provided by the metal and an acidic function provided by the zeolite acid sites [28, 29]. A common method for preparing bifunctional catalysts is a multi-step process consisting mainly of the deposition of a metal precursor compound over the support surface, either by impregnation, ion exchange or mechanical mixing; the drying and calcination of the catalysts; and the transformation of the precursor compound into the active metallic phase by reduction. It is well established that the metal dispersion and thus the metal/acid ratio depend markedly on the way the metal is introduced [30]. The catalysts listed in Table 1 were loaded with 1 wt% platinum via the ion-exchange technique using an amino complex of platinum, and were fully described in a previous work [29–31]. The CP814E catalyst was loaded with 1 wt% platinum by wet impregnation using tetraammine platinum (II) chloride, at room temperature with excess solution under constant stirring overnight, followed by evaporation at 70 °C and drying at 110 °C. A series of monometallic and bimetallic bifunctional catalysts was formed by loading 350HOA USY with different amounts of platinum, nickel and platinum–nickel by competitive ion exchange. The first series comprised monometallic bifunctional catalysts containing 0.5, 1, 1.5 and 2 wt% of platinum, the second was a series of monometallic bifunctional catalysts containing 0.5, 1, 1.5, 2, 10 and 20 wt% of nickel and the third series was of bimetallic bifunctional catalysts containing 1 wt% of platinum and 0.25, 0.5, 0.75 and 1 wt% of nickel. The monometallic catalysts were prepared using the ion-exchange method, while the method used for the bimetallic catalysts was simultaneous competitive ion exchange, which means that the platinum and nickel precursors, $Pt(NH_3)_4Cl_2H_2O$ and $Ni(NO_3)_26H_2O$ (Fischer Chemical, 98 % purity), respectively, were mixed with the zeolite at the liquid-to-solid ratio of 50 (1 g of zeolite to 50 ml of the metal complex solution), using ammonium hydroxide solution to adjust the pH of the suspension obtained at >9.

Composite catalysts

Two composite bifunctional catalysts consisting of beta and USA zeolites were prepared as follows: (1) the 1 wt% Pt/CP814E and 1 wt% Pt/USY-C catalysts were mixed physically at a mass ratio of 1:1; (2) CP814E and USY-C in their ammonium forms were mixed at a mass ratio of 1:1 and were loaded by 1 wt% Pt using the ion-exchange method.

Catalyst characterisation

Characterisation techniques used were ICP optical emission spectroscopy (ICP–OES), X-ray diffraction (XRD), ^{27}Al and ^{29}Si MAS-solid-state NMR, pyridine adsorption Fourier transform infrared spectroscopy, ammonia temperature programmed desorption (NH_3-TPD), scanning electron microscopy and energy dispersive X-ray analysis (SEM), transmission electron microscopy (TEM), X-ray photoelectron spectroscopy (XPS), hydrogen chemisorption of platinum, BET surface area analysis by nitrogen adsorption, and thermogravimetric analysis (TGA).

Catalyst loading and testing

It was fully described in a previous work [31].

Results and discussion

Deactivation behaviour of 320HOA USY zeolite

320HOA USY zeolite was supplied by Tosoh Corporation in the form of a white powder with particle sizes between 6 and 10 μm [15]. Figures 1 and 2 show the XRD pattern and SEM microgram of 320HOA USY, respectively, and Table 4 shows the in-house characterisation of 320HOA USY using various techniques.

USY zeolite structure was confirmed from the XRD pattern for a fresh sample with 2 wt% sodium content as supplied, 320HOA USY has a lower crystallinity but a

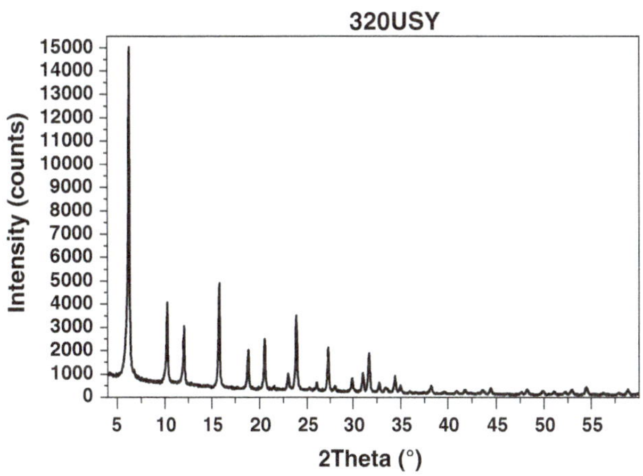

Fig. 1 XRD pattern of 320HOA USY

Fig. 2 SEM micrograph of 320HOA USY

higher unit cell size compared with CBV712 USY and CBV720 USY as explained in the previous work [31]. This can be related to the fact that 320HOA USY was produced via less steaming dealumination process and that no acid leaching was applied afterwards, which can be confirmed by the lower bulk Si/Al molar ratio. As a result, a reduced amount of unit cell size shrinking occurred. The rounded edged cubic crystallites, ranging in size between 1 and 3 µm, are combined with an amorphous phase which normally surrounds the zeolite Y crystals after the dealumination process, as is confirmed by the SEM images shown in Fig. 2. The ^{27}Al solid-state NMR for this USY catalyst, as shown in Fig. 3, indicated the presence of non-framework aluminium species as penta- and octa-coordinated aluminium, besides the tetrahedral aluminium framework. Four chemical environments were also obtained for Si, as shown in Fig. 4 for the ^{29}Si NMR solid state, which confirms that a shorter dealumination process was necessary to produce 320HOA USY. In addition, the Si/Al molar ratio for the framework was calculated to be 6.3, compared with 8.34 and 31.2 for CBV712 and CBV720 USY, respectively. Moreover, the ^{27}Al NMR peaks shapes and their corresponding chemical shifts were totally different from those for CBV712 and CBV720 USY, which may be explained by the different dealumination conditions that were applied to create these catalysts in the different structures. It can be concluded that most USY catalysts

Table 4 In-house characterisation of 320HOA USY

Elemental analysis (bulk)	Before ion exchange				After ion exchange with Pt precursor as tetraammine platinum (II) chloride $(Pt(NH_3)_4Cl_2).H_2O$			
	Si/Al mole ratio	2.90			Si/Al mole ratio	2.90	Pt wt%	0.83
XPS[a]	Si	Al	Si/Al	Pt	No Pt on the surface			
atomic conc %	24.42	10.89	2.24	0				
XRD	Crystallinity				Related to in-house as-synthesised Y			
	63.68							
NMR	Si/Al mole ratio of the framework				Four chemical environments of Si-NMR: Si(0Al), Si(1Al), Si(2Al) and Si(3Al)			
					Lewis acid sites existed: $Al(OSi)_4$, $(Al(H_2O)_5)^{3+}$ and $(Al(H_2O)_6)^{3+}$			
	6.30							
Ammonia TPD	NH_3 desorption through TPD/mmol g^{-1}				Ammonia adsorption calorimetry followed by temperature programmed desorption			
	0.30 ± 0.01							

Hydrogen chemisorption[b]	Metal dispersion	Metallic surface area (m^2 g^{-1} metal)	Metallic surface area (m^2 g^{-1} sample)	Crystallite size (nm)
	14.89	36.77	0.36	7.61

BET[b]	Surface area (m^2 g^{-1})	Pore volume (cm^3 g^{-1})		Pore size (Å)	Shows both micro- and mesoporosity	
	735.87	0.39		34.52		
TGA for coke content in spent catalyst (mass loss wt% in air) at different reaction temperatures	210 °C	Top	0.11	250 °C	Top	0.35
		Mid	0.14		Mid	0.31
		Bot	0.10		Bot	0.30
	230 °C	Top	0.24	270 °C	Top	0.10
		Mid	0.20		Mid	0.26
		Bot	0.13		Bot	0.15

[a] LPD Laboratory Services Ltd.

[b] MCA Services

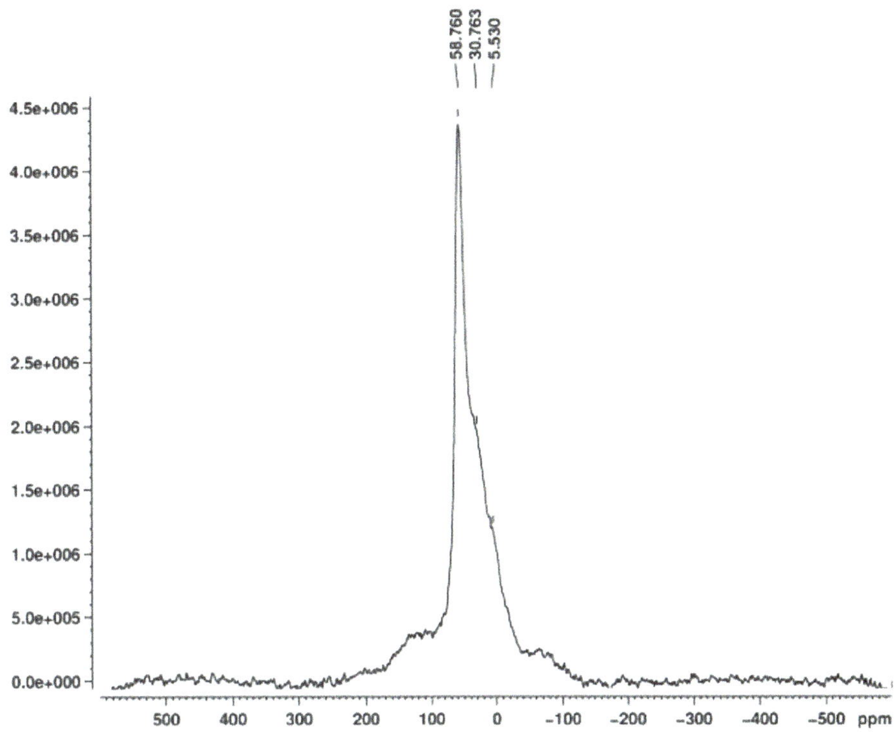

Fig. 3 ^{27}Al NMR spectra of 320HOA USY

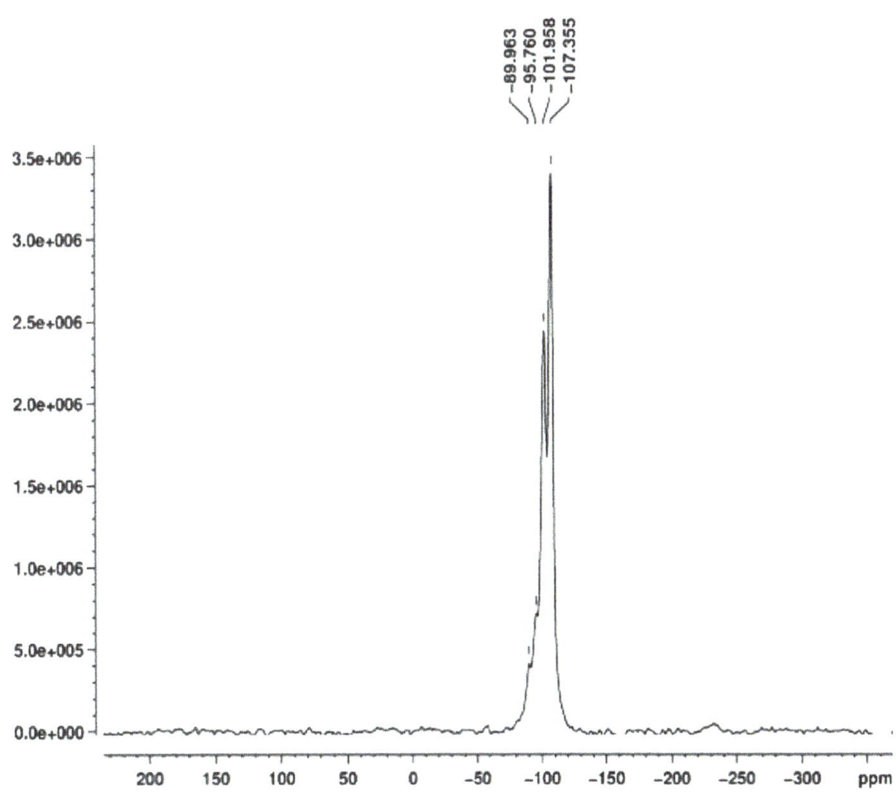

Fig. 4 ^{29}Si NMR spectra of 320HOA USY

have non-framework aluminium species; the nature and crystallographic locations within the zeolite cavities of USY catalysts are different from other catalysts, due to the dealumination conditions that have been applied to create them. Similar observations can be made for the ^{27}Al NMR peak shapes and the corresponding chemical shifts for USY-A compared with 320HOA USY as was described in [31]. They were all produced via steaming with no acid leaching, which is clearly reflected in their catalytic performance and deactivation behaviour.

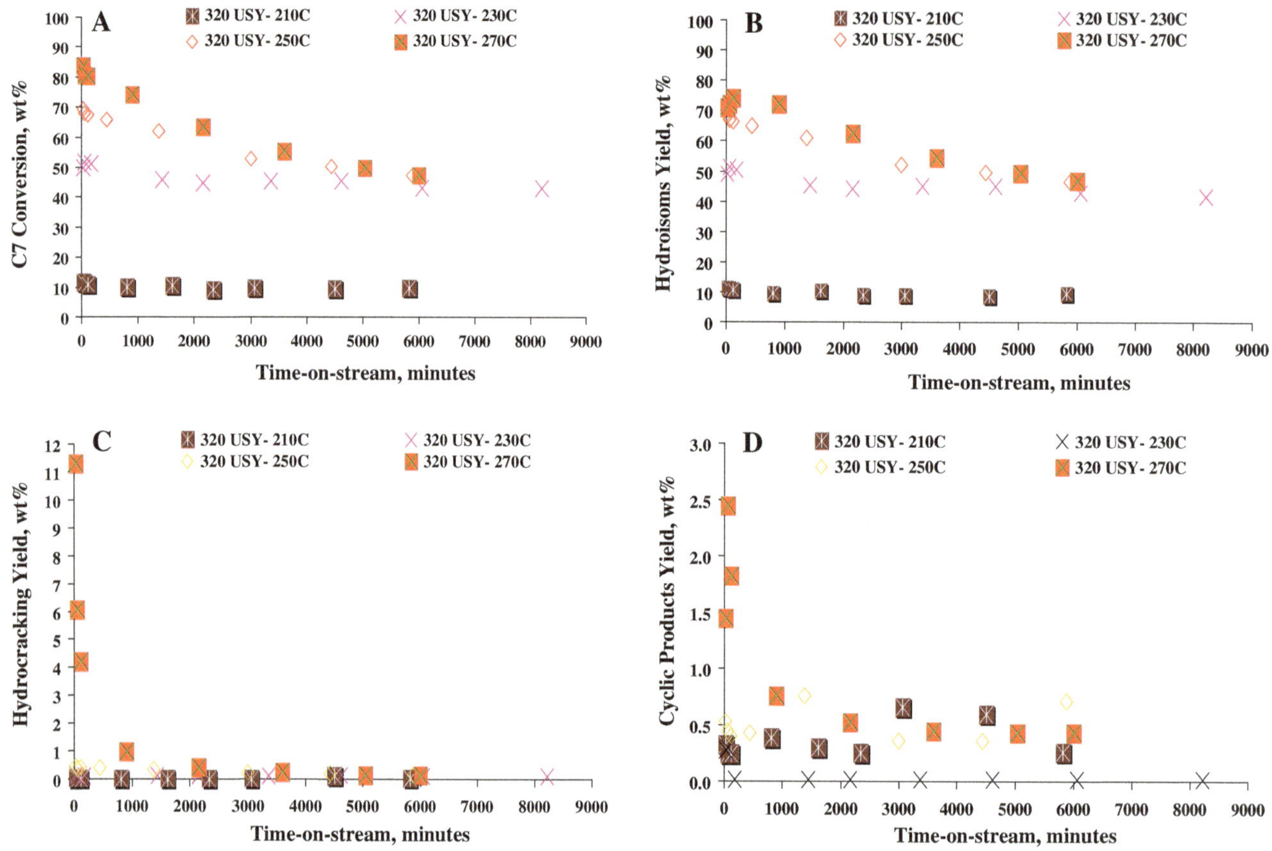

Fig. 5 Overall conversion (**a**), total yield of C7 isomers (mono + di + tribranched isomers) (**b**), total hydrocracking yield (**c**), total yield of the cyclic products (**d**) as a function of TOS for 320HOA USY at various reaction temperatures. WHSV = 5.13 h^{-1}; H_2/C7 (mol/mol) = 9; total pressure = 1 atm; temperature = 210, 230, 250 and 270 °C; 1 wt% Pt loading

The effect of TOS for the C7 conversion, hydroisomerisation, hydrocracking, and cyclic product yields at different reaction temperatures 210–270 °C are shown in Fig. 5a–d. Fresh catalyst was used for each run at each different temperature. Initially, the overall conversions were 12, 50, 70, and 84 wt% after 30 min following introduction of the feed to the reactor at the temperatures of 210, 230, 250, and 270 °C, respectively. At 210 °C, the catalyst lost about 1–2 wt% of its activity before levelling off and reaching a constant value after 60 h on stream, whereas the deactivation rate at 230 °C for the new activated catalyst was slightly faster, as can be seen in Fig. 5a. The overall conversion was 50 wt% after 30 min from the point at which the feed was introduced to the reactor, which then decreased with the TOS to 46 wt% after the first 24 h, and then decreased slowly within the deactivation rate range of 2–3 wt% every 24 h until the reaction terminated after about 6 days (140 h). A moderate drop in the catalyst activity occurred during the first few hours at the higher temperatures of 250 and 270 °C, after which a constant deactivating value was reached, ranging from 3 to 5 wt% every 24 h, until the reaction terminated. The initial selectivity towards the hydroisomerisation products was 98.5 and 84 % after 30 min following introduction of the feed to the reactor, which then increased throughout the TOS to 99 and 92 % after 120 min at 250 and 270 °C, respectively. The maximum isomer yield obtained was around 71 wt% at 270 °C and around 68 wt% at 250 °C, and selectivity towards cracked products was fairly high at the higher temperatures during the first few hours of the deactivation reaction. The highest selectivity for cracked products was 13.5 % 30 min after introducing the feed to the reactor at 270 °C, decreasing rapidly throughout the TOS and finally reaching 5 wt% after 120 min. The yield of cyclic products at the temperatures of 250 and 270 °C was also higher than at the lower temperatures.

It has been found that after the deactivation runs at temperatures of 210–270 °C, the hard coke content is lower than that found for the previously studied catalysts: CBV712USY, CBV720USY, and USY-A [31], which can be linked to the acidity properties of each catalyst. Thus, it can be seen that 320HOA USY has a lower acidity than the above catalysts resulting in a lower cracking activity and thus a lower coke formation. However, the negative effect of the presence of extra-framework Al species (EFAL) could be expected to be lower in 320HOA USY due to the pore

Table 5 The C7 conversion (wt%), product yield (wt%), and product selectivity (%) at different reaction conditions using the pressure rig for 1wt% Pt/320HOA USY

	Conversion (wt%)	Hydroisoms yield (wt%)	Hydrocrack yield (wt%)	Cyclic yield (wt%)	Hydroisoms selectivity (%)	Hydrocrack selectivity (%)	Cyclic selectivity (%)
WHSV (h^{-1})	T = 210 °C, P = 1 bar						
2.57	28.45	28.16	0.01	0.29	98.97	0.02	1.01
5.13	16.89	16.64	0.00	0.24	98.56	0.00	1.44
10.26	7.99	7.73	0.00	0.26	96.76	0.00	3.23
WHSV (h^{-1})	T = 210 °C, P = 8 bar						
2.57	3.54	3.25	0.00	0.29	91.73	0.00	8.27
5.13	4.27	3.98	0.00	0.29	93.14	0.00	6.86
10.26	0.52	0.27	0.00	0.25	52.70	0.00	47.30
WHSV (h^{-1})	T = 210 °C, P = 15 bar						
2.57	0.46	0.24	0.00	0.22	51.84	0.00	48.16
5.13	0.77	0.54	0.00	0.23	69.70	0.00	30.17
10.26	3.97	3.72	0.00	0.26	93.51	0.00	6.49
Pressure (bar)	T = 210 °C, WHSV = 5.13 h^{-1}						
1	16.89	16.64	0.00	0.24	98.56	0.00	1.44
8	4.27	3.98	0.00	0.29	93.14	0.00	6.86
15	0.77	0.54	0.00	0.23	69.70	0.00	30.17
WHSV (h^{-1})	T = 230 °C, P = 1 bar						
2.57	46.38	45.89	0.16	0.33	98.94	0.35	0.71
5.13	45.47	45.06	0.08	0.33	99.10	0.17	0.72
10.26	26.11	25.36	0.03	0.28	97.16	0.11	1.06
WHSV (h^{-1})	T = 230 °C, P = 8 bar						
2.57	32.35	32.00	0.04	0.31	98.91	0.13	0.97
5.13	19.90	19.56	0.03	0.31	98.29	0.16	1.55
10.26	9.08	8.80	0.00	0.28	96.94	0.00	3.05
WHSV (h^{-1})	T = 230 °C, P = 15 bar						
2.57	16.95	16.60	0.02	0.33	97.94	0.09	1.97
5.13	10.92	10.64	0.00	0.28	97.43	0.00	2.55
10.26	6.28	6.02	0.00	0.26	95.81	0.00	4.17
Pressure (bar)	T = 230 °C, WHSV = 5.13 h^{-1}						
1	45.47	45.06	0.08	0.33	99.10	0.17	0.72
8	19.90	19.56	0.03	0.31	98.29	0.16	1.55
15	10.92	10.64	0.00	0.28	97.43	0.00	2.55
WHSV (h^{-1})	T = 250 °C, P = 1 bar						
2.57	78.83	76.38	1.82	0.62	96.90	2.31	0.79
5.13	68.83	67.52	0.85	0.47	98.09	1.23	0.68
10.26	39.93	38.96	0.55	0.43	97.55	1.36	1.08
WHSV (h^{-1})	T = 250 °C, P = 8 bar						
2.57	70.07	68.91	0.83	0.32	98.35	1.19	0.46
5.13	52.62	51.99	0.66	0.31	98.81	1.25	0.58
10.26	30.51	30.05	0.19	0.27	98.50	0.61	0.89
WHSV (h^{-1})	T = 250 °C, P = 15 bar						
2.57	52.05	51.45	0.30	0.31	98.83	0.58	0.59
5.13	42.85	42.32	0.25	0.29	98.76	0.58	0.68
10.26	25.38	24.94	0.15	0.29	98.28	0.58	1.14

Table 5 continued

	Conversion (wt%)	Hydroisoms yield (wt%)	Hydrocrack yield (wt%)	Cyclic yield (wt%)	Hydroisoms selectivity (%)	Hydrocrack selectivity (%)	Cyclic selectivity (%)
Pressure (bar)	$T = 250$ °C, WHSV = 5.13 h^{-1}						
1	68.83	67.52	0.85	0.47	98.09	1.23	0.68
8	52.62	51.99	0.66	0.31	98.81	1.25	0.58
15	42.85	42.32	0.25	0.29	98.76	0.58	0.68
WHSV (h^{-1})	$T = 270$ °C, $P = 1$ bar						
2.57	78.02	71.93	4.18	1.91	92.19	5.36	2.45
5.13	72.53	69.68	1.88	0.97	96.07	2.59	1.34
10.26	41.78	40.09	0.96	0.73	95.94	2.31	1.75
WHSV (h^{-1})	$T = 270$ °C, $P = 8$ bar						
2.57	66.87	63.21	3.27	0.39	94.53	4.89	0.58
5.13	54.19	50.28	3.65	0.26	92.79	6.74	0.48
10.26	30.19	28.25	1.67	0.27	93.57	5.54	0.88
WHSV (h^{-1})	$T = 270$ °C, $P = 15$ bar						
2.57	58.01	54.98	2.71	0.32	94.78	4.68	0.54
5.13	43.24	40.59	2.37	0.29	93.86	5.47	0.66
10.26	29.63	27.47	1.90	0.26	92.69	6.42	0.89
Pressure (bar)	$T = 270$ °C, WHSV = 5.13 h^{-1}						
1	72.53	69.68	1.88	0.97	96.07	2.59	1.34
8	54.19	50.28	3.65	0.26	92.79	6.74	0.48
15	43.24	40.59	2.37	0.29	93.86	5.47	0.66

blockage effect, especially in the polymeric species [24], such that the coke content can be trapped within the pores since these will be smaller than those found for USY-A.

Effects of reaction pressure and contact time for 1 wt% Pt/320HOA USY

The effect of increasing the hydrogen partial pressure, and hence the overall reaction pressure and variation in contact time, representing the time taken by reactants to pass through the catalyst bed during a reaction, has been studied in this work at a range of temperatures from 210 to 270 °C for the hydroisomerisation of n-C7 over 1 wt% platinum-loaded 320HOA USY. A stainless-steel higher pressure unit was used, and the catalyst (1 wt% Pt/320HOA USY) was tested at temperatures ranging from 210 to 270 °C, WHSV of 2.57, 5.13, and 10.26 h^{-1}, and pressures of 1, 8 and 15 bar. At each temperature, products of the reaction were collected at the point where steady-state conditions were achieved; catalysts were calcined and then re-reduced before proceeding to the next temperature. Unfortunately, the reaction rig could not operate overnight at high pressures due to the necessary safety precautions and therefore it was not possible to fully study deactivation behaviours at these pressures.

Table 5 shows the results from the n-C7 conversion (wt%), product yield (wt%), and product selectivity (%) at the different reaction conditions using the pressure rig for the hydroisomerisation of n-C7 over 1 wt% Pt/320HOA USY. In general, it has been found that an increase in pressure results in a decrease in overall conversion and a slight decrease in selectivity to isomers with a noticeable increased selectivity to the cyclic products, particularly at the lower temperatures of 210 and 230 °C. There was no obvious cracking activity at the lower temperatures of 210 and 230 °C, and a constant WHSV was observed whilst changing reaction pressures. In contrast, at the highest temperature of 270 °C and WHSV of 5.13 h^{-1}, increased selectivity towards the cracked products was observed with increased reaction pressures, with comparable results being reported by Chao et al. [32] who investigated the effect of reaction pressure over the range 1–41.3 bar on mordenite catalysts with different Si/Al ratios and the zeolite Beta. In addition, they showed that pressure reduced the cracking yield for all catalysts as well as improved the mordenite catalysts' stability. Additionally, an increase in hydrogen pressure has been found to reduce the hydroisomerisation reaction rate [33], where an increase in the hydrogenation activity at higher hydrogen pressure results in the hydrogenation of intermediate olefins, due to a shorter intermediate olefin residence time inside the catalyst [34]. This minimises the cracking activity, since pressure reduces overall conversion but improves hydroisomerisation selectivity compared with that of cracking, which to some extent, was not the case for catalyst 1 wt% Pt/320HOA

Fig. 6 XRD pattern for 350HOA USY

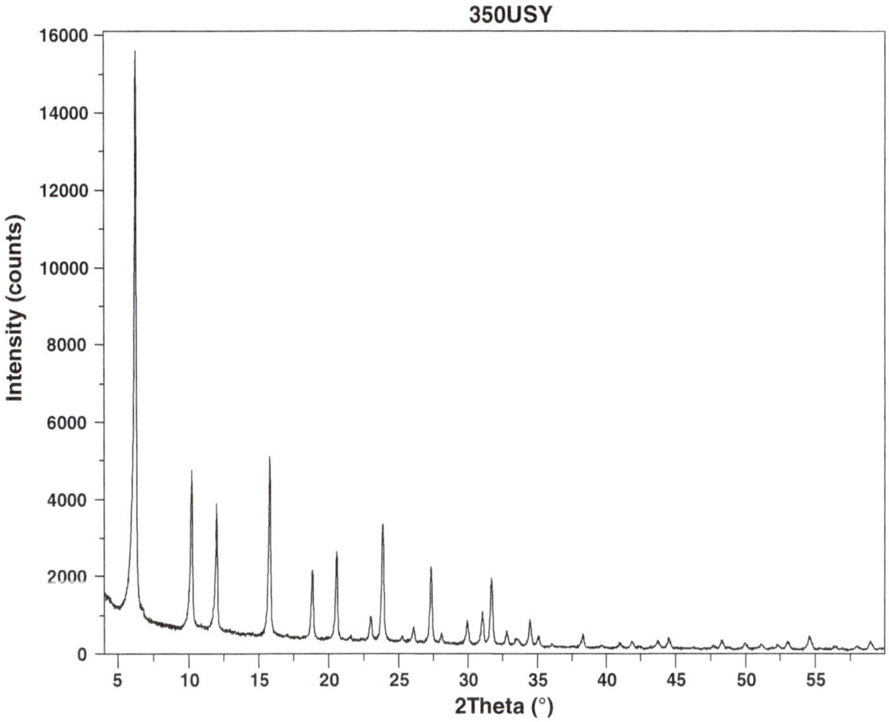

USY, whose small average pore diameter most likely limited the movement of larger molecules at high pressures, and therefore promoted cracking.

Table 5 illustrates that the conversion increased as the contact time increased (WHSV $\leq$ 10.26, 5.13, 2.57 h^{-1}) at constant temperature and pressure. However, increasing contact time between the reaction feed (n-C7) and the catalyst appears to cause a reduction in cyclic product yield, giving rise to an increase in cracking and hydroisomerisation selectivity.

Wang et al. [35] found that increasing the contact time results in an increased overall conversion of n-C7 at a constant temperature of 220 °C and in a reduced selectivity in hydroisomerisation versus cracking. Similar observations were made by Chica and Corma [36] when they tested the hydroisomerisation of nC_5, nC_6 and nC_7 over platinum-loaded zeolite supports. However, the decrease in selectivity to isomers was opposed by an increase in the overall yield of isomers at short contact times. It was also shown that with increasing contact times, the formation of multi-branched isomers also increases, leading to a lower mono/multi-branched isomer ratio. This behaviour could result from an initial formation of mono-branched isomers and their subsequent transformation into multi-branched isomers.

Deactivation behaviour of 350HOA USY zeolite

350HOA USY zeolite was supplied by Tosoh Corporation in the form of a white powder with particle sizes between 6 and 10 μm. The Figs. 6 and 7, respectively, show the XRD pattern and SEM microgram for 350HOA USY, and Table 6 the in-house characterisation for 350HOA USY using different techniques.

Fig. 7 SEM micrograph for 350HOA USY

Figures 8 and 9 show solid-state NMR spectra for ^{27}Al and ^{29}Si, respectively. ^{27}Al MAS NMR analysis revealed that the catalyst had EFAL species even though it was a dealuminated and partially acid-leached zeolite. Information regarding the Si atoms coordinated (via oxygen) with neighbouring T atoms (Si and Al) is provided by ^{29}Si MAS NMR, indicating two different chemical configurations of SiO_4, which were assigned to (4Si,0Al) and SiO_4 (3Si,1Al).

This confirms that 350HOA was produced by a dealumination process and acid leached to achieve a high Si/Al

Table 6 Some in-house characterisations of 350HOA USY

Elemental analysis (bulk)	Before ion exchange				After ion exchange with Pt precursor as tetraammine platinum (II) chloride ($Pt(NH_3)_4Cl_2).H_2O$			
	Si/Al mole ratio	5.55			Si/Al mole ratio	5.55	Pt wt%	0.90
XPS[a]	Si	Al	Si/Al	Pt	Pt is likely to be metallic			
Atomic Conc %	28.32	5.69	4.98	0.07				
XRD	Crystallinity %				Related to in-house as-synthesised Y			
	66							
NMR	Si/Al mole ratio for the framework				Two chemical environments for Si-NMR: Si(0Al) and Si(1Al)			
	12.38				Lewis acid sites are present			
Ammonia TPD	NH_3 desorption through TPD/ mmol g^{-1}				Ammonia adsorption calorimetry followed by temperature programmed desorption			
	0.76 ± 0.01							
BET[b]	Surface area (m^2 g^{-1})	Pore volume (cm^3 g^{-1})		Pore size (Å)	Shows both micro and mesoporosity			
	872.97	0.47		26.10				
TGA for coke content in spent catalyst (mass loss wt% in air) after different reaction temperatures	210 °C	Top	0.10	250 °C	Top	0.26		
		Mid	0.13		Mid	0.37		
		Bot	0.13		Bot	0.28		
	230 °C	Top	0.52	270 °C	Top	1.20		
		Mid	0.76		Mid	1.12		
		Bot	0.71		Bot	1.29		

LPD Laboratory Services Ltd.

[b] MCA Services

mole ratio; thus a high surface area was obtained, as there were fewer EFAL species. This would result in a mesoporous channel network, due to the larger pore volume. Moreover, it should be noted that the dealumination process required to produce 350HOA USY was less severe than for the CBV720USY catalyst [31], which has a higher Si/Al mole ratio in the framework than 350HOA USY. Nevertheless, both catalysts have mesoporous systems, due to larger pore volumes and surface areas, but less acidity and thus lower catalytic activity compared to just-dealuminated catalysts.

The effects of TOS on the C7 conversion, hydroisomerisation, hydrocracking, and cyclic product yields at different reaction temperatures 210–270 °C are shown in Fig. 10a–d, respectively. Fresh catalyst was used for each run at the different reaction temperatures. Initially, the overall conversion was 17, 56, 75, and 84 wt% after 30 min from the point when the feed was introduced to the reactor at the temperatures of 210, 230, 250, and 270 °C, respectively. At 210 °C, the catalyst lost about 1–2 wt% of its activity and then reached a pseudo-stable state for the initial 60 min on stream. The deactivation rate at 230 °C for the new activated catalyst was slightly faster, with the overall conversion being 56 wt% after 30 min after which it decreased with the TOS to 51 wt% after 27 h, and then decreased slowly within the deactivation rate range of 2–3 wt% every 24 h until the reaction finally terminated after about 6 days (140 h). Rapid initial deactivation occurred during the first few hours at the higher temperatures of 250 and 270 °C, followed by a constant deactivating value being reached, ranging from 3 to 5 wt%, every 24 h until the reaction terminated.

The initial selectivity towards the hydroisomerisation products was 98 and 85 % after the initial 30 min, and then increased throughout the TOS to 99 and 94 % after 120 min at 250 and 270 °C, respectively. Thus, the selectivity towards cracked products was fairly high at higher temperatures during the first few hours of the deactivation reaction, with the highest cracked product selectivity being 14 % 30 min after introducing the feed to the reactor at 270 °C; however, this decreased rapidly throughout the TOS to 4 % after 120 min. Additionally, the yield of cyclic products at the temperatures of 210 and 230 °C was higher than that at the higher temperatures.

The TOS results provide evidence that the higher activity of 350HOA USY compared to CBV720 USY [31] is related to the number of acid sites. Furthermore, moderately slow deactivation for both catalysts was observed, reaching the pseudo-stable state in the initial few minutes at the lower temperatures of 210 and 230 °C. The n_{Pt}/n_A ratio for 350HOA USY was calculated as 0.06, which is in the suggested range of $0.03 \leq n_{Pt}/n_A \leq 0.17$ for a well-

Fig. 8 ^{27}Al NMR spectra for 350HOA USY

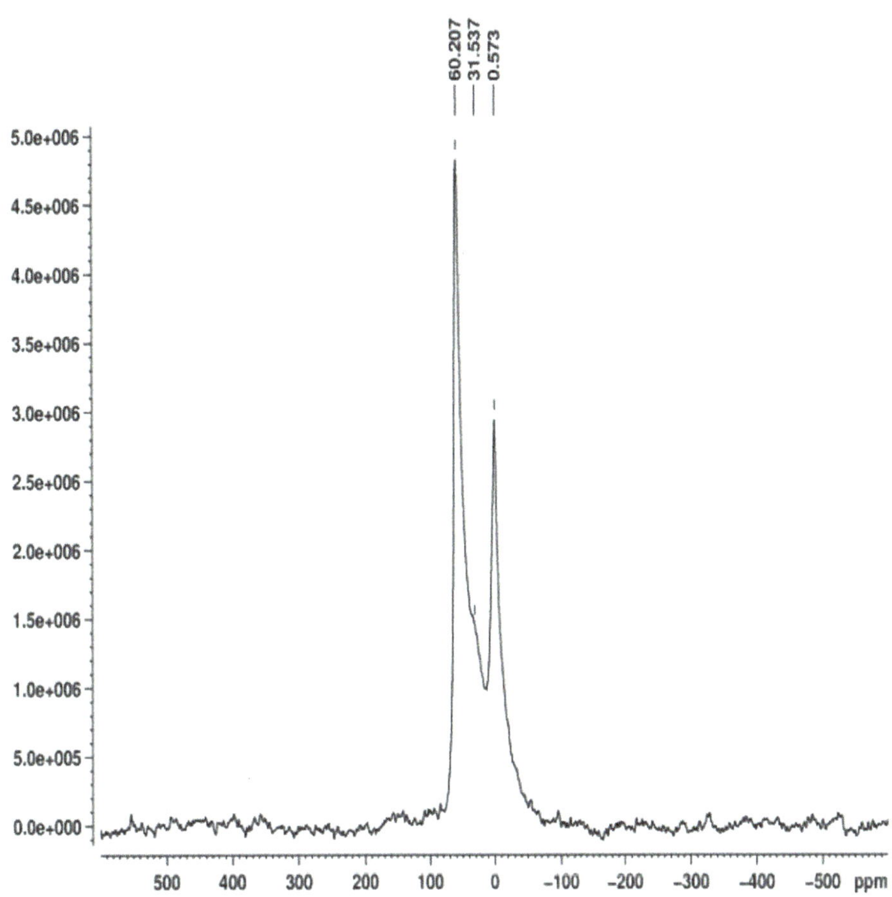

balanced bifunctional catalyst, in conjunction with 0.12 for CBV720 USY, which allowed deactivation to occur at reasonably low temperatures with lower coke content. Although the mesoporosity texture was similar for both catalysts, the average coke content was higher for 350HOA USY, which may be due to its higher acidity, as the framework Si/Al was lower, resulting in cracking activity at temperatures of 250 and 270 °C.

Extraction of EFAL in USY zeolites by AHFS post-synthetic modification

Ammonium hexafluorosilicate solution was used to remove the EFAL from the USY samples: USY-A, USA-C, 320HOA, and from the acid-leached 350HOA, due to the samples having different silicon-to-aluminium ratios, and thus different numbers of aluminium atoms as non-framework on their crystal unit cells. As with dealumination, after the process the aluminium removed from the framework remains on the crystal and migrates to the surface in an extension that is dependent on the partial pressure of the steam used. This can be confirmed by Si/Al ratio results on the above USYs surfaces by XPS analysis as shown in Tables 4 and 6 and it can be seen that the Si/Al ratios on the surfaces are much lower for zeolites that have only been dealuminated, such as USY-A, USY-C, and 320HOA than for the acid-leached zeolites such as CBV712, CBV720 and 350HOA. The resulting AHFS modified zeolite samples were characterised by XRD, MAS NMR, ammonia TPD, ICP, XPS and BET, with the effects of the treatments on the total amount and type of EFAL, the relationship between amount and nature of acid sites, and the deactivation behaviour discussed below.

Optimisation of chelating extraction parameters

The USY sample 320HOA containing 2wt% sodium with the ammonium form as the starting material has been fine tuned using a chelate compound (ammonium hexafluorosilicate $(NH_4)_2SiF_6$) to provide the best condition to remove only the non-framework aluminium whilst maintaining the aluminium framework and original zeolite crystallinity—as these are unaffected by this chelating agent. Four main factors were considered when optimising the chemical treatments: (1) the AHFS and USY mixture volume, (2) the treatment duration time, (3) the zeolite to AHFS weight ratio, and (4) the temperature used during the chemical treatment. Table 7 lists the resulting modified sample characteristics for each factor and illustrates, for the AHFS modified samples, that the increase in mixture volume

Fig. 9 ^{29}Si NMR spectra for 350HOA USY

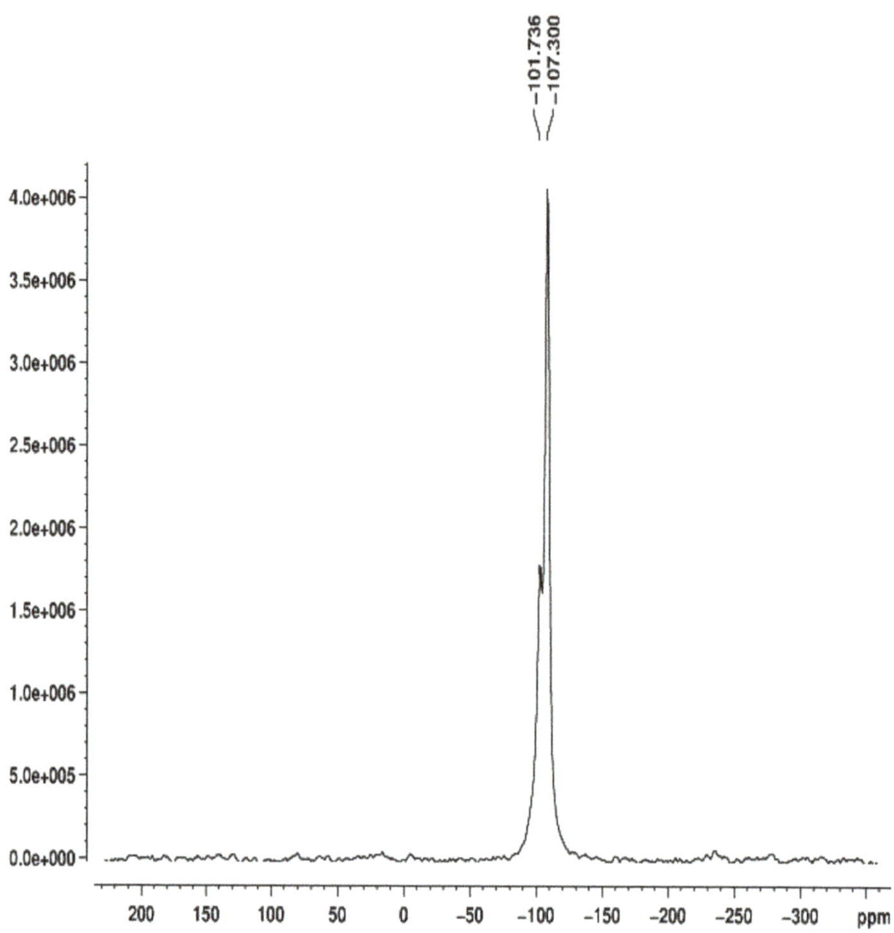

caused the total aluminium content in the bulk to decrease from 9.73 wt% before the modification to between 4.83 and 5.51 wt%. Additionally, both the Si/Al mole ratio in the framework and the crystallinity were maintained in the ranges of 6.45–6.75 mol ratio and 57–68 %, respectively. There was no significant effect on the mixing time factors, such that the Si/Al mole ratio in the framework and the crystallinity were maintained in the ranges of 6.45–6.75 and 57–66 %, respectively. The zeolite to AHFS mass ratio was optimised in the range of 6.67–1.67, so the framework structure had been attacked by the AHFS at a ratio of 1.67 mol, leading to a Si/Al mole ratio in the framework which increased to 11.10.

A temperature range of 25–100 °C was used during the chemical treatment and it was found that the resulting modified samples had the total aluminium content in the range of 4.79–5.56 wt%, and crystallinity in the range of 57–64 %. As well as this, the framework was unaffected as the temperature rose, leaving the Si/Al in the framework in the range of 6.43–6.53 mol ratio. It can be seen that the total Si content in the modified samples is higher than that detected for the original sample 320HOA before the acid chelating treatment, which can explained by the existence of the silicon fluoride containing species on the outer surfaces of the modified zeolites, an effect that has been reported in the literature [27, 37]. The detected residual fluoride species was removed from the structure by applying an extra ammonium ion-exchange step.

Table 8 shows the unit cell for the AHFS modified samples, the starting material sample (320HOA), the total aluminium number Al_{total} (based on the elemental analysis), the framework aluminium number Al_F (measured using ^{29}Si MAS NMR $(Si/Al)_F$, and on the basis of 192 T atoms (Si or Al) per unit cell with the composition of $H_x(AlO_2)_x(SiO_2)_{192-x}$) and the EFAL number Al_{nF}, (the difference between the Al_{total} and Al_F) [26, 27, 38, 39]. To some extent, the value for $Al_{F/u.c}$ was greater than the value for the total $Al_{total/u.c}$ as shown for the samples designated as F9, F10, and F22 in the table, so this difference reflects the errors in analyses for some AHFS modified samples which may have resulted from the existence of silicon fluoride containing species during the chemical treatment. The Al_{nF} extraction efficiency percentage was calculated for each resulting sample by dividing the difference between Al_{nF} for the modified sample and Al_{nF} for the starting material, which is equal to 23 atoms for 320HOA.

It is clear from Table 8 that the most effective extraction of EFAL by AHFS chelating agent was for the samples F4,

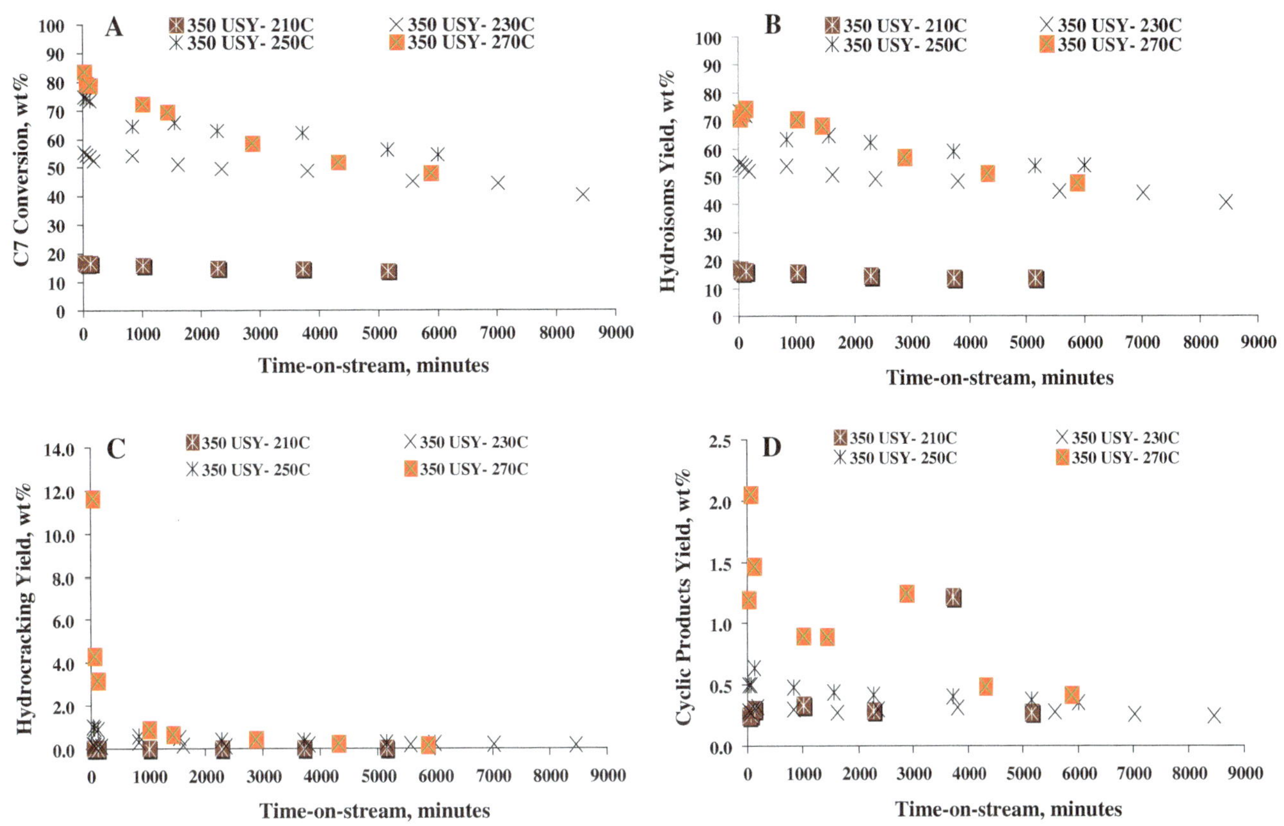

Fig. 10 Overall conversion (**a**), total yield of C7 isomers (mono + di + tribranched isomers) (**b**), total hydrocracking yield (**c**), and total yield of the cyclic products (**d**) as a function of TOS for 350HOA USY at various reaction temperatures. WHSV = 5.13 h^{-1}; H_2/C7 (molar) = 9; total pressure = 1 atm; temperature = 210, 230, 250 and 270 °C; 1 wt% Pt loading

F5, F6, F7, and F11. The results for the starting material (320HOA) are arranged in the table according to the chemical treatment conditions applied, where the EFAL formed represents approximately 47 % of the total aluminium. Furthermore, the obtained $(Si/Al)_T$ as bulk (measured by the chemical analysis) is almost in agreement with $(Si/Al)_F$ measured by NMR analysis for those samples (F4, F5, F6, F7, and F11) which has 100 % extraction efficiency. The lowest extraction efficiency was found to be 46 % for the modified sample F8, which was prepared by the chemical treatment using the highest mass ratio (6.67 wt/wt) of zeolite to AHFS, which can be considered as the mildest condition, with moderate extraction efficiency observed for the modified samples F1, F2, F3, and F12 in which partial extraction occurred, leaving the framework aluminium unaffected, and maintaining the framework $(Si/Al)_F$ ratio as prior to modification. This resulted in reasonable crystallinity values in the range of 61–66 % compared with the original crystallinity of 64 % for the starting material.

Catalysts USY-A, USY-C, 320HOA USY, and 320HOA USY were chemically treated by AHFS to remove the EFAL as these have a detrimental effect on catalytic and transport properties, and consequently adversely affects the coke formation and deactivation behaviour. The EFAL was removed from the catalysts in their ammonium form using the chelating agent AHFS under the same chemical treatment conditions as sample F2. This involved 50 ml of a solution mixture, stirred for 1 h at room temperature, with a zeolite to AHFS mass ratio of 3.33 wt/wt. From Table 9, it can be seen that samples F20 and F21, the modified samples of catalysts USY-A and USY-C, respectively, still have EFAL which has not been extracted by AHFS, and that framework Al atoms were attacked. In contrast, F2 was partially extracted but was not attacked by the AHFS. It can be interpreted that F2 was produced by the AHFS treatment of 320HOA USY which in its ammonium form contains approximately 2 wt% sodium, and thus the presence of sodium inhibits aluminium extraction [27]. Table 9 shows the in-house characterisation for the original and modified catalysts using different techniques, and it can be seen that the $(Si/Al)_F$ increased for modified samples F20 and F21, confirming that the AHFS has dealuminated the framework Al atoms, and subsequently decreased their crystallinity. It was also observed that the zeolites prepared with AHFS retained less than 0.6 wt% fluorine based on the ion chromatography analysis, which was not seen by XPS analysis measurements for the

Table 7 The chemical treatment parameters of AHFS and the resulting samples characteristics

	The chemical treatment conditions					Elemental analysis ICP			MAS NMR	XRD
	AHFSwt (g)	320USYwt (g)	Mixing time (h)	Temp (°C)	Vol (ml)	Si (wt%)	Al (wt%)	Bulk (Si/Al) (mol/mol)	Framework (Si/Al) (mol/mol)	Crystallinity (%)
320 USY	–	–	–	–	–	29.36	9.73	2.9	6.41	63.68
Optimising the mixture volume										
F1	0.6	2	1	25	25	30.01	5.51	5.23	6.71	63.57
F2	0.6	2	1	25	50	30.71	5.22	5.65	6.45	57.27
F3	0.6	2	1	25	75	31.18	5.33	5.62	6.75	66.47
F4	0.6	2	1	25	100	32.37	4.83	6.44	6.23	68.46
Optimising the mixing time										
F2	0.6	2	1	25	50	30.71	5.22	5.65	6.45	57.27
F5	0.6	2	3.5	25	50	33.12	4.75	6.70	6.75	66.28
F6	0.6	2	7	25	50	33.73	4.81	6.74	6.60	62.91
F7	0.6	2	14	25	50	33.72	4.6	7.04	6.59	66.69
Optimising the AHFS weight										
F8	0.3	2	1	25	50	30.94	7.37	4.03	6.51	60.57
F2	0.6	2	1	25	50	30.71	5.22	5.65	6.45	57.27
F9	1.2	2	1	25	50	37.68	1.68	21.55	11.10	67.11
Optimising the mixture temperature										
F2	0.6	2	1	25	50	30.71	5.22	5.65	6.45	57.27
F10	0.6	2	1	50	50	33.37	4.58	7.00	6.51	63.53
F11	0.6	2	1	75	50	32.5	4.79	6.52	6.43	64.47
F12	0.6	2	1	100	50	32.16	5.56	5.56	6.53	61.13

samples surface. In addition, the XPS measurements revealed that the surface atomic concentration for the aluminium atoms decreased for the AHFS modified samples; thus, it can observed that the $(Si/Al)_F$ mole ratio measured by NMR, (Si/Al) measured by the elemental analysis, and Si/Al measured by XPS present a closer match to each other, particularly in the case of sample F2 as opposed to samples F21 and F22 in which the Al_F was attacked by the AHFS. In terms of the obtained texture of the samples treated with AHFS, some mesoporosity developed within the crystal as the pore volume increased and pore size decreased, thus the total surface area was consequently increased for all of the modified samples.

Figure 11a–d shows the effect of TOS on the n-C7 conversion, hydroisomerisation, hydrocracking and cyclic product yields at the reaction temperature of 230 °C for the samples F2, F20 and F21, which are the AHFS modified samples for the catalysts 320HOA USY, USY-A, and USY-C, respectively. Figure 11a shows that the initial overall conversion was 63 wt% after 30 min from introducing the feed to the reactor at the temperature of 230 °C for F21 compared to 50 wt% for 320HOA USY before the modification, and that the catalyst lost about 1–2 wt% of its activity before reaching a constant "level-off" value after 60 h on stream. The deactivation rate for the F2 catalyst was slightly faster than the parent catalyst along the TOS to be in the range of 3–4 wt% and then reached a constant "level-off" value until the reaction terminated after about 4 days (5,580 min).

Figure 11b shows that the initial selectivity towards the hydroisomerisation products was 98 % for F2 compared to 99 % for 320HOA USY before the modification and kept almost the same along the deactivation run. As demonstrated in Fig. 11c and d, the selectivity towards cracked and cyclic products was fairly high for F2 in comparison with 320HOA USY which may be due to the n_{Pt}/n_A ratio being disturbed as a consequence of increasing the number of acid sites for F2.

Figure 12a shows that the initial overall conversion was 60 wt% after 30 min from the point when the feed was introduced to the reactor at the temperature of 230 °C for F20 compared to 70 wt% for USY-A before the modification [31], which is then decreasing with the TOS to 43 wt% after 780 min and decreasing slowly within the deactivation rate ranging between 1 and 3 wt% every 24 h until the reaction terminated after almost 4 days. In comparison with USY-A, which was completely deactivated in the initial few minutes, F20 deactivated much more steadily with TOS, which may be explained by the effect of pore blockage by the polymeric EFAL species being overcome even in the presence of some EFAL due to further dealumination occurring during the acid leaching

Table 8 Al content of AHFS modified samples and extraction efficiency

	Al_{total}/u.c	Al_F/u.c	Al_{nF}/u.c	Al_{nF} %	Extraction efficiency (%)
320USY	49	26	23	47	–
F1	31	25	6	19	75
F2	29	26	3	11	87
F3	29	25	4	15	82
F4	26	26	0	0	100
F5	25	25	0	0	100
F6	25	25	0	0	100
F7	24	24	0	0	100
F8	38	26	13	33	46
F9	9	16	*	*	–
F10	24	26	*	*	–
F11	26	26	0	0	100
F12	29	26	4	13	84
USY-A	50	24	26	52	**
F20	24	18	6	25	**
USY-C	51	22	29	57	**
F21	24	15	9	38	**
350USY	29	14	15	52	**
F22	5	13	*	*	–

* This difference reflects the errors in the analyses. ** Al_F attacked by AHFS

via the AHFS chelating agent. Figure 12b illustrates that the initial selectivity towards the hydroisomerisation products was 97 % for F20 compared to 92 % for USY-A before the modification and was maintained fairly steadily along the deactivation run. As shown in Fig. 12c, d, the selectivity towards cracked and cyclic products was fairly high for F20 compared with USY-A, which may be due to the n_{Pt}/n_A ratio being disturbed as a consequence of increasing the number of the acid sites (n_A) for F20, and also due to the presence of some EFAL species.

From the Fig. 13a, the initial overall conversion was 36 wt% after 30 min from introducing the feed to the reactor at the temperatures of 230 °C for F21 compared to 46 wt% for USY-C before the modification, and then decreasing with the TOS to 32 wt% after 1,000 min compared to 43 wt% after the same period of time, and decreasing slowly within the deactivation rate ranging between 1 and 3 wt% every 24 h until the reaction was terminated after almost 4 days. Figure 13b shows that the initial selectivity towards the hydroisomerisation products was 99 % for F21, which was the same as USY-C before the modification, and kept almost same along the deactivation run. As shown in Fig. 13c and d, the selectivity towards cracked and cyclic products was fairly similar for F21 compared with USY-C.

The influence of nickel on the stability of Pt/350HOA USY in the hydroisomerisation of n-C7

The catalytic properties of a bifunctional catalyst depend on the balance between its acidity and its hydrogenating functions. With respect to its selectivity and stability, increasing the proportion of the hydrogenation function at constant acidity leads to an improved catalyst performance, and for this reason, a series of monometallic and bimetallic bifunctional catalysts was formed by loading the H-USY (350HOA USY, with a global Si/Al = 5 mol/mol, supplied by Tosoh Corporation) with different amounts of platinum, nickel and platinum–nickel by competitive ion exchange, then investigating the effects of nickel as a second metal to Pt on the stability of the Pt/350HOA USY catalyst in the hydroisomerisation of n-C7 at 230 °C and 1 atm using H_2/C7 = 9 and WHSV = 5.13 h^{-1}. The first series comprised monometallic bifunctional catalysts containing 0.5, 1, 1.5 and 2 wt% of platinum supported on H-USY (350HOA USY), the second was a series of monometallic bifunctional catalysts containing 0.5, 1, 1.5, 2, 10 and 20 wt% of nickel, also supported on 350HOA USY, and the third series was of bimetallic bifunctional catalysts containing 1 wt% of platinum and 0.25, 0.5, 0.75 and 1 wt% of nickel, again supported on 350HOA USY. The monometallic catalysts were prepared using the ion-exchange method, while the method used for the bimetallic catalysts was simultaneous competitive ion exchange, which means that the platinum and nickel precursors, $Pt(NH_3)_4Cl_2H_2O$ and $Ni(NO_3)_26H_2O$, respectively, were mixed together with the zeolite, at the liquid-to-solid ratio of 50 (1 g of zeolite to 50 ml of the metal complex solution), using ammonium hydroxide solution to adjust the pH of the suspension obtained at >9 in order to enhance the uptake and homogenous distribution of the metals within the zeolite grain.

Table 10 shows some characteristics of the mono- and bimetallic bifunctional catalysts so formulated using characterisation techniques such as ICP, XRD, NMR and TGA. Some results were also obtained using XPS and TEM. The quantitative uptake of the loaded metals was verified by elemental analysis using inductively coupled plasma spectroscopy for the hydrofluoric acid digested zeolite samples. It was observed that the uptake of metals by the zeolite support was less than the calculated value. It has been reported that the USY zeolite in its ammonium form had greater ion-exchange efficiency than the zeolite H-USY, due to the greater reduction in the pH of the reaction medium in the latter case [40]. The pH was adjusted for this reason, even though the desired molar ratio for the bimetallic catalyst Pt–Ni/USY was obtained. Using X-ray diffraction, it was found that all the diffraction patterns of the metal-loaded catalysts were identical to that of the parent sample of 350HOA USY. Moreover, no

Table 9 In-house characterisation of original catalysts and those modified by AHFS

	AHFSwt (g)	USYwt (g)	Mixing time (h)	Temp (°C)	Vol (ml)	Si (wt%)	Al (wt%)	F (wt%)	Pt (wt%)	Bulk Si/Al, molar ratio	Framework Si/Al, molar ratio	Crystal (%)	NH_3 TPD (mmol g^{-1})	(Si/Al), Atomic Conc %(XPS)	Pt, Atomic conc % (XPS)	Surface area (m^2/g)	Pore volume (cm^3/g)	Pore size (Å)
320USY	–	–	–	–	–	29.36	9.73	–	0.83	2.90	6.41	63.68	0.3	2.24	0	735.87	0.39	34.52
F2	0.6	2	1	25	50	30.71	5.22	0.56	0.94	5.65	6.45	57.27	1.05	5.32	0	818.98	0.45	31.43
USY-A	–	–	–	–	–	26.75	9.71	–	1.01	2.87	7.11	99.16	0.86	2.22	0.03	755.67	0.42	27.99
F20	0.6	2	1	25	50	33.57	4.53	0.03	0.87	7.12	9.51	70.17	1.11	10.23	0.03	774.25	0.42	26.66
USY-C	–	–	–	–	–	28.51	9.80	–	0.99	2.79	7.84	98.35	0.60	1.25	0.03	519.16	0.37	41.84
F21	0.6	2	1	25	50	35.84	4.81	0.05	0.86	7.16	11.93	63.38	1.50	3.36	0	720.30	0.43	28.78
350USY	–	–	–	–	–	34.04	5.89	–	0.90	5.55	12.38	66.39	0.76	4.98	0.07	872.97	0.47	26.10
F22	0.6	2	1	25	50	39.39	1.06	0.05	0.84	36.19	13.69	72.58	0.49	16.16	0	917.30	0.46	22.12

diffraction peaks were assigned to Pt or Ni, which may indicate high dispersion of these metals. The crystallinity results obtained were in the range of 60–78, versus 66 % for the parent material. The TEM images for the 1 wt% Pt–1 wt% Ni, 2 wt% Pt and spent 2 wt% Pt/350USY catalysts are shown in Fig. 14. The black dots appearing on the support matrix are assumed to be Pt or Pt–Ni particles. The average size of the visible metal particles was found to be in the range of 3–7 nm for the Pt–Ni bimetallic catalyst and in the range of 3–30 nm for the monometallic samples, as measured manually and averaged [41], indicating that the presence of nickel and platinum in the bimetallic catalysts enhanced greatly the formation of metal particles and reduced their sizes. It was also found that the presence of platinum improved the nickel reduction in comparison to the presence of nickel only [40]. As in the faujasite zeolite structure (zeolite Y), the cavities had a maximum diameter of 1.3 nm, which means that the observed metal particles were located outside the zeolite cavities. Nevertheless, these metallic particles could have been located within the mesopore channels which were created during the hydrothermal dealumination of the Y zeolite, or at the outer surface of the catalyst grains, as has been reported [40]. It has also been suggested that there are likely to be smaller metal particles inside the cavities which are not detectable by TEM [42, 43]. It was observed that the average metallic particle size in the case of the 2 wt% Pt/350USY was unchanged by the deactivation reaction, which may indicate that no metal sintering occurred; however, this conclusion cannot be assured, as TEM was able to study only limited regions of the samples and consequently the results should be interpreted as mainly local [43].

The Figs. 15, 16, and 17 show the effect of TOS on the n-C7 conversion, hydroisomerisation, hydrocracking and cyclic product yields at the reaction temperature of 230 °C for the first, second and third series of monometallic and bimetallic bifunctional catalysts containing platinum and nickel, supported on H-USY (350HOA USY). Figure 15a indicates that the initial overall conversion values were 47, 56, 57, and 51 wt% for the monometallic catalysts containing Pt of 0.5, 1, 1.5, and 2 wt%, respectively, after 30 min from introducing the feed to the reactor at the temperature of 230 °C. All the catalysts underwent small decreases in conversion with TOS up to 12 h, after which there was a significant decrease in conversion for the catalyst containing 0.5 wt% Pt. The catalyst containing 2 wt% Pt was found to have the smaller drop in conversion, while that containing 0.5 wt% Pt underwent greatest fall in activity during the TOS. The higher sustainability of the catalyst containing 2 wt% Pt can be explained by the formation of catalytically active metallic particles and better balance between metal and acid sites, as shown in Table 10.

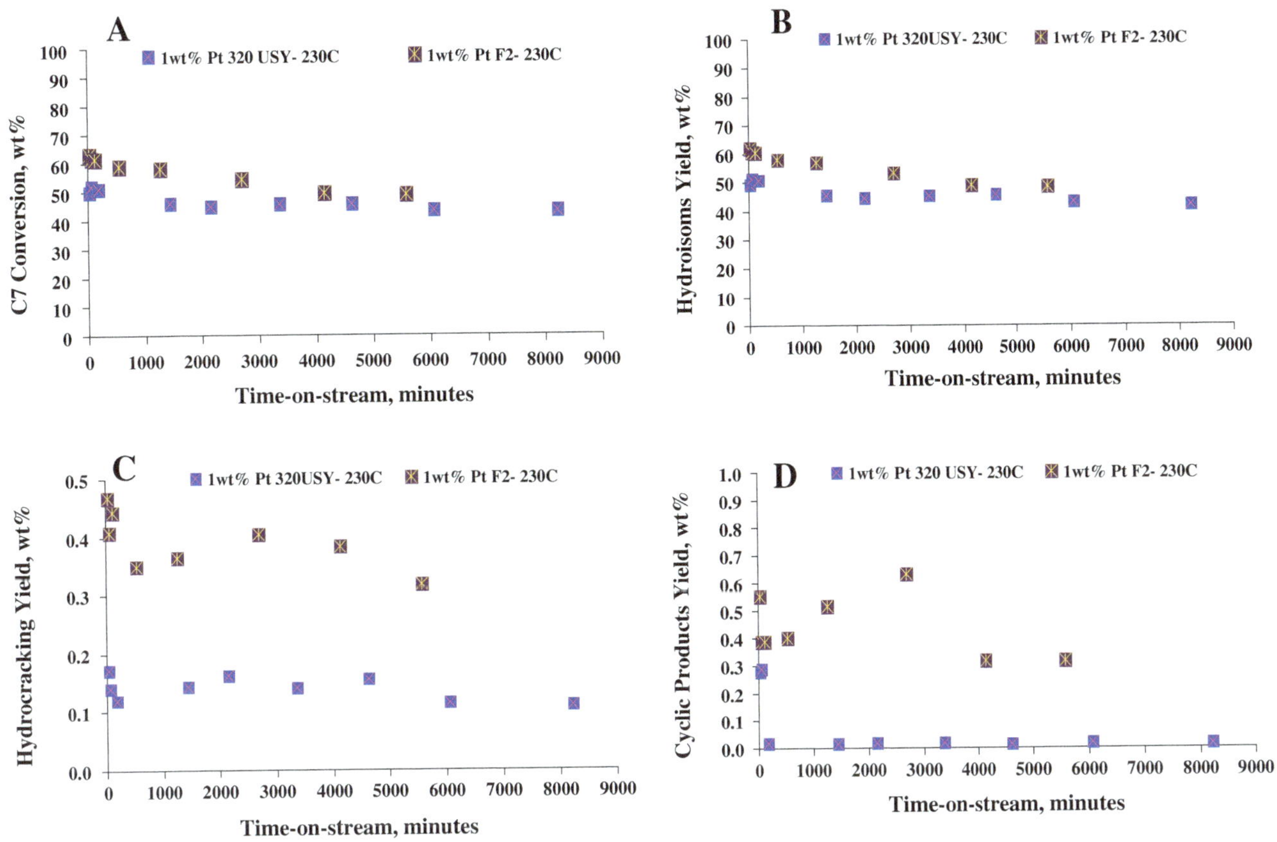

Fig. 11 Overall conversion (**a**), total yield of C7 isomers (mono + di + tribranched isomers) (**b**), total hydrocracking yield (**c**), and total yield of the cyclic products (**d**) as a function of TOS for F2 catalyst at various reaction temperatures. WHSV = 5.13 h^{-1}; H_2/C7 (mol/mol) = 9; total pressure = 1 atm; temperature = 230 °C; 1 wt% Pt loading

Figure 15b shows that the initial selectivity towards the hydroisomerisation products ranged between 98 and 99 % for all the catalysts, remaining almost unchanged throughout the deactivation run. Furthermore, as shown in the Fig. 15c and d, the selectivity towards cracked and cyclic products respectively was fairly in the range of 1 and 2 % for all the catalysts even if their n_{Pt}/n_A ratio is quite different. Nevertheless, n_{Pt}/n_A ratio was found to be within the optimum balance limits of 0.03 and 0.17 ($0.03 \leq n_{Pt}/n_A \leq 0.17$) for the most catalysts, which may reflect the great dehydrogenation–hydrogenation function capacity of platinum even at the lower contents [40, 41, 44]. The average hard coke contents were found to be in the range of 0.1–0.2 wt% for all the catalysts except that containing 1 wt% platinum, for which the coke content was about 0.7 wt%, which may due to the effect of the longer TOS when compared with the other catalysts.

Figure 16a shows that 30 min after introducing the feed into the reactor at 230 °C, the initial overall conversion rates were 52 and 59 wt%, respectively, for the monometallic catalysts containing 10 and 20 wt% Ni. After 180 min on stream, there was a large drop in activity to 30 and 45 wt%, then a sustained and rapid deactivation rate until a pseudo-stable state was reached at activity levels of 10 and 16 wt%, respectively, persisting until the reactions terminated. Deactivation was also found to be very rapid for the monometallic catalysts containing Ni at 0.5, 1, 1.5 and 2 wt%. This behaviour may be explained in terms of the poor hydrogenating capacity of Ni compared to Pt [44], making it more difficult for the Ni atoms to be fully reduced, so that not enough Ni metallic sites were available to sustain the reaction, since the hydroisomerisation reaction includes a dehydrogenation step [40, 42]. Therefore, selectivity to form the C7 isomers was lower than for the catalysts containing Pt, while higher cracked and cyclic product yields were obtained, which means that higher coke content was formed, possibly blocking and poisoning the active sites of the catalysts during the deactivation reaction. In addition, the low stability of catalysts with higher nickel content may be explained by the tendency of Ni to polymerise olefins and hence to form coke [42].

Figure 17a shows that the initial overall conversion rates were 66, 72, 66 and 67 wt% for the bimetallic bifunctional catalysts containing 1 wt% Pt and 0.25, 0.5, 0.75 and 1 wt% Ni, respectively, 30 min after introducing the feed into the reactor. All of these catalysts showed small decreases in conversion with TOS up to 12 h, after which there was a continuous moderate decrease in conversion

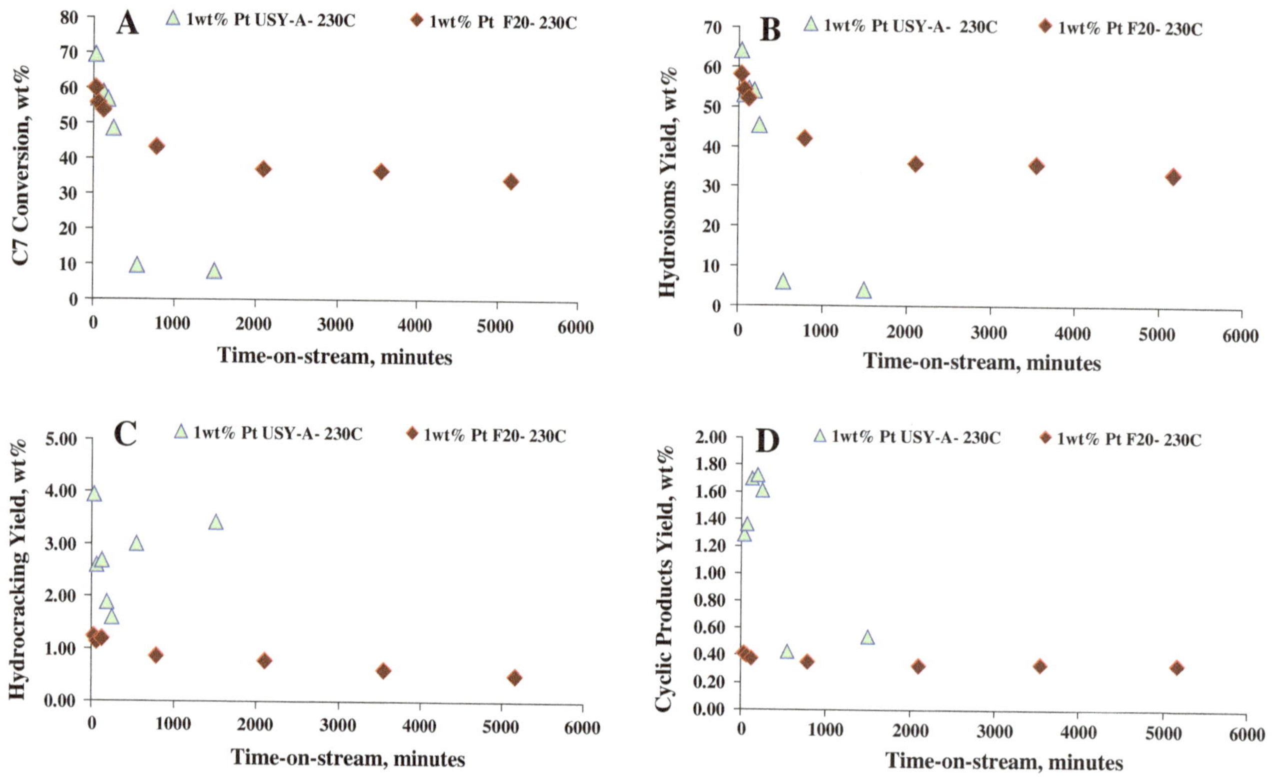

Fig. 12 Overall conversion (**a**), total yield of C7 isomers (mono + di + tribranched isomers) (**b**), total hydrocracking yield (**c**), and total yield of the cyclic products (**d**) as a function of TOS for F20 catalyst at various reaction temperatures. WHSV = 5.13 h^{-1}; $H_2/C7$ (mol/mol) = 9; total pressure = 1 atm; temperature = 230 °C; 1 wt% Pt loading

until the reactions terminated. Figure 17b indicates that initial selectivity towards the hydroisomerisation products ranged between 98 and 99 % for all of the catalysts, remaining almost constant throughout the deactivation run. Furthermore, as shown in the Fig. 17c and d, selectivity towards cracked and cyclic products respectively was broadly in the range of 1–2 % for all of the catalysts. On the other hand, the maximum hard coke content was found in the bimetallic catalyst containing 1 wt% Pt and 1 wt% Ni, as shown in the Table 10.

It can be concluded that a small addition of Pt motivated a substantial increase in hydrogenating activity when compared with the catalysts containing only Ni, as reported in the literature [30]. It also appears that the presence of platinum enhanced the reduction of the nickel cations, increasing the dispersion of the two metals as the nickel particles provided support for the Pt atoms, indicating that the Pt atoms were inserted into the nickel crystallite [40, 42]. On the other hand, it was found that increasing Ni loadings inhibited both the activity and selectivity of the catalyst, which may be interpreted as indicating that the formation of larger bimetallic Ni–Pt particles at higher Ni loadings resulted in some Ni content not being reduced, thus contributing to pore blockage of the catalyst and reducing the availability of catalyst acid sites accordingly [44].

Effect of platinum loading method on CP814E beta

Bifunctional zeolite catalysts typically have two types of active centres, the hydrogenation/dehydrogenation sites provided by the metal (such as platinum) for the formation of olefinic intermediates, and the acidic site for isomerisation and cracking. In the case of Pt-loaded zeolite catalysts in the ammonium form, the method most commonly used to introduce platinum into the pores and obtain a homogeneous filling in the crystallites of the zeolite is cationic exchange, using platinum tetraammine $[Pt(NH_3)_4]^{2+}$ complexes in an aqueous solution in the presence of a competing cation, generally an ammonium ion, NH_4^+. A second method to introduce platinum in the pores is incipient wetness impregnation by which the platinum tetraammine will randomly be deposited in the zeolite pores whereas only one platinum tetraammine complex can be associated with an aluminium site, and thus an atomic dispersion of the complexes inside the pores is expected with the ion-exchange method. Therefore, more dispersed platinum and consequently smaller particles using the ion-exchange method would be produced than via the impregnation method [27, 29, 45]. The purpose of this study was to examine the effects of platinum loading of beta zeolite by ion-exchange and impregnation methods, examining the

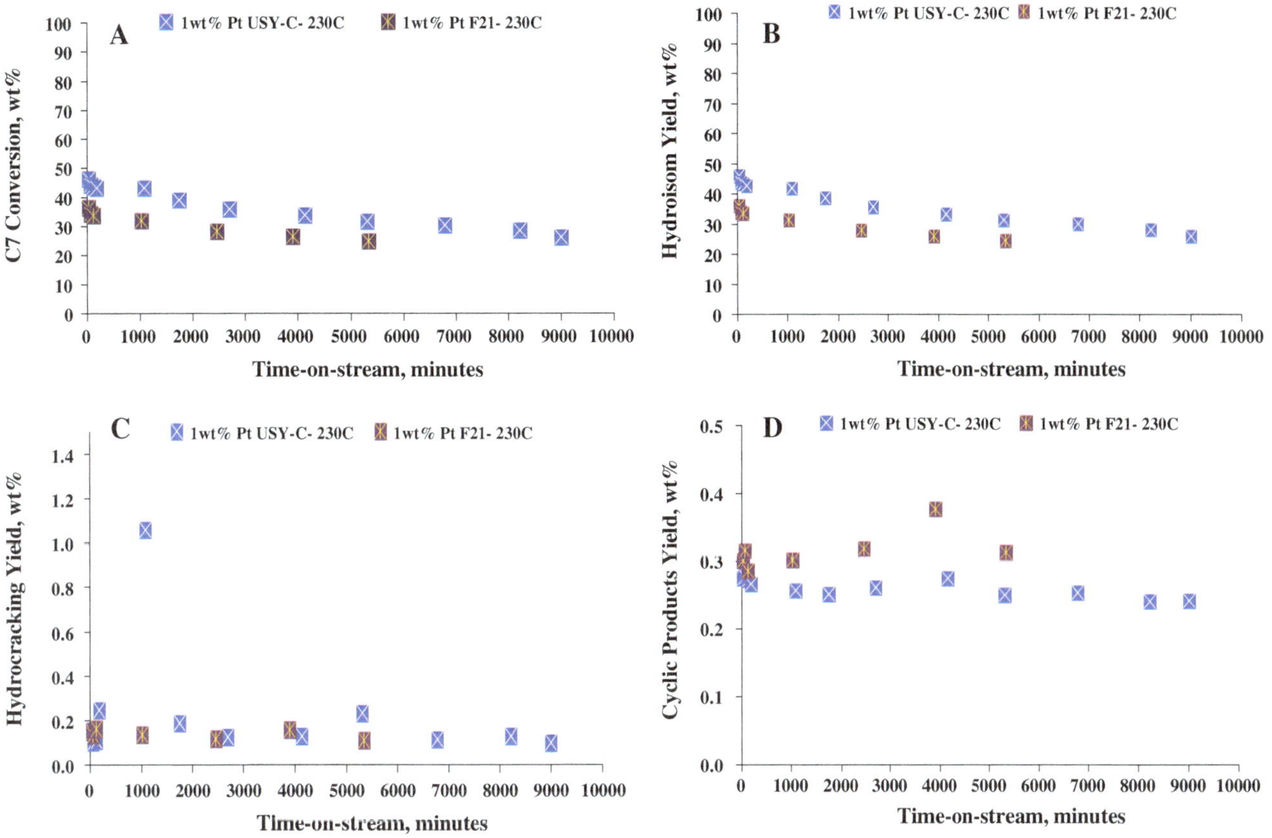

Fig. 13 Overall conversion (**a**), total yield of C7 isomers (mono + di + tribranched isomers) (**b**), total hydrocracking yield (**c**), and total yield of the cyclic products (**d**) as a function of TOS for F21 catalyst at various reaction temperatures. WHSV = 5.13 h^{-1}; H_2/C7 (mol/mol) = 9; total pressure = 1 atm; temperature = 230 °C; 1 wt% Pt loading

TOS stability of 1wt% Pt-loaded CP814 beta, where its deactivation behaviour was studied in the previous work [31], in the hydroisomerisation of n-C7 at constant reaction conditions and the effect of pH on the ion-exchange process.

Figure 18a–d shows the effects of TOS for the C7 conversion, hydroisomerisation, hydrocracking, and cyclic product yields for the catalyst with 1 wt% Pt/CP814E in which the platinum was loaded by one of three procedures: (1) preparation of the ion-exchange system using an aqueous solution of tetraammine platinum (II) chloride complex mixed with the ammonium-beta zeolite CP814E and the addition of the competing cation NH_4^+ using ammonia solution at pH > 9, (2) preparation of the ion-exchange system using an aqueous solution of tetraammine platinum (II) chloride complex mixed with the ammonium-beta zeolite CP814E without adding ammonia solution such that the pH = 4–5, and (3) preparation by wet impregnation using an aqueous solution of tetraammine platinum (II) chloride complex mixed with the ammonium-beta zeolite CP814E without adding ammonia solution, such that the pH = 4–5. As Fig. 18a shows, initially the overall conversion of C7 for the three catalysts was 77, 88, and 90 wt% after 30 min from the point when the feed was introduced to the reactor at 230 °C, prepared by the three procedures described above. The catalysts lost about 1–2 wt% of their activity before reaching a pseudo stable state for the initial 60 min on stream, and then slightly decreased with a further increase in time until the reactions were finally terminated. On the other hand, the selectivity towards hydroisomerisation products was 99, 85, and 85 % after 30 min and then increased with time to 99, 89, and 89 % for the three catalysts, respectively. Furthermore, the greatest selectivity towards cracked products was 15 % and slightly decreased with TOS to approximately 9 % for the catalysts which were prepared by the ion-exchange method with pH = 4–5 and by the wet impregnation method. The selectivity towards the cyclic products was very low for the three catalysts along the deactivation runs.

The variation in C1/C3 mass ratio with TOS for the three catalysts at constant reaction conditions is shown in Fig. 19, with both the acid and metal functions being initially active and deactivated rapidly as indicated by the C1/C3 ratio. This could be explained by a change occurring in the balance between the two functions with TOS due to coke depositions for both functions. Furthermore, in all the

Table 10 Different metal loadings wt% on 350HOA USY

	Al (wt%)	Si (wt%)	Bulk (Si/Al) (mol/mol)	Framework (Si/Al) (mol/mol)	Crystallinity (%)	Actual Ni (wt%)	Actual Pt (wt%)	Theoretical Ni (wt%)	Theoretical Pt (wt%)	Ni/Pt (mol/mol)	n_{Pt}/n_A	Hard coke (wt%)
350HOA USY	5.89	34.04	5.55	12.38	66	–	–	–	–	–	–	–
Pt/350HOA USY												
0.5 %Pt	–	–	–	–	–	–	0.36	–	0.50	–	0.02	0.19
1 %Pt	5.89	34.04	5.55	12.38	66	–	0.90	–	1.00	–	0.06	0.66
1.5 % Pt	–	–	–	–	–	–	1.17	–	1.50	–	0.08	0.13
2 % Pt	–	–	–	–	–	–	1.46	–	2.00	–	0.10	0.14
Pt-Ni/350HOA USY												
1 % Pt–0.25 %Ni	5.29	27.50	5.00	–	60	0.22	0.90	0.25	1.00	0.81	0.11	0.47
1 % Pt–0.5 % Ni	5.29	27.60	5.01	–	63	0.44	0.91	0.50	1.00	1.61	0.16	0.62
1 % Pt–0.75 %	5.31	27.65	5.00	–	61	0.68	0.92	0.75	1.00	2.46	0.21	0.43
1 % Pt–1 % Ni	5.31	27.97	5.06	11.57	60	0.92	0.91	1.00	1.00	3.36	0.27	1.01
Ni/350HOA USY												
0.5 % Ni	–	–	–	–	–	0.44	–	0.50	–	–	0.10	–
1 % Ni	–	–	–	–	–	0.88	–	1.00	–	–	0.20	–
1.5 % Ni	–	–	–	–	–	1.28	–	1.50	–	–	0.29	–
2 % Ni	–	–	–	11.03	–	1.86	–	2.00	–	–	0.42	–
10 % Ni	4.74	27.74	5.62	10.86	73	7.48	–	10.00	–	–	1.68	–
20 % Ni	4.28	24.83	5.57	10.58	–	11.96	–	20.00	–	–	2.68	–

three catalysts, the metal function deactivated faster than the acid function; however, this was not severe enough to cause a drop in the n-C7 conversion. The worst metal/acid balance was found for the catalysts that were prepared by ion exchange with the pH = 4–5 and by the wet impregnation method, as indicated by the lower values of C1/C3, and as such higher cracking activity was observed for those two catalysts. Moreover, the two catalysts demonstrated poor metal dispersion resulting in some of the metal being deposited on the outside of the crystals and not within the zeolite pores, as compared to the catalyst that was prepared by ion exchange with the pH > 9.

Effect of CP814E composites

The zeolite-based composite materials consisting of two distinct phases with entirely different pore structure and acidity have been used extensively both as catalysts, and to minimise the formation of coking, thus increasing the lifetime of the catalyst. Beta zeolite can also be used as an additive to FCC cracking catalysts because it favours reactions that improve the octane number of gasoline since the feedstocks consist of bulky poly-nuclear molecules which cannot diffuse inside the zeolite pores. Composite catalysts made of a zeolitic and a non-zeolitic active oxide/poly-oxide as a matrix can be used to crack the bulky feed molecules into smaller molecules which can diffuse inside the zeolite pores providing a two-phase system of the composite that operates synergistically. This synergism is a result of the open structure of the matrix that allows easy intra-particle diffusion and favours the primary cracking of the bulky feed molecules, as a result of the balance between the acidic sites and the metallic sites due to the dispersion of Pt throughout the matrix [27, 46, 47].

In this work, TOS experiments were carried out on the hydroisomerisation of n-C7 using hybrid catalysts consisting of CP814E and USY-C at identical reaction conditions, and thus the relation between zeolite pore structure and its resistance to coke deactivation is examined for the composite catalysts formed. Two composite bifunctional catalysts were prepared as follows: (1) the catalysts 1 wt% Pt/CP814E and 1 wt% Pt/USY-C were mixed physically at mass ratio of 1:1. (2) CP814E and USY-C in their ammonium forms were mixed at mass ratio of 1:1, and were loaded by 1 wt% Pt using the ion-exchange method.

The effect of TOS on the C7 conversion, hydroisomerisation, hydrocracking, and cyclic product yields at reaction temperature of 230 °C are shown in the Fig. 20a–d,

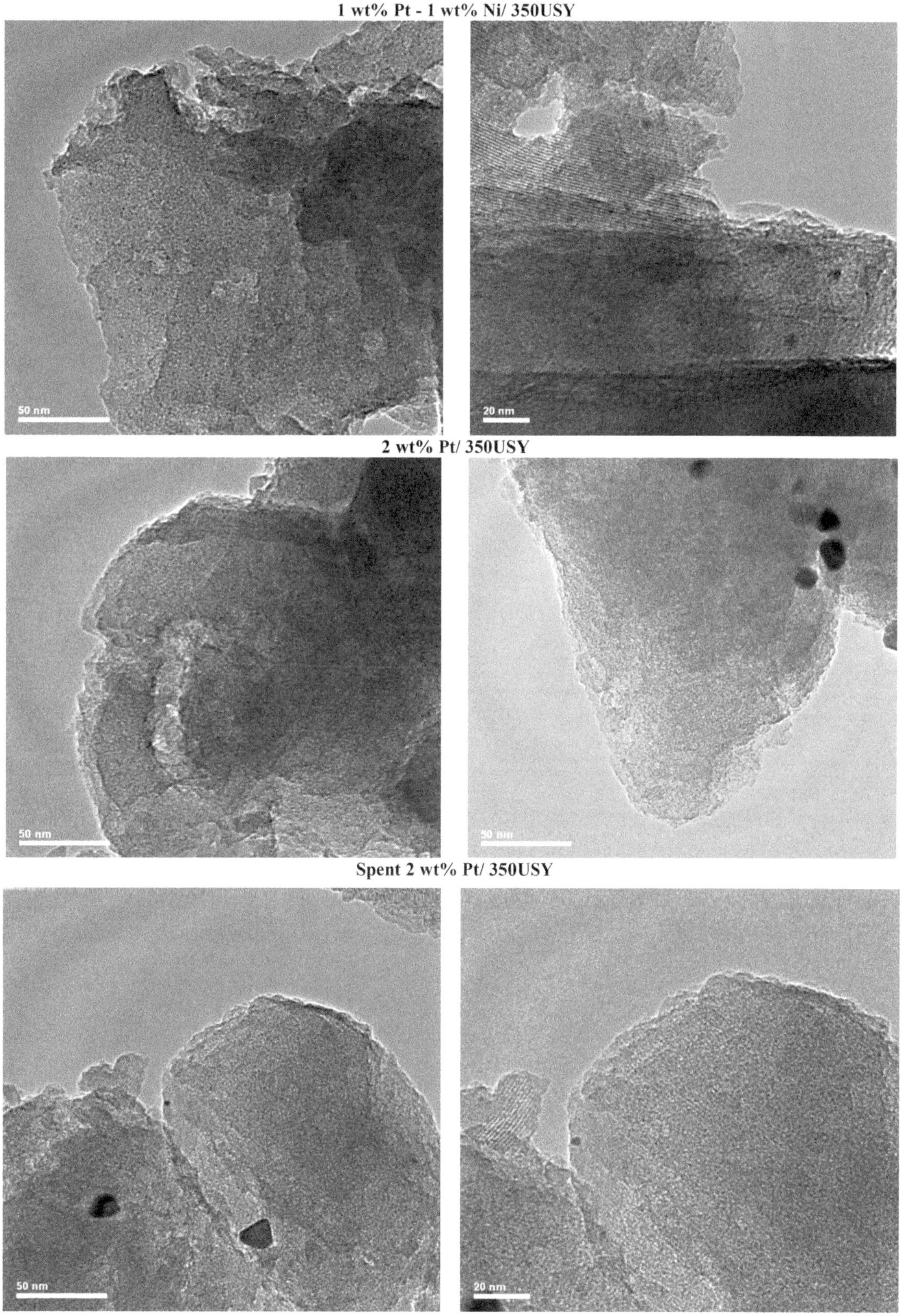

Fig. 14 TEM images of 1 wt% Pt–1wt% Ni, 2 wt% Pt and spent 2 wt% Pt/350USY

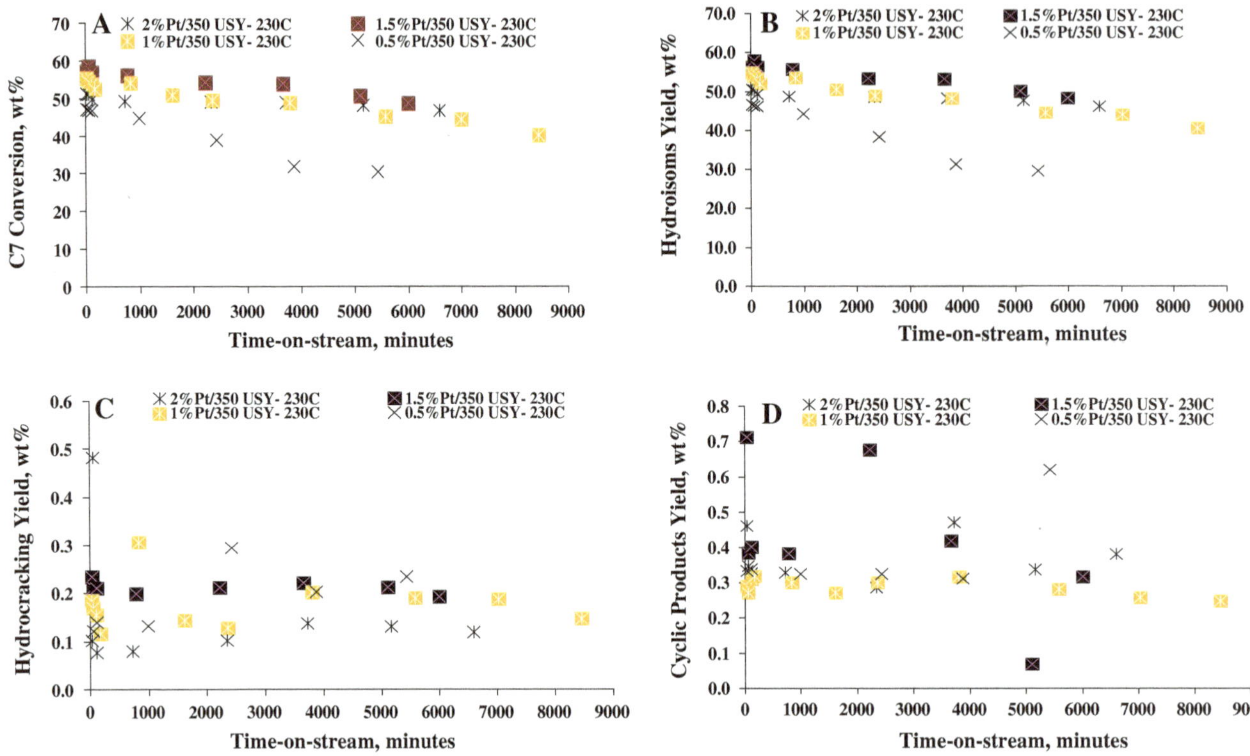

Fig. 15 Overall conversion (**a**), total yield of C7 isomers (mono + di + tribranched isomers) (**b**), total hydrocracking yield (**c**), and total yield of the cyclic products (**d**) as a function of TOS for the monometallic catalysts containing Pt at various reaction temperatures. WHSV = 5.13 h^{-1}; H_2/C7 (mol/mol) = 9; total pressure = 1 atm; temperature = 230 °C; 1 wt% Pt loading

respectively, for the catalysts: 1 wt% Pt/CP814E, 1 wt% Pt/USY-C, 1 wt% Pt/USY-CP814E physically mixed, and 1 wt% Pt/USY-CP814E physically mixed and loaded by Pt together by the ion-exchange method with ammonia added to adjust the pH to be more than 9. As Fig. 20a shows, initially the overall conversion of C7 for the 1 wt% Pt/USY-CP814E physically mixed was 66 wt%, which is in between that of the corresponding individual catalysts, compared with 77 and 46 wt% for the catalysts 1 wt% Pt/CP814E and 1 wt% Pt/USY-C, respectively. The conversion then decreased abruptly at the initial stages of the reaction, and finally reached a relatively steady value, after which the conversion decreased steeply with the TOS until the reaction was terminated after 183 h. Thus, this composite catalyst demonstrated a better TOS activity than 1 wt% Pt/USY-C catalyst, but slightly worse than 1 wt% Pt/CP814E. Even its selectivity toward the C7 isomers and cracked products was very similar to both catalysts along the deactivation run.

On the other hand, the composite catalyst 1 wt% Pt/USY-CP814E, physically mixed and loaded with Pt using ion exchange and subsequently deactivated as with the 1 wt% Pt/CP814E catalyst, offered slightly higher activity towards the cracked products along the deactivation run, with the initial conversion of 78 wt% decreasing slightly with TOS to 67 wt% with the reaction terminating after 83 h. The selectivity towards the C7 isomers and cracked products was 96 and 3 %, respectively, and the composite catalyst demonstrated a better TOS activity than the 1 wt% Pt/USY-C catalyst, and a similar deactivation pattern to the 1 wt% Pt/CP814E even though the selectivity towards the cracked products was higher. This indicates that the 1 wt% Pt/USY-C catalyst was the dominant component in the formulated composite catalyst of 1 wt% Pt/USY-CP814E and provided a better TOS stability and higher activity. On the other hand, the 1 wt% Pt/CP814E catalyst was the dominant component for the composite catalyst 1 wt% Pt/USY-CP814E loaded with Pt using ion exchange. Figure 21 shows the deposited hard coke contents over the top, middle, and bottom zones of the individual and composite catalysts. It can be seen that the lowest hard coke content was found for the 1 wt% Pt/USY-C and the highest for the 1 wt% Pt/CP814E catalyst, with the composite catalyst having a hard coke content part way between that of the corresponding individual catalysts. The total acidity for ammonia TPD for the 1 wt% Pt/CP814E, 1 wt% Pt/USY-C, and 1 wt% Pt/USY-CP814E physically mixed was found to be 0.63, 0.60 and 1.88 mmol g^{-1}, respectively. Thus, it can be seen that even the composite catalyst has the highest acidity which is almost double compared with the corresponding individual catalysts. It has moderate hard coke content and fairly good TOS catalytic stability.

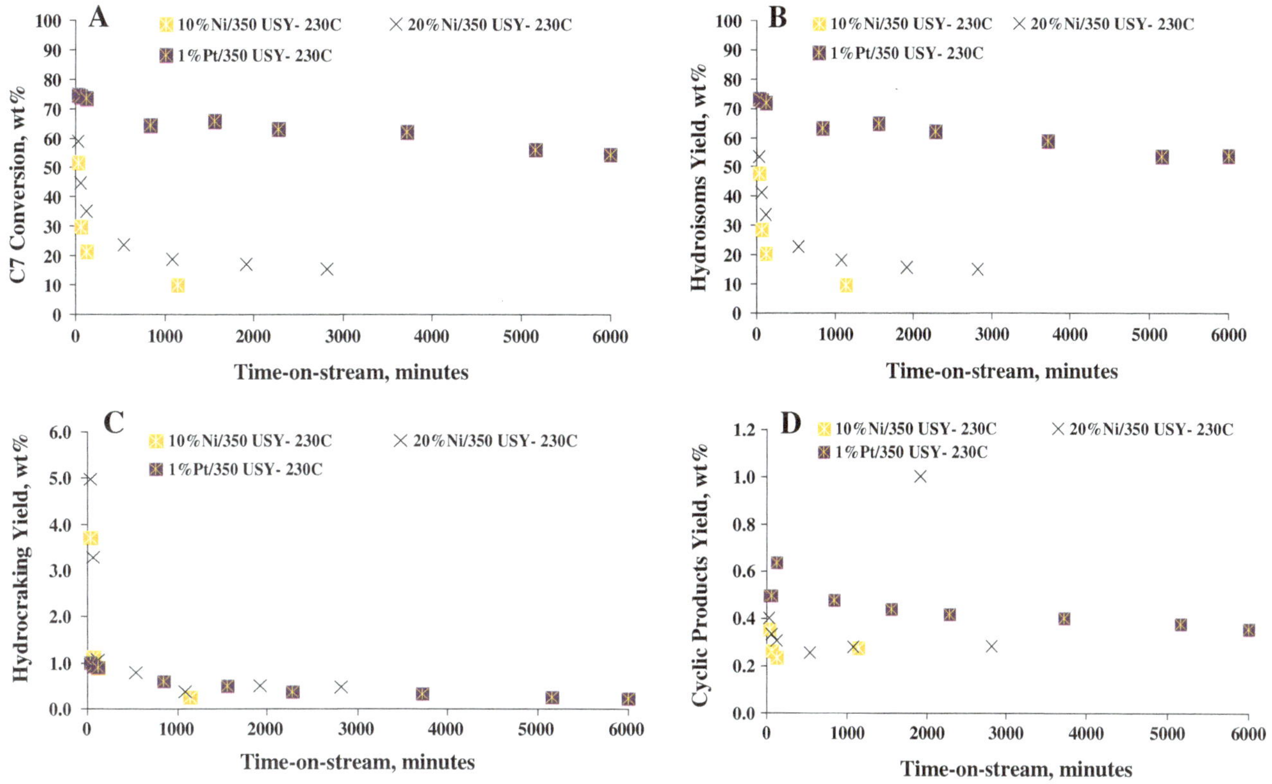

Fig. 16 Overall conversion (**a**), total yield of C7 isomers (mono + di + tribranched isomers) (**b**), total hydrocracking yield (**c**), and total yield of the cyclic products (**d**) as a function of TOS for the monometallic catalysts containing Ni at different reaction temperatures. WHSV = 5.13 h^{-1}; H_2/C7 (mol/mol) = 9; total pressure = 1 atm; temperature = 230 °C; 1 wt% Pt loading

Main characteristics of in-house mordenite zeolites

Mordenite zeolite was synthesised using the molar gel composition of (6 Na_2O: Al_2O_3: 30 SiO_2: 780 H_2O), which is the synthesis recipe of mordenite obtained from a verified synthesis handbook [21]. The mordenite phase with the highest purity and crystallinity was optimised against certain synthesis parameters while keeping this composition ratio constant. Three factors were optimised: the ageing time of the synthesis gel, the crystallisation time and the crystallisation temperature. Table 11 shows the phase crystallinity, the bulk Si/Al ratio and the framework Si/Al ratio of the mordenite obtained by optimising these three factors. It can be seen that the optimum synthesis conditions were when the ageing time was 12 h, crystallisation time was 24 h and crystallisation temperature was 180 °C. Under these conditions, the mordenite zeolite phase obtained had a crystallinity of 82 %, bulk Si/Al ratio of 8.44 (mol/mol) and framework Si/Al ratio of 9.02 (mol/mol). This procedure was repeated more than 10 times and reproducible results were obtained in terms of the yield, crystallinity, bulk and framework Si/Al ratio, as well as of the crystal morphology and size.

Figures 22 and 23 show the SEM microgram and XRD pattern of this in-house mordenite catalyst. The former shows that the catalyst contained large crystallites of 15–20 μm whose elliptical shape was different from that of commercial catalysts, and it has been previously reported that synthetic mordenite presents several morphologies, as a function of the crystallisation conditions [48]. The XRD patterns show a pure phase of mordenite consistent with that found previously [49]. However, it had a lower crystallinity than CBV21A [31], which also has a higher bulk Si/Al ratio. Figure 24a and b shows respectively the ^{27}Al and ^{29}Si solid-state NMR spectra of this in-house mordenite catalyst. It can be seen that the catalyst had only one ^{27}Al peak at about 56 ppm, which was assigned to tetrahedrally coordinated aluminium, and there were no additional EFAL peaks. In the ^{29}Si MAS NMR spectrum, two peaks at 112 and 106 ppm were assigned to Si(0Al) and Si(1Al) configurations, respectively.

Post-synthesis modification of in-house mordenite zeolite

Post-synthesis treatment by steam or acid attack, often referred to as dealumination, can remove aluminium from the zeolite framework. Steaming, which is performed at temperatures above 450 °C, extracts aluminium atoms from the framework and deposits them on the external

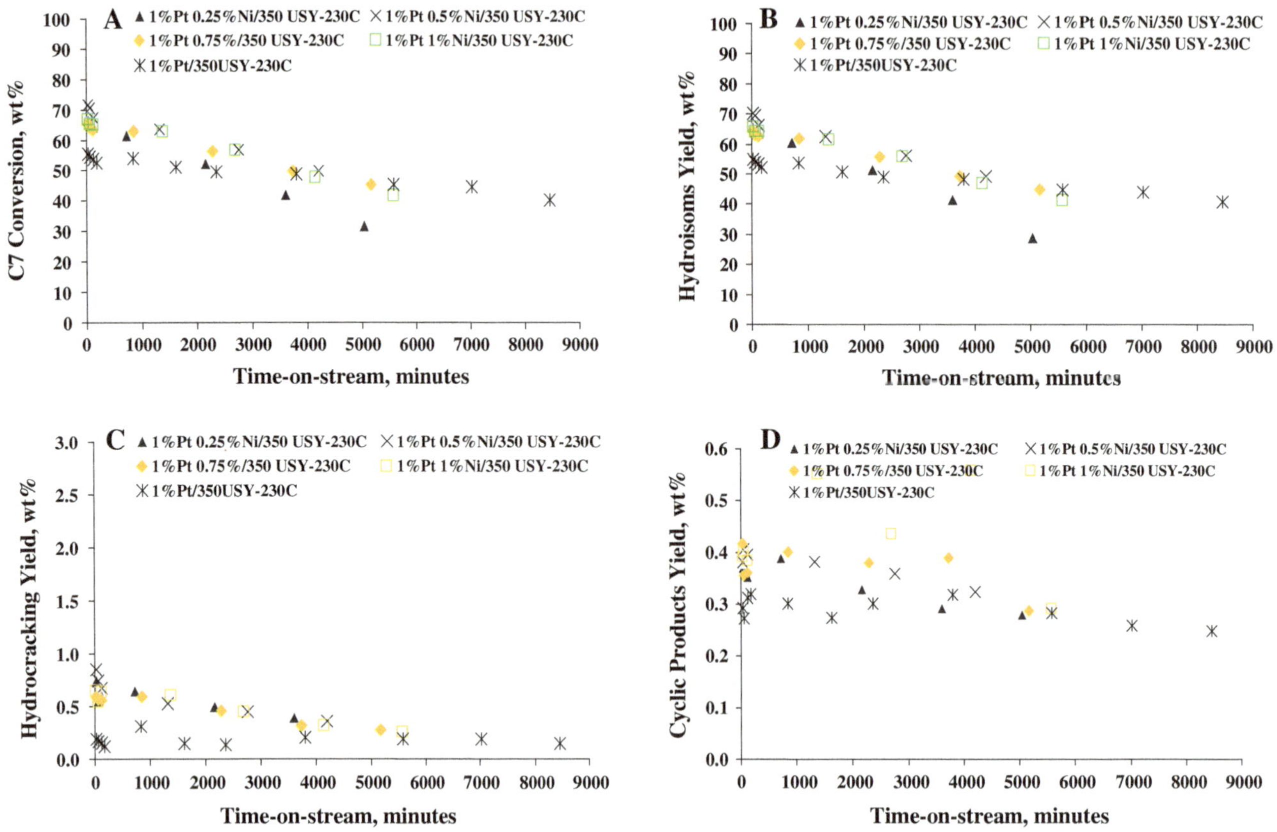

Fig. 17 Overall conversion (**a**), total yield of C7 isomers (mono + di + tribranched isomers) (**b**), total hydrocracking yield (**c**), and total yield of the cyclic products (**d**) as a function of TOS for the bimetallic catalysts containing Pt–Ni at different reaction temperatures. WHSV = 5.13 h^{-1}; H_2/C7 (mol/mol) = 9; total pressure = 1 atm; temperature = 230 °C; 1 wt% Pt loading

surface of the zeolite. An additional acid treatment can also be employed to completely remove aluminium atoms from the zeolite pore system. Increasing the framework Si/Al ratio by reducing the aluminium content in this way can result in the isolation of the remaining aluminium atoms, which in turn increases the strength of the acid sites and decreases the total number of active sites. This serves to improve both thermal and hydrothermal stability, while also generating mesopores, thus making the crystallite more accessible and reducing the severity of the diffusion limitations. Shortening the diffusion path length is of paramount importance for increasing the activity and stability of mordenite, as it is a one-dimensional, non-mesoporous zeolite. A small number of blockages can deactivate large parts of the mordenite structure [18, 50–52].

Two series of dealuminated mordenite catalysts were prepared by acid leaching and steaming applied to the in-house synthesised mordenite in its ammonium form. Table 12 shows the different acid leaching conditions that were applied to produce different catalysts with differing Si/Al ratios (acidity) based on the severity of acid leaching. In addition, three dealuminated mordenite catalysts were produced via the steaming dealumination technique at temperatures of 500, 600 and 700 °C. Table 13 shows the main characteristics of the acid-leached and steamed catalysts.

From Tables 5 and 6, it can be seen that the highest Si/Al molar ratios obtained for the framework and bulk were 19.13 and 42.76 (mol/mol), respectively, by acid leaching using nitric acid at a concentration of 10 molar, a temperature of 120 °C and a leaching time of 87 h (3.6 days), as in condition 15 in Table 12. A similar Si/Al ratio for the framework was obtained by increasing the acid leaching time to 168 h (7 days) at the same mixing temperature, as in condition 16, the most severe that was applied. It is generally expected that the bulk Si/Al molar ratio should remain the same even after steaming modification and that the steamed mordenite should have the same bulk Si/Al molar ratio as synthesised mordenite, but the framework Si/Al molar ratio was higher and increased as the steaming temperature increased from 500 to 700 °C, as shown in Table 13, which means that the framework aluminium atoms were dealuminated by steam to become EFAL.

Figures 25 and 26 show the SEM micrograms of steamed mordenites and the XRD patterns of acid-leached and steamed mordenites, respectively. From the SEM images, it can be seen that the steamed catalysts contained

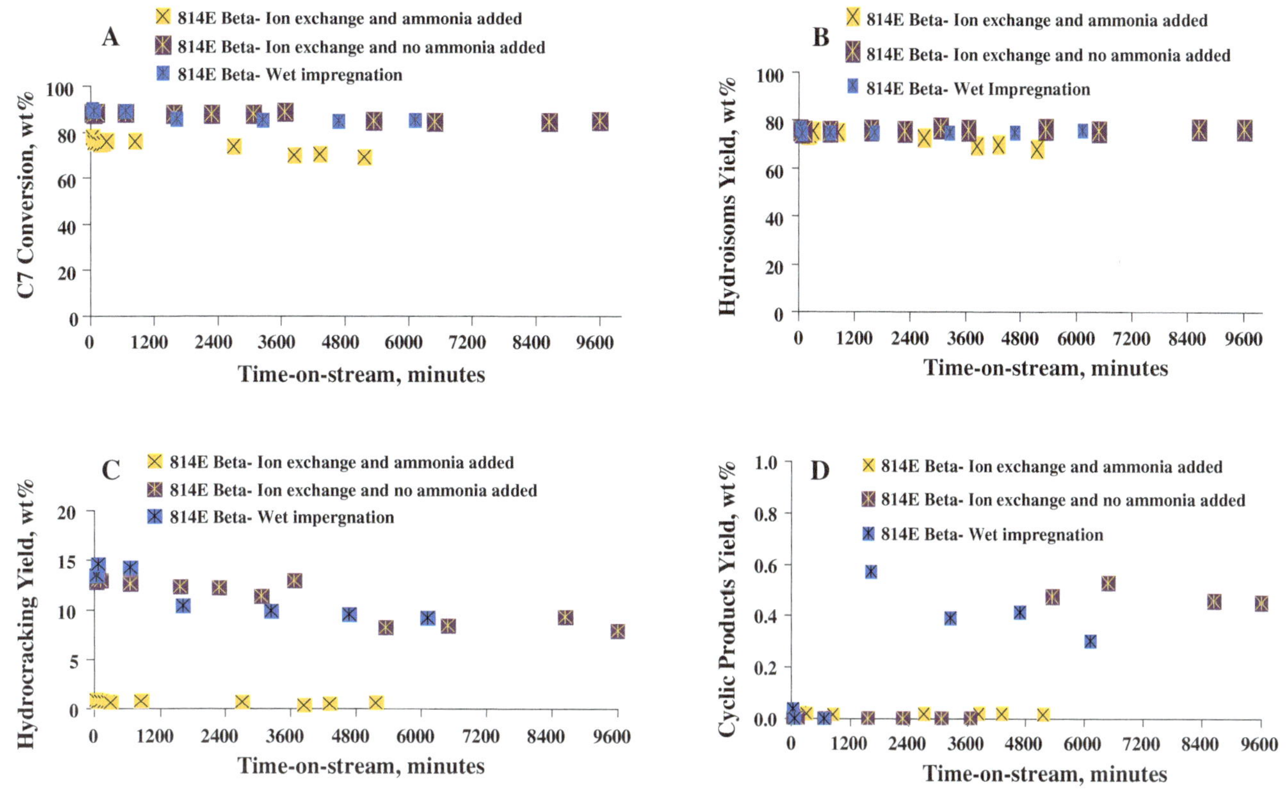

Fig. 18 Overall conversion (**a**), total yield of C7 isomers (mono + di + tribranched isomers) (**b**), total hydrocracking yield (**c**), and total yield of the cyclic products (**d**) as a function of TOS for CP814E. WHSV = 5.13 h^{-1}; H_2/C7 (mol/mol) = 9; total pressure = 1 atm; temperature = 230 °C; 1 wt% Pt loaded by different methods

Fig. 19 Variation of the C1/C3 mass ratio as a function of TOS of CP814E

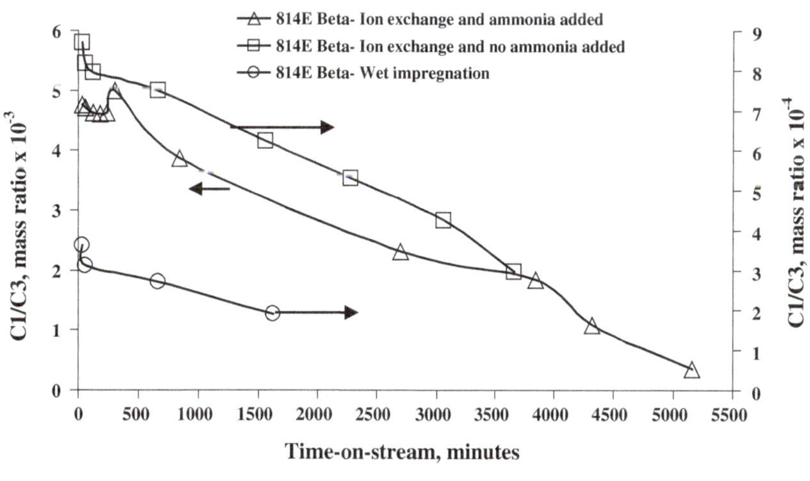

large and small crystallites of different sizes with some smaller debris and agglomerates whereas the acid-leached catalyst kept almost its original large crystal shape with some holes on the crystal surface. The XRD patterns confirm the pure mordenite phase for the steamed and acid-leached samples, with less overall crystallinity than as-synthesised mordenite.

Figures 27 and 28 show the ^{27}Al and ^{29}Si solid-state NMR spectra of acid-leached and steamed samples. It can be seen that the acid-leached catalysts had two ^{27}Al peaks, at about 56 and 0 ppm, assigned to a tetrahedrally coordinated aluminium atom in the framework and a smaller quantity of octahedrally coordinated aluminium as EFAL. Conversely, the steamed samples showed a very large quantity of EFAL at unique resonance spectra for octahedral and tetrahedral aluminium atoms at about 1 and 3 ppm, respectively, alongside in-framework aluminium in tetrahedral coordination. The quantity of EFAL also

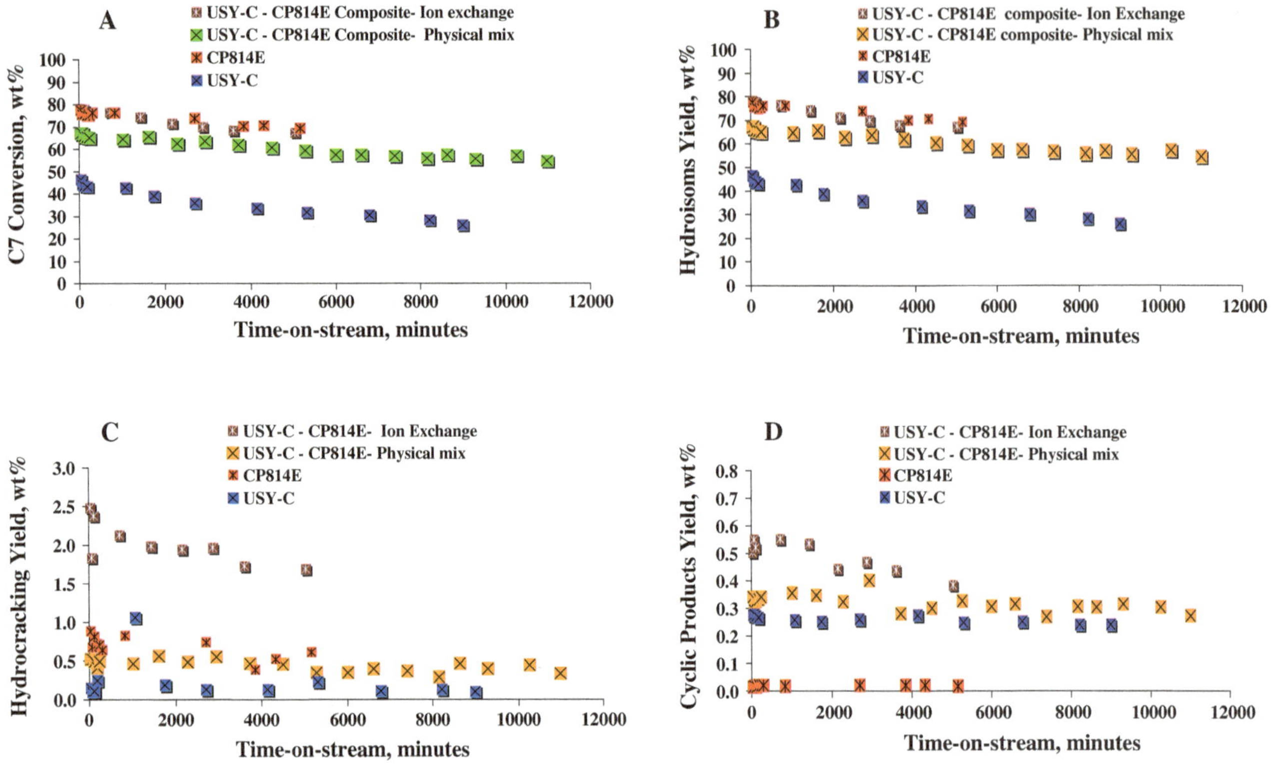

Fig. 20 Overall conversion (**a**), total yield of C7 isomers (mono + di + tribranched isomers) (**b**), total hydrocracking yield (**c**), and total yield of the cyclic products (**d**) as a function of TOS for USY and CP814E composites. WHSV = 5.13 h^{-1}; H_2/C7 (mol/mol) = 9; total pressure = 1 atm; temperature = 230 °C; 1 wt% Pt loading

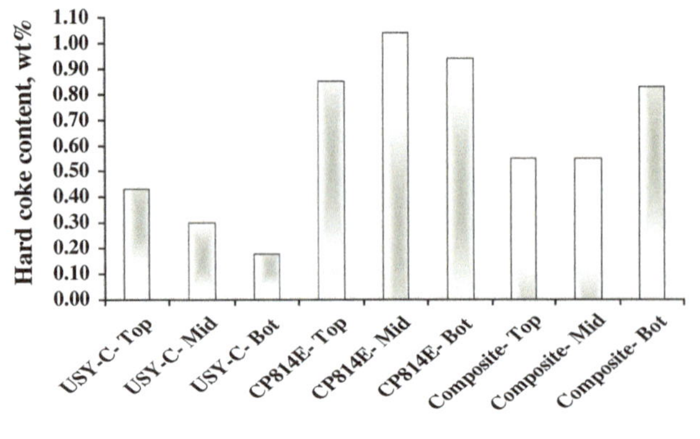

Fig. 21 The hard coke content over the aged CP814E, USY-C, and USY-CP814E physically mixed

increased as the steaming temperature increased. In the ^{29}Si MAS NMR spectra, two peaks at about 114 and 107 ppm were assigned respectively to Si(0Al) and Si(1Al) configurations, and these chemical shifts differ slightly from the sample produced by steaming. In addition, silicon atoms are different crystallographically as indicated by the different peak shape, depending upon the steaming temperature that was employed. However, the strong increase in the Si(0Al) signal intensity as the steaming temperature increased indicates that the replacement of Al by Si in the framework was more effective [20].

Table 13 and Fig. 29 show that the dealumination treatments by acid leaching and steaming resulted in an increase in mesoporosity, as illustrated with N_2-physisorption as a sharp increase in adsorption at high relative pressures, indicating the formation of large mesopores and a corresponding increase in the external surface area. Additionally, an average pore size of 27 Å was observed in the treated samples, compared with 18 Å for as-synthesised mordenite.

TOS behaviour during C7 hydroisomerisation conversion was examined for the following commercial and in-house mordenite catalysts: CBV10A, CBV21A, 640HOA, 660HOA, 690HOA [31], acid-leached mordenite and in-house mordenite steamed at 600 °C. All of the above catalysts were loaded with 1 wt% Pt by the ion-exchange method. Three very acidic commercial catalysts—1 wt% Pt/CBV10A, 1 wt% Pt/640HOA and 1 wt% Pt/660HOA—and the acid-leached mordenite deactivated very rapidly during the initial few seconds, due to a rapid coking that may have poisoned the acid sites of these catalysts and changed the effective acid site density needed for an ideal bifunctional conversion. As a consequence, no TOS results were reported for these catalysts. Figure 30a–d shows the effects of TOS at a reaction temperature of 230 °C on C7 conversion, hydroisomerisation, hydrocracking and cyclic

Table 11 In-house mordenite synthesis parameters and the resulting sample characteristics

Ageing time (h)	Crystallisation time (h)	Crystallisation temp (°C)	XRD crystallinity (%)	ICP Si/Al (mol/mol)	ICP Na (wt%)	NMR Si/Al (mol/mol)
1	1 day	180	43	9.50	3.63	8.03
3			64	9.70	3.99	9.13
4			75	8.36	3.26	8.97
6			77	8.13	3.15	8.97
8			75	7.83	3.08	8.98
10			75	7.92	3.07	8.98
12			82	8.44	3.14	9.02
15			80	7.99	3.23	8.97
24			49	8.92	3.09	8.62
12	12	180	Amorphous	–	–	–
	20		76	–	–	8.71
	22		75	–	–	8.89
	23		71	–	–	8.92
	24		82	8.44	3.14	9.02
	25		76	–	–	8.63
	26		72	–	–	8.63
	28		70	–	–	8.65
	36		Amorphous	–	–	–
12	1 day	170	Amorphous	–	–	–
		175	32	–	–	7.67
		180	82	8.44	3.14	9.02
		185	72	–	–	8.80
		190	71	–	–	8.63

Fig. 22 SEM micrograph of in-house and as-synthesised mordenite

product yields respectively for the following catalysts: 1 wt% Pt/CBV21A, 1 wt% Pt/690HOA and in-house mordenite steamed at 600 °C.

Initially, the overall conversions were 12, 41 and 52 wt% after 60 min from the point when the feed was introduced to the reactor at 230 °C, for the 1 wt% Pt/CBV21A, 1 wt% Pt/690HOA and steamed mordenite catalysts, respectively. A rapid deactivation occurred for the 1 wt% Pt/CBV21A catalyst with high cracking activity, resulting in selectivity towards the cracked products at

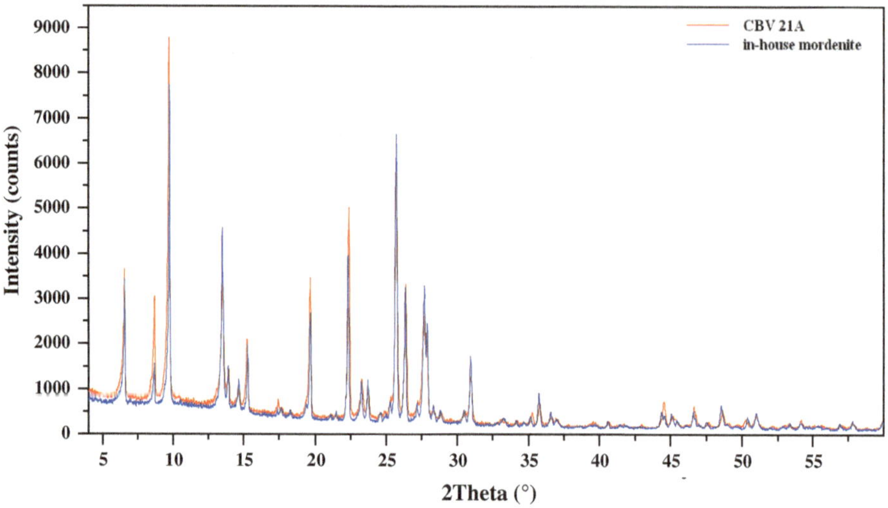

Fig. 23 XRD patterns of in-house and as-synthesised mordenite

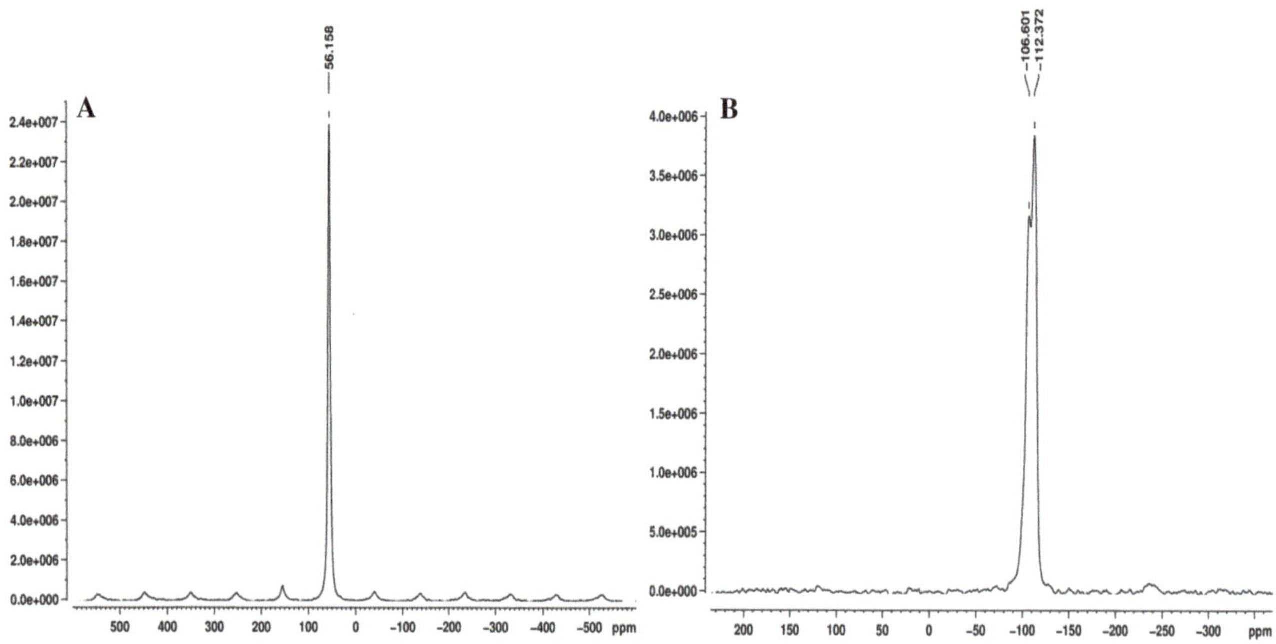

Fig. 24 ^{27}Al (a) and ^{29}Si (b) MAS NMR spectra of in-house and as-synthesised mordenite

24 %, compared with 71 % for the isomer products, and thereafter the catalyst was almost deactivated after 240 min on stream. On the other hand, greater TOS stability was shown by the 1 wt% Pt/690HOA and steamed mordenite catalysts. However, a shorter coking rate was observed for the steamed mordenite, so it preserved its initial activity for almost 24 h, after which it deactivated at the rate of 8–10 wt% every day before achieving a pseudo-stable state at the end of the reaction. In contrast, the 1 wt% Pt/690HOA catalyst deactivated continuously from the point when the feed was introduced to the reactor at the slower rate of 3–5 wt% every day, before achieving a pseudo-stable state at the end of the reaction. In addition, selectivity towards C7 isomers and cyclic products for both catalysts was in the range of 97–98 and 2–3 %, respectively.

Faster deactivation and the mono-dimensional pore system of 1 wt% Pt/CBV21, coupled with a higher density of strong acid sites and as-synthesised zeolite, may have caused the lower activity, which may also be attributable to platinum crystallites blocking the pore openings, as previously reported [35, 53]. There are side pockets in the pore topology of mordenite that match the 12MR and 8MR crossing. The higher number of Brønsted acid sites in the cavities also increases the sensitivity of the 8 MR pores to coking and they can easily fill up and become blocked by carbonaceous deposits [54]. The average lifetime of the carbocations on the surface can be lengthened by strong

Table 12 Acid leaching conditions applied and the main characteristics of samples obtained

Condition	NH_4-MOR (wt)	Acid type	Acid conc. (molar)	Acid, Vol. (ml)	Leaching time (h)	Temp (°C)	XRD crystallinity (%)	NMR Si/Al (mol/mol)	ICP Si/Al (mol/mol)
1	1	HCl	6	50	12	25	71	9.53	–
2	1	HCl	6	50	24	25	73	9.63	–
3	1	HCl	6	100	24	25	74	9.69	–
4	1	HCl	Conc	100	24	120	70	15.32	–
5	1	HNO_3	10	100	16	25	73	9.24	–
6	1	HNO_3	6	50	24	25	70	9.32	–
7	1	HNO_3	Conc	100	24	25	70	9.36	–
8	1	HNO_3	Conc	50	24	25	72	9.39	–
9	1	HNO_3	Conc	100	24	120	78	15.30	–
10	1	HNO_3	6	100	24	120	73	16.09	–
11	1	HNO_3	6	100	48	120	75	16.78	–
12	2	HNO_3	10	100	24	120	70	17.10	–
13	1	HNO_3	10	100	16.5	120	77	17.88	–
14	1	HNO_3	10	100	48	120	74	18.75	–
15	1	HNO_3	10	100	87	120	74	19.13	42.76
16	1	HNO_3	10	100	168	120	76	19.25	–

Table 13 The main characteristics of acid-leached and steamed mordenite samples

Bulk (Si/Al) mol/mol	Framework (Si/Al) (mol/mol)	Crystallinity (%)	Bulk Pt (wt%)	Acidity (mmol g^{-1})	Surface area (m^2 g^{-1})	Pore volume (cm^3 g^{-1})	Pore size (Å)
As-synthesised mordenite							
8.44	9.02	82	–	1.24	406.95	0.17	18.02
Acid-leached mordenite							
42.76	19.13	74	–	0.48	478.85	0.23	26.57
Mordenite steamed at 500 °C							
8.97	20.38	66	–	0.30	419.52	0.21	26.97
Mordenite steamed at 600 °C							
9.06	28.84	64	0.58	0.22	415.00	0.20	26.12
Mordenite steamed at 700 °C							
8.89	∞	67	–	–	404.93	0.20	26.68

acidity and narrow pore diameter, while the diffusion of the branched products can be slowed down. Each of these factors will help to crack the tertiary carbocations that have formed during isomerisation prior to desorption and will assist in the re-adsorption and cracking of the branched paraffins before they leave the pores and enter the gas stream [55].

There are various possible explanations for the slower deactivation of 1 wt% Pt/690HOA and steamed mordenite. First, it may be due to the medium acid strength of the catalysts and their reduced acid site density. Alternatively, it may be due to the balance between acid and metal active sites. It could also be ascribed to the existence of mesopores and to the slight increase in the pore volume, which may be beneficial to the diffusion of bulky moieties. Finally, it may be due to post-synthesised mordenite having smaller crystallites. The last two characteristics can be seen as reducing the resistance of the product molecules to formation and diffusion, resulting in less cracking and deactivation of the catalyst caused either by blockages at the site or by plugging of channels. The method of dealumination applied to the final lattice Si/Al ratio influences catalyst activity. For example, steam dealumination leaves the lattice with aluminium removed as EFAL, whereas samples that are practically free of EFAL can result from acid leaching, so EFAL may have a negative effect by blocking pores and preventing the Pt from being well dispersed during the ion-exchange process. This may explain the deactivation of the 1 wt% Pt/690HOA steamed mordenite catalysts.

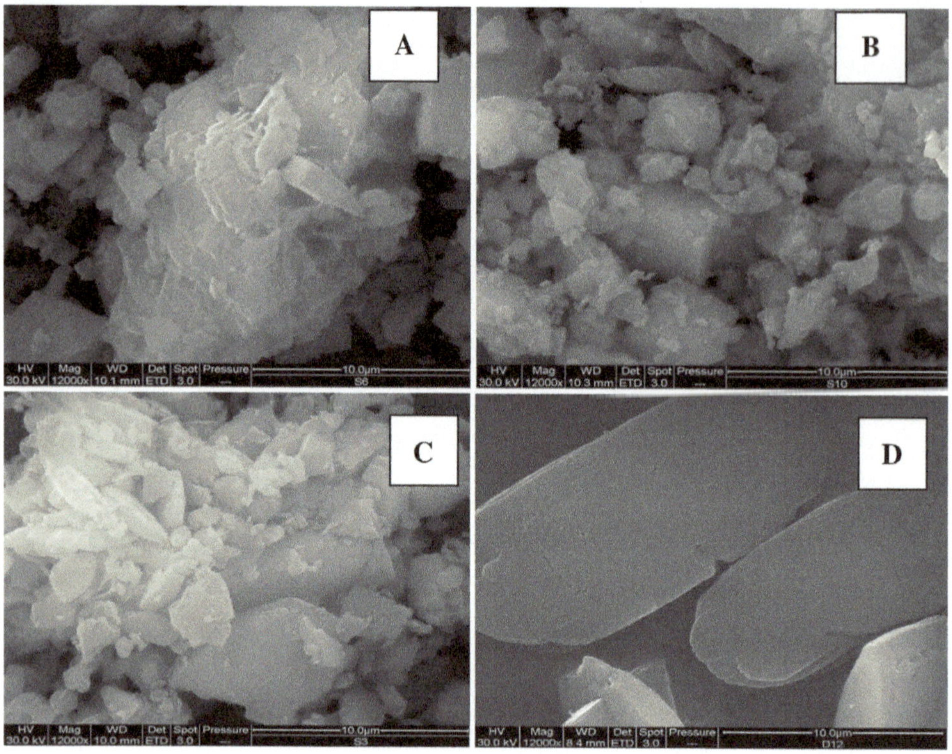

Fig. 25 SEM micrographs of mordenites steamed at 500 °C (**a**), 600 °C (**b**), 700 °C (**c**) and acid leached (**d**)

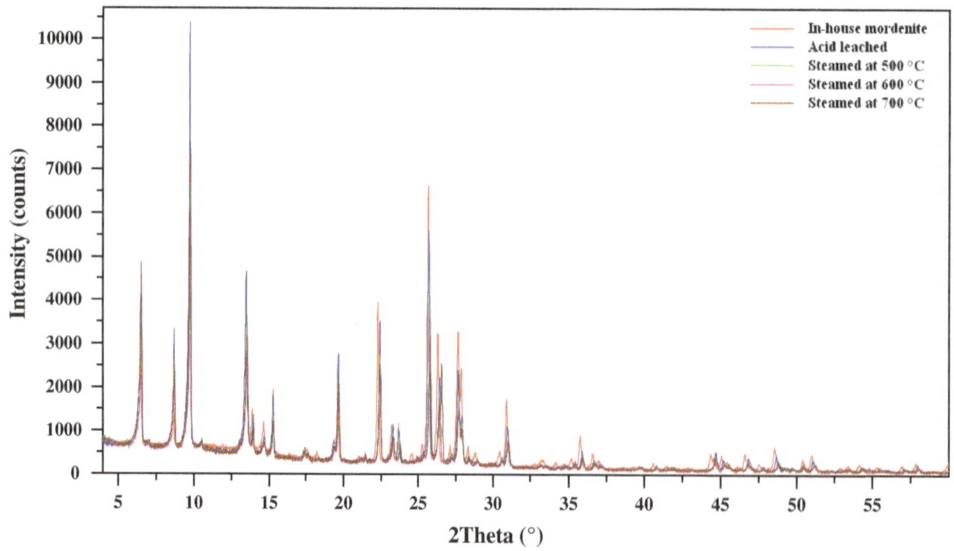

Fig. 26 XRD patterns of acid-leached and steamed mordenites

Deactivation behaviour of two selected commercial mesoporous aluminosilicates and a γ-alumina-based catalyst

TOS stability during the hydroisomerisation of n-C7 under identical reaction conditions was examined using two selected commercial mesoporous aluminosilicates—Al-SBA-15 and Al-MCM-48—and a γ-alumina catalyst, which were loaded with 1 wt% platinum by the ion-exchange technique and tested using the atmospheric pressure glass fixed-bed flow reactor system. The two mesoporous aluminosilicates, Al-SBA-15 and Al-MCM-48, containing 8 and 3 wt% aluminium, respectively, were supplied by Claytec (USA) [20] as white powders in their proton form. Their main surface texture features and the methods by which the alumination and protonation were

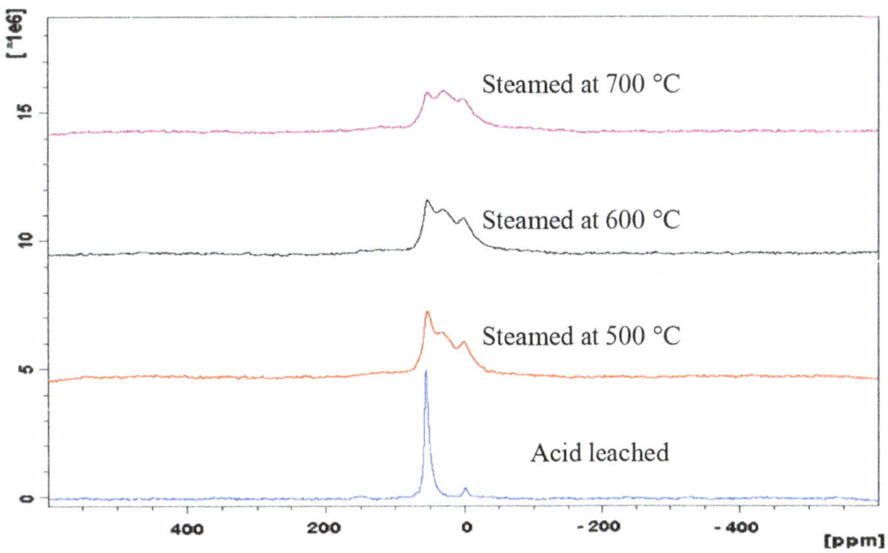

Fig. 27 ^{27}Al NMR of acid-leached and steamed samples

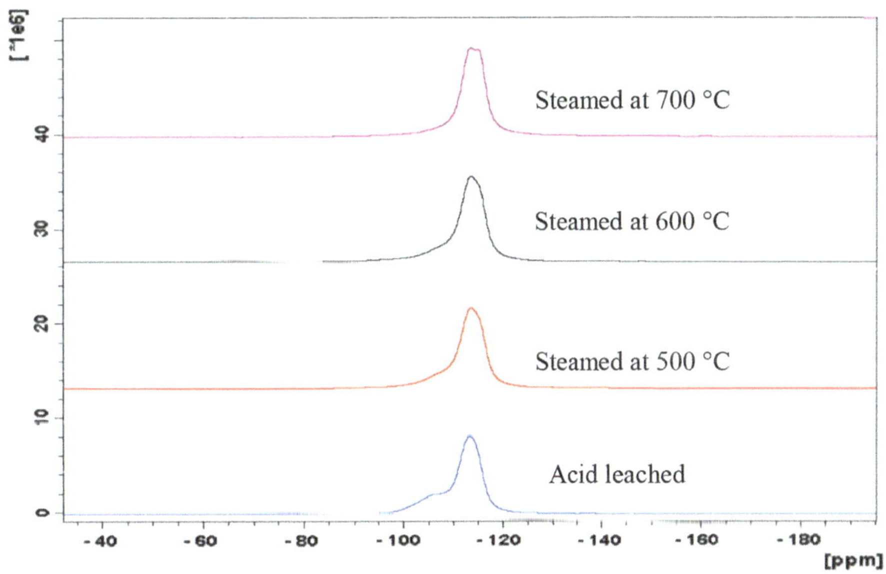

Fig. 28 ^{29}Si NMR of acid-leached and steamed samples

preformed, as specified by the supplier, are shown in Table 14.

Figure 31 shows the SEM image of Al-SBA-15 and Al-MCM-48 catalysts. It can be observed that both samples consist of non-uniform and irregular-shaped particles varying in size from 0.5 to 2 μm. XRD patterns of the Al-SBA-15 and Al-MCM-48 catalysts are shown in Fig. 32. The Al-SBA-15 sample shows well-resolved patterns with a prominent diffraction peak (100) and two additional diffraction peaks indexed to (110) and (200) reflections, which match well with the pattern reported for SBA-15 [56], indicating that the Al-SBA-15 product prepared by post-alumination had a well-ordered hexagonal mesostructure, whereas the XRD pattern of the Al-MCM-48 sample was identical to the one reported by Beck et al. [57]. Four diffraction peaks were also clearly observed and indexed to (211), (220), (420) and (332), indicating that the Al-MCM-48 product prepared by post-alumination had a well-ordered cubic mesostructure. Table 15 shows some in-house characterisations of the Al-SBA-15, Al-MCM-48 and γ-alumina catalysts using various different techniques.

Solid-state MAS NMR spectroscopy of ^{27}Al indicated the presence of tetrahedrally coordinated aluminium atoms in the Al-SBA-15 (A) and Al-MCM-48 lattices. It also revealed extra-framework octahedral and pentahedral Al in the pores of both molecular sieves, which may have been formed during hydrothermal crystallisation and/or by dealumination during calcination [58], as shown in Fig. 33. Therefore, it can be confirmed that not all Al atoms were inserted in the frameworks of SBA-15 and MCM-48.

In the ^{29}Si MAS NMR spectra, two peaks at 110 and 102 ppm were assigned to Si(0Al) and Si(1Al)

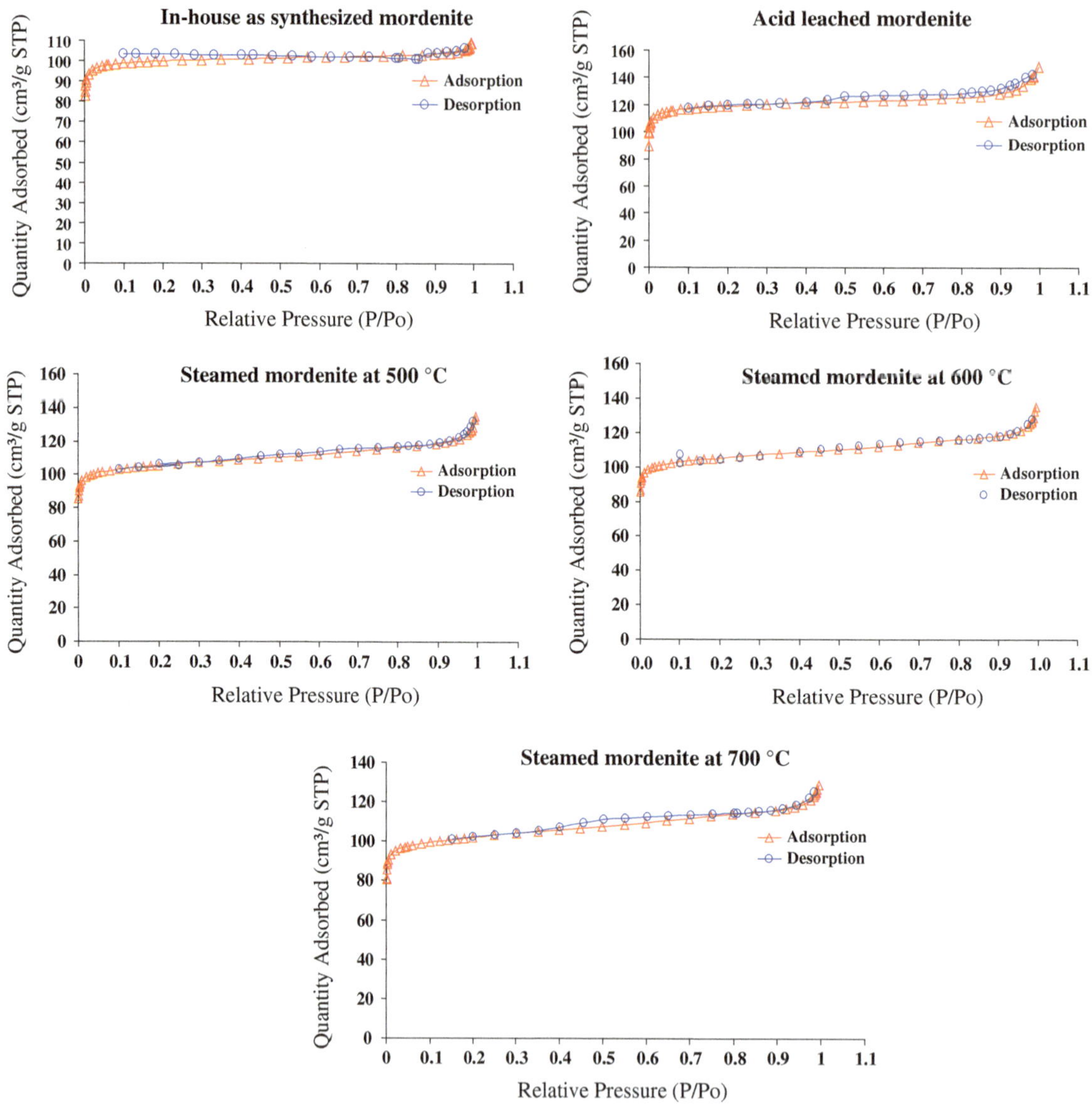

Fig. 29 N_2 adsorption and desorption isotherms of acid-leached and steamed samples

configurations respectively for Al-SBA-15, and three peaks at 110, 101, and 92 ppm were assigned to Si(0Al), Si(1Al) and Si(2Al) configurations respectively for Al-MCM-48, as shown in Fig. 34. This confirmed that some Si atoms were substituted by Al atoms to generate the mesoporous aluminosilicate structure for both catalysts.

The Al-SBA-15 and Al-MCM-48 catalysts were also characterised using nitrogen adsorption at −196 °C, as shown in Fig. 35, to confirm that these materials had retained their mesostructures after the post-alumination process to substitute the silicon with aluminium. Figure 35 shows the nitrogen adsorption isotherms for both catalysts, which are type IV according to the IUPAC classification [59]. The isotherm of the mesoporous molecular sieve exhibits a sharp inflection characteristic of capillary condensation within uniform pores, where the P/P_0 position of the inflection point is related to the diameter of the pore. The sharp rise near 0.4 P/P_0 corresponds to condensation in the primary mesopores. As the size of the primary mesopores increases, the capillary condensation step shifts to higher relative pressures and hysteresis becomes more pronounced. The existence of a hysteresis loop in the isotherms indicates the presence of mesopores and its shape is related to their shape: roughly, a vertical loop indicates cylindrical mesopores, whereas a horizontal one indicates inkbottle-shaped mesopores [51].

As shown in Tables 14 and 15 alongside Fig. 35, the two commercial catalysts exhibited high pore size and pore volume and retained their mesostructures, even after the post-alumination process. Table 14 shows that Al-SBA-15 had a much lower surface area than expected for mesoporous materials (>1,000 $m^2\ g^{-1}$), which can be explained

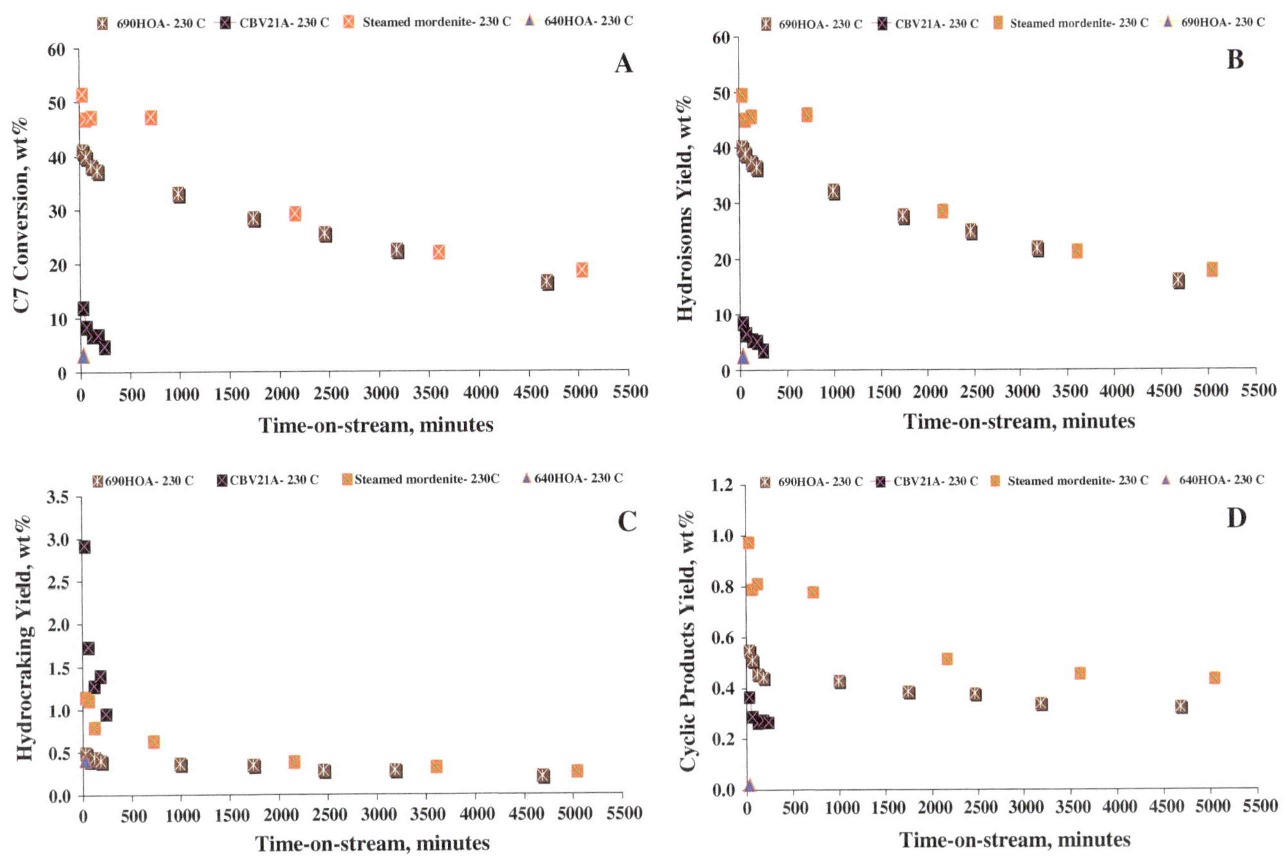

Fig. 30 Overall conversion (**a**), total yield of C7 isomers (mono + di + tribranched isomers) (**b**), total hydrocracking yield (**c**) and total yield of cyclic products (**d**) as a function of TOS for commercial and in-house steamed mordenite catalysts. WHSV = 5.13 h^{-1}; $H_2/C7$ (mol/mol) = 9; total pressure = 1 atm; temperature = 230 °C; 1 wt% Pt loading

in two ways. First, the presence of EFAL on the mesostructure may partially block the mesopores and subsequently hinder nitrogen adsorption. The existence of EFAL was revealed by NMR, as shown in Fig. 33a. Alternatively, the Al-SBA-15 mesostructure may have partially collapsed, so some residual amorphous material was generated after the post-alumination process. Despite the appearance of the lower surface area, the Al-SBA-15 catalyst retained its general mesostructure, as revealed by the XRD and nitrogen adsorption isotherms shown in Figs. 32 and 35 respectively, while its large pore size and pore volume were comparable with those of Al-MCM-48.

The deactivation behaviour of the 1 wt% Pt/Al-SBA-15 and 1 wt% Pt/Al-MCM-48 catalysts was investigated during the catalytic conversion of n-C7 at reaction temperatures of 210, 230, 250 and 270 °C and at atmospheric pressure, using the glass fixed-bed flow reactor system loaded with 1 g of each catalyst, as described in "Results and discussion" section. The constant mole ratio of hydrogen to n-C7 was 9 and WHSV was 5.13 h^{-1}. At temperatures below 300 °C, very low n-C7 conversion was found, and so 370 °C was chosen as the optimum temperature for studying the activity decay of these catalysts.

Figure 36a–d illustrates the effects of TOS on the C7 conversion, hydroisomerisation, hydrocracking and cyclic product yields at reaction temperature of 370 °C for the two catalysts loaded with 1 wt% Pt by the ion-exchange method. The γ-alumina support supplied by Alfa-Aesar (Johnson–Matthey) was also loaded with 1 wt% Pt using the wet impregnation method and tested under the same reaction conditions.

Initially, the respective overall conversions for 1 wt% Pt/Al-SBA-15, 1 wt% Pt/Al-MCM-48 and 1 wt% Pt/γ-Al_2O_3 were 51, 39 and 24 wt%, 30 min after the introduction of feed to the reactor at 370 °C. The 1 wt% Pt/Al-SBA-15 catalyst deactivated slightly with TOS, reaching a conversion of 25 wt% (<26 wt% of the initial catalytic activity) after about 2 days, then remained in a pseudo-stable state until the reaction terminated after 4 days. The 1 wt% Pt/Al-MCM-48 started with less activity than the 1 wt% Pt/Al-SBA-15, but with greater TOS stability, deactivating slightly to reach a conversion of 30 wt% (<9 wt% of the initial catalytic activity) after 2 days, then maintained a pseudo-stable state until the reaction terminated after 3 days. Deactivation was rapid for the 1 wt% Pt/γ-Al_2O_3 catalyst, with high cyclisation activity under the conditions of the investigation, resulting in selectivity towards the cyclic products at 55 %, compared with 40 % for the isomer products; thereafter, the catalyst was almost deactivated after 180 min on stream. The 1 wt% Pt/Al-SBA-15 and 1 wt% Pt/Al-MCM-48 catalysts also exhibited fairly high cyclisation selectivity, of 20 and 27 wt%, respectively, decreasing with TOS to 7 and 9 wt%, respectively, by the end of the reactions. The cracking selectivity of the 1 wt% Pt/Al-SBA-15, 1 wt% Pt/Al-MCM-48 and 1 wt% Pt/γ-Al_2O_3 catalysts was 1, 3 and 10 wt%, respectively. The maximum isomer yields for 1 wt% Pt/Al-SBA-15, 1 wt% Pt/Al-MCM-48 and 1 wt% Pt/γ-Al_2O_3 were found to be 40, 27 and 10 wt%, respectively, decreasing with TOS in the same way as conversion.

As the major reactions were hydroisomerisation and cyclisation, negligible cracking was observed for the

Table 14 Main surface texture features and the alumination and protonation methods of Al-SBA-15 and Al-MCM-48 as supplied by Claytec (USA) [20]

Sample	BET surface area ($m^2\ g^{-1}$)	Total pore volume @ $P/P_0 = 0.98$ ($cm^3\ g^{-1}$)	BJH pore size (nm)	Alumination and protonation methods
8 wt% Al-SBA-15	408	0.75	8.6	Calcination of the as-made pure silica mesophase at 600 °C to remove surfactant, followed by grafting reaction of the pure silica mesophase with aluminium i-propoxide to achieve alumination, followed by another calcination at 550 °C to form the protonated derivative
3 wt% Al-MCM-48	1,217	1.4	2.4	Sodium aluminate was used for Al incorporation during framework assembly, followed by ammonium ion exchange in the presence of surfactant, then calcination at 600 °C to form the protonated derivative

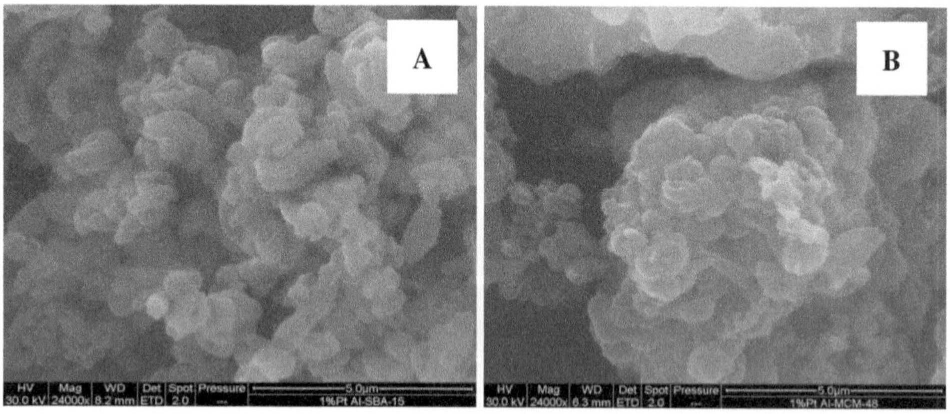

Fig. 31 SEM micrographs of Al-SBA-15 (**a**) and Al-MCM-48 (**b**)

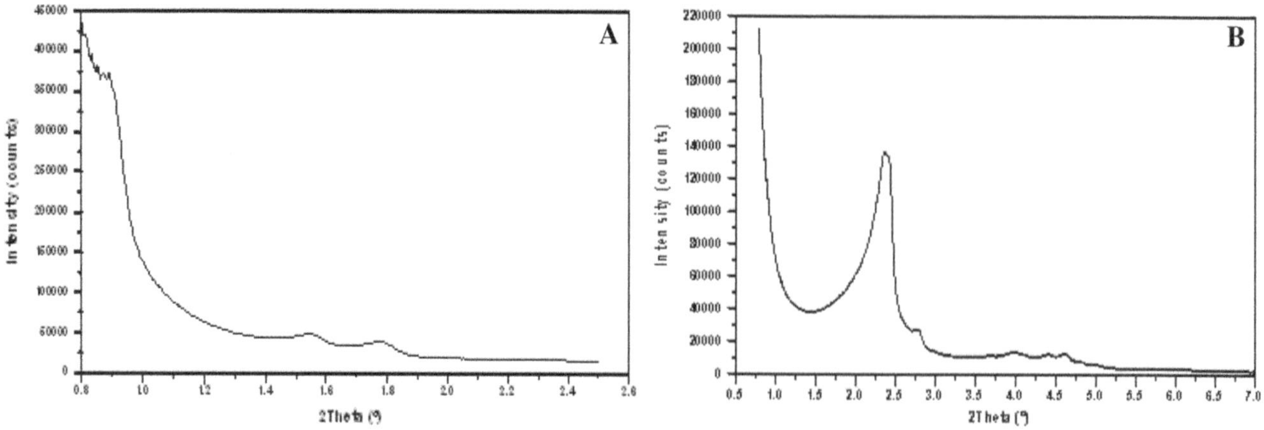

Fig. 32 XRD patterns of Al-SBA-15 (**a**) and Al-MCM-48 (**b**)

Table 15 Some characterisations of Al-SBA-15 and Al-MCM-48 catalysts

Bulk (Si/Al) (mol/mol)	Framework (Si/Al) (mol/mol)	Bulk Pt (wt%)	Si/Al atomic conc % (XPS)	Pt atomic conc % (XPS)	Acidity (mmol g^{-1})	Surface area (m^2 g^{-1})	Pore volume (cm^3 g^{-1})	Pore size (Å)
Al-SBA-15								
11.05	6.39	0.95	10.62	0.04	0.29	–	–	–
Al-MCM-48								
32.65	9.33	0.95	–	–	–	1,035	1.33	34.36
γ-alumina								
–	–	–	–	–	0.19	211	0.59	90.49

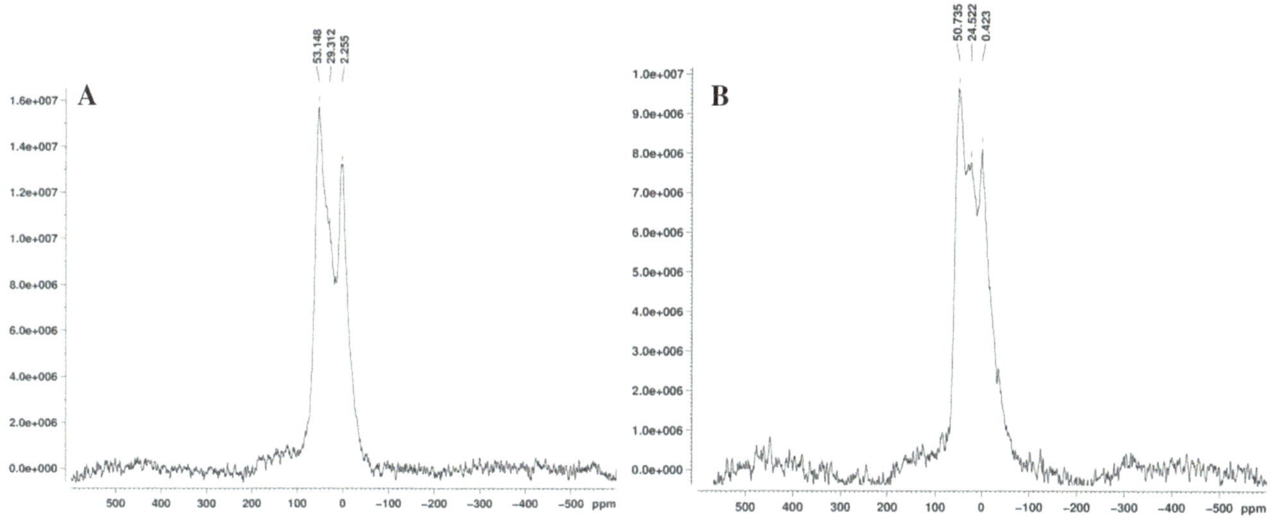

Fig. 33 ^{27}Al MAS NMR spectra of Al-SBA-15 (**a**) and Al-MCM-48 (**b**)

1 wt% Pt/Al-SBA-15 and 1 wt% Pt/Al-MCM-48. Small amounts of alkenes were also detected in the products. Although the deactivation rate was significant during the first few minutes, the activity of the catalyst became reasonably stable after about 2 days. A certain amount of initial deactivation has been reported by other workers during the isomerisation of *n*-alkanes over Pt catalysts, which is expected under atmospheric pressure conditions [60]. However, it was shown that a great quantity of oligomers (cyclic products) is produced from propene at low temperatures on MCM-41-type catalysts and that the pore size of the mesoporous material can make a difference when the aim is to oligomerise larger olefins or to produce large lubricant molecules in general [12].

It is reported that the very high surface area of ordered mesoporous materials makes it easier to obtain highly dispersed noble metal catalysts [12, 61]. However, because these catalysts have fewer acid sites, as shown in Table 15, the turnover frequency (TOF) of n-C7 hydroisomerisation at 370 °C over these sites would be expected to be low, which means in turn that there would be a relatively low level of interaction between the low acid sites in the large pore walls of these catalyst and the iso-heptane molecules generated by dehydrogenation on the platinum sites. Thus, these alkenes will be oligomerated to produce the cyclic products. It can be concluded that the acid and metal sites are not sufficiently well balanced to produce ideal bifunctional catalysts for n-C7 conversion. The oligomers formed may be attributed to the formation of carbonaceous species on the catalyst surface, which contribute to the poisoning and/or blocking of the active centres. Thus, it is clear that good isomerisation activity is obtained when the hydrogenation activity of the metal (Pt) and the Brønsted acid activity of the bifunctional catalyst are in balance. In comparison to the 1 wt% Pt/γ-Al_2O_3 catalyst, the 1 wt% Pt/Al-SBA-15 and 1 wt% Pt/Al-MCM-48 produced smaller quantities of cracked products, probably due to their effectively greater Pt dispersion.

The maximum coke content for the 1 wt% Pt/Al-SBA-15 catalyst was found to be 0.49, 0.72, and 0.70 wt% in the top, middle and bottom reactor zones, respectively, while the minimum coke content for the 1 wt% Pt/γ-Al_2O_3 catalyst was found to be 0.36 wt%, as revealed by TGA analysis. It can be clearly seen that the middle and bottom

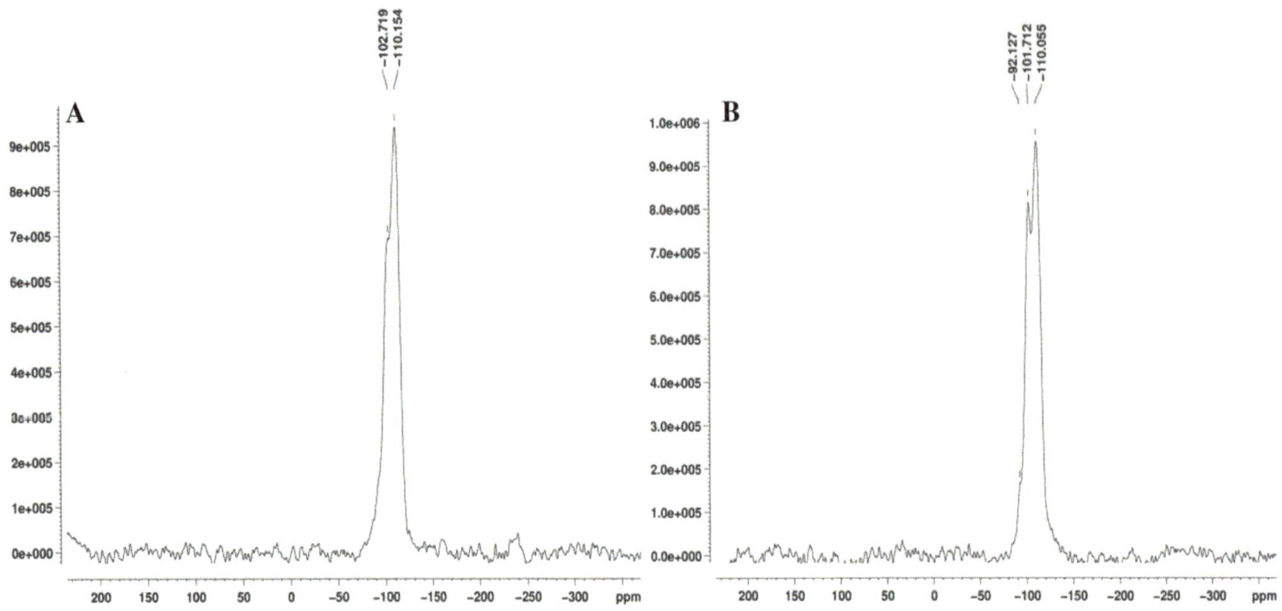

Fig. 34 ^{29}Si MAS NMR spectra of Al-SBA-15 (**a**) and Al-MCM-48 (**b**)

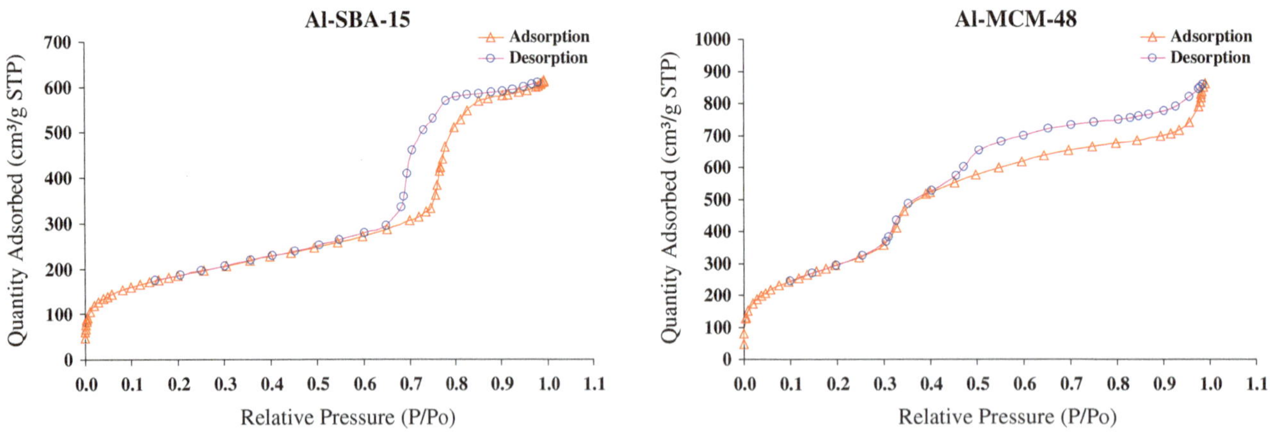

Fig. 35 N_2 adsorption and desorption isotherms of Al-SBA-15 and Al-MCM-48

zones of 1 wt% Pt/Al-SBA-15 had the highest hard coke content in the catalyst bed, whereas there was slightly less coke in the top zone. This suggests that the middle and bottom zones of the catalyst bed contributed the most activity, where the reaction took place. In addition, the lower coke content in the top zone is to some extent indicative of the low acidity of this catalyst and the fact that no diffusion limitations were encountered with it, since such limitations would result in coking and would cause the top section of the bed to be coke rich.

Conclusions

There are many conclusions which can be drawn from this investigation of the deactivation of a range of commercial bifunctional nanoporous-based catalysts during the hydroisomerisation of normal heptane (n-C7) under identical reaction conditions. The main cause of deactivation of acid zeolite catalysts is generally coking, as coke compounds become trapped in the zeolite pores and can poison the acid sites or block their access:

- It has been shown that many different parameters control the stability of USY zeolites in the hydroisomerisation of n-C7. Not only reaction conditions but also catalyst characteristics play the crucial roles.
- For most of the catalysts studied, total conversion increased dramatically as reaction temperature increased in the range of 210–270 °C. However, the total selectivity of C7 isomers at lower temperatures was very high throughout the deactivation reaction at

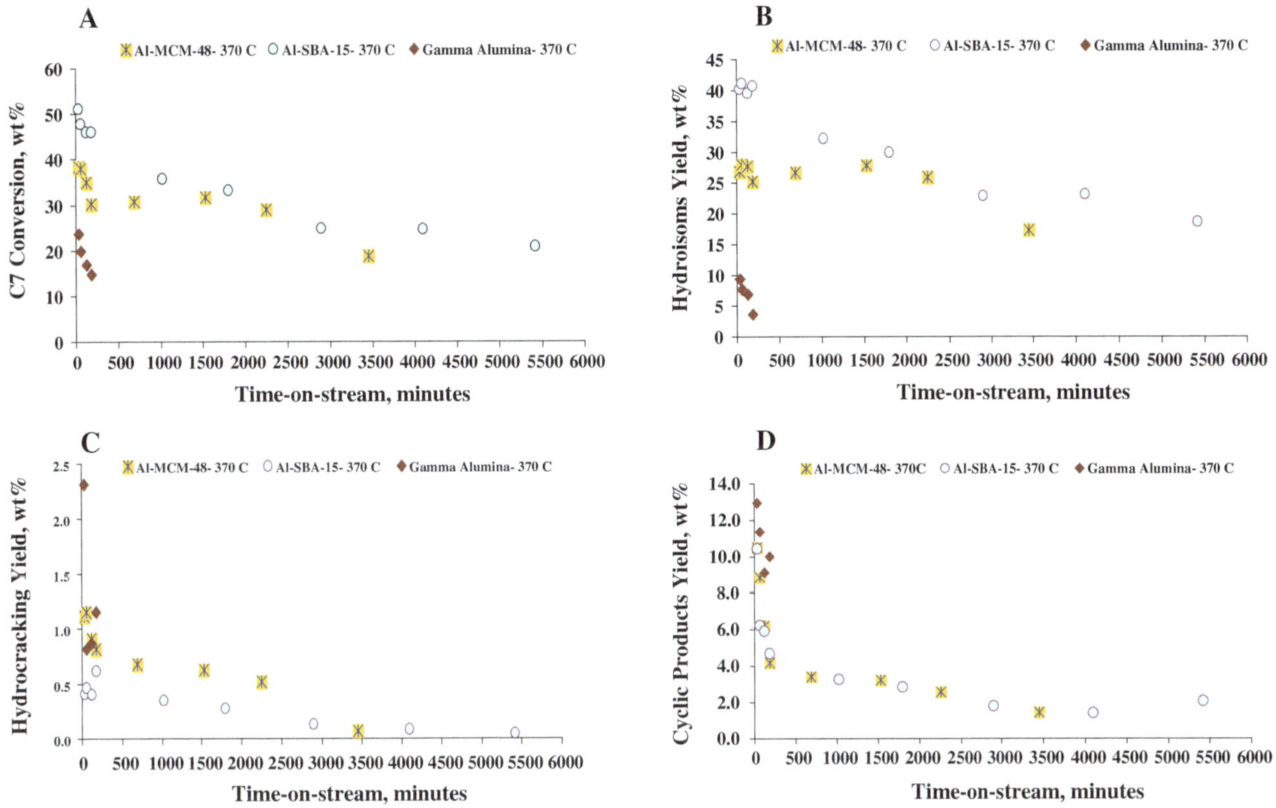

Fig. 36 Overall conversion (**a**), total yield of C7 isomers (mono + di + tribranched isomers) (**b**), total hydrocracking yield (**c**) and total yield of the cyclic products (**d**) as a function of TOS for the Al-SBA-15, Al-MCM-48 and γ-alumina catalysts. WHSV = 5.13 h^{-1}; H_2/C7 (mol/mol) = 9; total pressure = 1 atm; temperature = 370 °C; 1 wt% Pt loading

210 and 230 °C, taking advantage of the thermodynamic equilibrium.

- The selectivity of cracked products increased at the higher temperatures of 250 and 270 °C during the first few hours of the deactivation reaction and then decreased rapidly throughout the TOS. The selectivity of cyclic products was slightly higher at the lower temperatures of 210 and 230 °C.
- Increasing pressure resulted in decreased overall conversion and slightly decreased selectivity to isomers with noticeable increased selectivity to the cyclic products, particularly at the lower temperatures of 210 and 230 °C.
- There was no obvious cracking activity at the lower temperatures of 210 and 230 °C, and constant WHSV as the reaction pressure changed, and in contrast there was increased selectivity towards the cracked product as the reaction pressure increased at the highest temperature of 270 °C.
- The total conversion increased as the contact time increased at constant temperature and pressure. However, increasing the contact time between the n-C7 feed and the catalyst seems to cause an increase in cracking selectivity.
- Zeolite Y commercial catalysts that were generated by steaming at different levels of severity behaved very differently to each other and to acid-leached commercial catalysts.
- USY zeolite pore structures which can inhibit coke formation in conjunction with careful dealumination to decrease the number of acid sites are necessary for the development of time-stable reforming zeolite-based catalysts.
- There is a clear correlation between the obtained yield of the cracked products, which relates to the existence of the stronger acid sites and to the hard coke. Maximum hard coke values were found at 270 °C.
- The balance between the number of Pt atoms and the number of acid sites plays an important role in determining the activity, selectivity and stability of the bifunctional catalyst. The greater the value of n_{Pt}/n_A, the slower the coke formation and the higher the catalyst stability.
- Lower platinum dispersion may have resulted in the formation of a considerably greater quantity of coke.
- Lower acidity, hence less cracking activity, resulted in the catalyst being more stable against deactivation by coking, which is the case for zeolites with high Si/Al ratios.

- High surface area and larger pore volume, due to the more severe dealumination process and acid leaching, reduced the hard coke content, thus slowing the deactivation rate.
- The AHFS acid-leached zeolite had a larger pore volume (smaller pore size and larger surface area), due to the alleviation of the obstruction by EFAL of the mesopore system, thus slowing the formation of coke and providing good catalytic stability.
- The higher Pt loading, up to 2 wt%, improved catalytic stability, due to a better balance and closeness of the catalytic functions.
- Pt–Ni bimetallic catalysts were better than monometallic catalysts containing only Ni because they had higher activity, better selectivity for C7 isomers and more stability. The presence of platinum in the bimetallic catalysts improved the formation of metal particles to a great extent and reduced their sizes, while enhancing the reduction of nickel.
- The best method to introduce the platinum into the beta zeolite pores was found to be the ion-exchange method at pH > 9 which achieved a better metal/acid balance, lower cracking activity, and relatively good metal dispersion.
- The metal functions deactivated faster than the acid functions for all the catalysts loaded with platinum, irrespective of the method used. However, this was not sufficiently severe to cause a drop in n-C7 conversion due to the unique channel system and pore structure of beta zeolite.
- The TOS behaviour of USY zeolites can be enhanced by forming composite catalysts of USY and beta zeolite, providing a unique pore structure and acidity to improve their TOS stability.
- The metal function deactivated faster than the acid function in all the composite catalysts but did not cause a severe drop in n-C7 conversion, indicating that the composite zeolite pore structure plays a very important role in the conversion, and an appropriate combination of the pore structure and acidity showed superior TOS behaviour.
- Although the composite catalyst offers a highest acidity compared to the corresponding individual catalysts; coking is moderate and as such the TOS stability is reasonable.
- Synthesis parameters greatly affect mordenite morphology in terms of the size and aspect ratio of the mordenite crystals, so a range of mordenite morphologies were obtained for a variety of catalysts, both in-house and commercial. Each catalyst had different diffusion and porosity properties; therefore, performance differed among them.
- The framework aluminium was removed to extra-framework positions as a result of steam dealumination as well as giving the final in-framework Si/Al ratio. On the other hand, acid leaching resulted in samples that were practically free of EFAL, which in turn led to the formation of aluminium gradients along the zeolite crystallites. A more uniform aluminium distribution was achieved by steam treatment.
- The main factors controlling catalyst stability and selectivity were pore topology, acid strength, acid site density and location of the acid sites. The crucial feature for coke formation was the pore topology of the catalyst.
- The synthesis method of mesoporous materials and the post-alumination process applied significantly affect the number and strength of acid sites and thereby their catalytic activity, selectivity and stability.
- As the mesoporous molecular sieves present very high total surface area and large pore size, this may depress the TOF, causing the olefins (alkenes) produced during the dehydrogenation of alkanes on the metal sites to have less interaction with the acid sites on which they can be hydrogenated to give their relevant isomers. Thus, the alkenes will be oligomerated to produce the cyclic products, which may be attributed to the formation of carbonaceous species on the catalyst surface, contributing to the poisoning and/or blocking of the active centres.

Acknowledgments I would like to thank Dr. Patrick Hill (Manchester University) for his help in carrying out the SEM analysis, Dr. Vladimir Zholobenk (Keele University) who kindly provided the FTIR technique, Dr. Christopher Muryn (Manchester University) for his help in performing the XRD analyses, Professor Michael Anderson and Dr. Barbara Gore (Manchester University), who kindly provided the NMR technique, Mr. Martin Jennings (Manchester University) for his help with the elemental analysis and Mr. David Gordon (Manchester University) for his cooperation in making the glass reactor.

References

1. Guisnet M, Ribeiro FR (2011) Deactivation and regeneration of zeolite catalysts. Imperial College Press, London
2. Bartholomew CH (2003) Catalyst deactivation and regeneration. In: Kirk-Othmer encyclopedia of chemical technology. Wiley, London
3. Zhang W, Smirniotis PG (1999) On the exceptional time-on-stream stability of HZSM-12 zeolite: relation between zeolite pore structure and activity. Catal Lett 60:223–228

4. Stöcker M (2005) Gas phase catalysis by zeolites. Microporous Mesoporous Mater 82:257–292
5. Gopal S, Smirniotis PG (2004) Factors affecting isomer yield for *n*-heptane hydroisomerization over as-synthesized and dealuminated zeolite catalysts loaded with platinum. J Catal 225:278–287
6. Kuznetsov PN (2003) Study of *n*-octane hydrocracking and hydroisomerization over Pt/HY zeolites using the reactors of different configurations. J Catal 218:12–23
7. Villegas JI, Kumar N, Heikkila T, Smieskova A, Hudec P, Salmi T, Murzin DY (2005) A highly stable and selective Pt-modified mordenite catalyst for the skeletal isomerization of n butane. Appl Catal 284:223–230
8. McDaniel CW, Maher PK (1968) New ultrastable form of faujasite. In: Molecular sieves. London: Soc Chem Ind, pp 86–194
9. Van Laak ANC, Sagala SL, Zečević J, Friedrich H, de Jongh PE, de Jong KP (2010) Mesoporous mordenites obtained by sequential acid and alkaline treatments—catalysts for cumene production with enhanced accessibility. J Catal 276:170–180
10. Moushey DL (2006) Formation of mesoporosity in zeolite and mesoporous molecular sieve structures through use of carbon as a secondary templating agent. Dissertation, University of Cincinnati, USA
11. Logar NZ, Kaucic V (2006) Nanoporous materials: from catalysis and hydrogen storage to wastewater treatment. Acta Chim Slov 53:117–135
12. Corma A (1997) From microporous to mesoporous molecular sieve materials and their use in catalysis. Chem Rev 97:2373–2419
13. Pinnavaia TJ, Liu Y (2002) Aluminosilicate mesostructures with improved acidity and hydrothermal stability. J Mater Chem 12:3179–3190
14. Zeolyst International, http://www.zeolyst.com/our-products/standard-zeolite-powders/zeolite-y.aspx. Accessed 29 May 2011
15. Tosoh Corporation, http://www.tosoh.com/Products/tcdzeo.htm#hzc. Accessed 26 June 2011
16. Rawlence DJ, Earl, GJ (1993) Materials having novel structure factors for catalysis of hydrocarbon transformation and development of selective processes. BRITE-EURAM 4633, 5th Meeting Report, Elf Solaize
17. Abudawood R, Alotaibi F, Garforth A (2011) Hydroisomerization of *n*-heptane over Pt-loaded USY zeolites. Effect of steaming, dealumination and the resulting structure on catalytic properties. Ind Eng Chem Res 50:9918–9924
18. Abudawood R (2010) Hydroisomerization of alkanes over metal-loaded zeolite catalysts. Dissertation, University of Manchester
19. Alotaibi F (2011) Comparative study of the time-on-stream stability of bifunctional nanoporous-based catalysts in *n*-heptane hydroisomerisation. Dissertation, University of Manchester
20. Claytec, Inc., http://claytecinc.com/products. Accessed 29 May 2011
21. Robson H, Lillerud KP (2001) Verified Syntheses of zeolitic materials, 2nd edn. Elsevier, Amsterdam
22. Weitkamp J, Puppe L (1999) Catalysis and zeolites. Springer, Berlin
23. Wang QL, Torrealba M, Giannetto G, Guisnet M, Perot G, Cahoreau M, Caisso J (1990) Dealumination of Y zeolite with ammonium hexafluorosilicate: a SIM-XPS study of the aluminium distribution. Zeolites 10:703–706
24. Wang QL, Giannetto G, Torreaaba M, Perot G, Kappenstein C, Guisnet M (1991) Dealumination of zeolites: II. Kinetic study of the dealumination by hydrothermal treatment of a NH4NaY zeolite. J Catal 130:459–470
25. Wang QL, Giannetto G, Guisnet M (1991) Dealumination of zeolites: III. Effect of extra-framework aluminum species on the activity, selectivity, and stability of Y zeolites in *n*-heptane cracking. J Catal 130:471–482
26. Remy MJ, Stanica D, Poncelet G, Feijen EJP, Grobet PJ, Martens JA, Jacobs PA (1996) Dealuminated H–Y zeolites: relation between physicochemical properties and catalytic activity in heptane and decane isomerization. J Phys Chem 100:12440–12447
27. Van Bekkum H, Flanigen EM, Jacobs PA, Jansen JC (2001) Introduction to zeolite science and practice, 2nd edn. Elsevier, Amsterdam, pp 261–288
28. Roussel ML, emberton J, Guisnet M, Cseri T, Benazzi E (2003) Mechanisms of *n*-decane hydrocracking on a sulfided NiW on silica–alumina catalyst. J Catal 218: 427–437
29. Guerin M, Kappenstein C, Alvarez F, Giannetto G, Guisnet M (1988) Preparation of PtHY catalysts: influence on the catalytic properties of the complexes used as platinum precursors. Appl Catal 45:325–333
30. Martins A, Silva JM, Ribeiro FR, Ribeiro MF (2006) Hydroisomerization of *n*-hexane over Pt–Ni/HBEA using catalysts prepared by different methods. Catal Lett 109(1–2):83–87
31. Alotaibi F, Abudawood R, Al-Megren H, Al-Kinany H, Garforth A (2013) The time-on-stream stability of some selected bifunctional nanoporous-based catalysts in *n*-heptane hydroisomerisation. J Appl Petrochem Res (submitted)
32. Chao K, Lin C, Lin C, Wu H, Tseng C, Chen S (2000) *n*-Heptane hydroconversion on platinum-loaded mordenite and beta zeolites: the effect of reaction pressure. Appl Catal A 203:211–220
33. Tran MT, Gnep NS, Szabo G, Guisnet M (1998) Isomerization of *n*-butane over H-mordenites under nitrogen and hydrogen: influence of the acid site density. J Catal 174:185–190
34. Guisnet M, Fouche V (1991) Isomerization of *n*-hexane on platinum dealuminated mordenite catalysts: I. Influence of the silicon-to-aluminium ratio of the zeolite. Appl Catal 71:283–293
35. Wang ZB, Kamo A, Yoneda T, Komatsu T, Yashima T (1997) Isomerization of *n*-heptane over Pt-loaded zeolite β catalysts. Appl Catal 159:119–132
36. Chica A, Corma A (1999) Hydroisomerization of pentane, hexane, and heptane for improving the octane number of gasoline. J Catal 187:167–176
37. Lónyi F, Lunsford JH (1992) The development of strong acidity in hexafluorosilicate-modified Y-type zeolites. J Catal 136:566–577
38. Cruz JM, Corma A, Fornes V (1989) Framework and extra framework aluminum distribution in $(NH_4)_2F_6Si$-dealuminated Y-zeolites—relevance to cracking catalysts. Appl Catal 50:287–293
39. Gola A, Rebours B, Milazzo E, Lynch J, Benazzi E, Lacombe S, Delevoye L, Fernandez C (2000) Effect of leaching agent in the dealumination of stabilized Y zeolites. Microporous Mesoporous Mater 40:73–83
40. Jordao MH, Valencia E, Vicemario S, Dilson C (1999) Characterization of Pt/HUSY and Pt-Ni/HUSY catalysts by transmission electron microscopy. Mate Res 2:219–223
41. Karthikeyan D, Lingappan N, Sivasankar B (2008) Hydroisomerization of *n*-octane over bifunctional Ni-Pd/HY zeolite catalysts. Ind Eng Chem Res 47:6538–6546
42. Barsi FV, Cardoso D (2009) Bimetallic Pt–Ni catalysts supported on USY zeolite for *n*-hexane isomerization. Braz J Chem Eng 26–2:353–360
43. Roldán R, Beale AM, Sánchez-Sánchez M, Romero-Salguero FJ, Jiménez-Sanchidrián C, Gómez JP, Sankar G (2008) Effect of the impregnation order on the nature of metal particles of bi-functional Pt/Pd-supported zeolite beta materials and on their catalytic activity for the hydroisomerization of alkanes. J Catal 254:12–26
44. Eswaramoorthi I, Lingappan N (2003) Ni–Pt/H–Y zeolite catalysts for hydroisomerization of *n*-hexane and *n*-heptane. Catal Lett 87(3–4):133–142
45. Kampers FWH, Engelen CWR, van Hoof JHC, Konigsberger DC (1990) Influence of preparation method on the metal cluster size

of platinum/ZSM-5 catalysts as studied with EXAFS. J Phys Chem 94(23):8574–8578
46. Changlei J, Bo M, Xiwen Z, Fengxiang L, Zhizhi Z, Bo Q (2008) Catalytic performance of Pt/HY-β in *n*-octane hydroisomerization. Petrol Sci 6:299–305
47. Smirniotis PG, Davydov L, Ruckenstein E (1999) Composite zeolite-based catalysts and sorbents. Catal Rev 41:43–113
48. Hamidi F, Bengueddach A, Di Renzo F, Fajula F (2003) Control of crystal size and morphology of mordenite. Catal Lett 87:149–152
49. Treacy MMJ, Higgins JB (2001) Collection of simulated XRD powder patterns for zeolites. Elsevier, Amsterdam
50. Janssen AH, Koster AJ, de Jong KP (2001) Three-dimensional transmission electron microscopy observations of mesopores in dealuminated zeolite Y. Angew Chem Int Ed 40:1102–1104
51. Van Donk S, Janssen AH, Bitter JH, de Jong KP (2003) Generation, characterization, and impact of mesopores in zeolite catalysts. Catal Rev 45:297–319
52. Janssen AH, Koster AJ, de Jong KP (2002) On the shape of the mesopores in zeolite Y: a three-dimensional transmission electron microscopy study combined with texture analysis. J Phys Chem 106:11905–11909
53. Lee JK, Rhee HK (1997) Characteristics of Pt/H-beta and Pt/H-mordenite catalysts for the isomerization of *n*-hexane. Catal Today 38:235–242
54. Van Donk S (2002) Adsorption, diffusion and reaction studies of hydrocarbons on zeolite catalysts. PhD Dissertation, Utrecht University, Netherlands
55. Chica A, Corma A, Miguel PJ (2001) Isomerization of C5–C7 *n*-alkanes on unidirectional large pore zeolites: activity, selectivity and adsorption features. Catal Today 65:101–110
56. Nie C, Huang L, Zhao D, Li Q (2001) Performance of Pt/Al-SBA-15 catalysts in hydroisomerization of *n*-dodecane. Catal Lett 71:117–125
57. Beck JS, Vartuli JC, Roth WJ, Leonowicz ME, Kresge CT, Schmitt KD, Chu CTW, Olson DH, Sheppard EW (1992) A new family of mesoporous molecular sieves prepared with liquid crystal templates. J Am Chem Soc 114:10834–10843
58. Eimer GA, Pierella LB, Monti GA, Anunziata OA (2002) Synthesis and characterization of Al-MCM-41 and Al-MCM-48 mesoporous materials. Catal Lett 78:65–75
59. Gregg SJ, Sing KSW (1982) Adsorption, surface area and porosity. Academic Press, London
60. Chaudhari K, Das TK, Chandwadkar AJ, Sivasanker S (1999) Mesoporous aluminosilicate of the MCM-41 type: its catalytic activity in *n*-hexane isomerization. J Catal 186:81–90
61. Chaudhari K (2000) Synthesis, characterization and catalytic properties of mesoporous molecular sieves. PhD Thesis, Catalysis Division, National Chemical Laboratory, Pune

Permissions

The contributors of this book come from diverse backgrounds, making this book a truly international effort. This book will bring forth new frontiers with its revolutionizing research information and detailed analysis of the nascent developments around the world.

We would like to thank all the contributing authors for lending their expertise to make the book truly unique. They have played a crucial role in the development of this book. Without their invaluable contributions this book wouldn't have been possible. They have made vital efforts to compile up to date information on the varied aspects of this subject to make this book a valuable addition to the collection of many professionals and students.

This book was conceptualized with the vision of imparting up-to-date information and advanced data in this field. To ensure the same, a matchless editorial board was set up. Every individual on the board went through rigorous rounds of assessment to prove their worth. After which they invested a large part of their time researching and compiling the most relevant data for our readers. Conferences and sessions were held from time to time between the editorial board and the contributing authors to present the data in the most comprehensible form. The editorial team has worked tirelessly to provide valuable and valid information to help people across the globe.

Every chapter published in this book has been scrutinized by our experts. Their significance has been extensively debated. The topics covered herein carry significant findings which will fuel the growth of the discipline. They may even be implemented as practical applications or may be referred to as a beginning point for another development. Chapters in this book were first published by Springer; hereby published with permission under the Creative Commons Attribution License or equivalent.

The editorial board has been involved in producing this book since its inception. They have spent rigorous hours researching and exploring the diverse topics which have resulted in the successful publishing of this book. They have passed on their knowledge of decades through this book. To expedite this challenging task, the publisher supported the team at every step. A small team of assistant editors was also appointed to further simplify the editing procedure and attain best results for the readers.

Our editorial team has been hand-picked from every corner of the world. Their multi-ethnicity adds dynamic inputs to the discussions which result in innovative outcomes. These outcomes are then further discussed with the researchers and contributors who give their valuable feedback and opinion regarding the same. The feedback is then collaborated with the researches and they are edited in a comprehensive manner to aid the understanding of the subject.

Apart from the editorial board, the designing team has also invested a significant amount of their time in understanding the subject and creating the most relevant covers. They scrutinized every image to scout for the most suitable representation of the subject and create an appropriate cover for the book.

The publishing team has been involved in this book since its early stages. They were actively engaged in every process, be it collecting the data, connecting with the contributors or procuring relevant information. The team has been an ardent support to the editorial, designing and production team. Their endless efforts to recruit the best for this project, has resulted in the accomplishment of this book. They are a veteran in the field of academics and their pool of knowledge is as vast as their experience in printing. Their expertise and guidance has proved useful at every step. Their uncompromising quality standards have made this book an exceptional effort. Their encouragement from time to time has been an inspiration for everyone.

The publisher and the editorial board hope that this book will prove to be a valuable piece of knowledge for researchers, students, practitioners and scholars across the globe.

List of Contributors

Y. Huang, H. Chen and T. Xiao
Guangzhou Boxenergy Technology Ltd, Guangzhou Hi-Tech Development Zone, Guangzhou, People's Republic of China

J. Su
Environmental Engineering College, Shandong University, No. 27 Shanda Nanlu, Jinan, People's Republic of China

P. N. Nelson
University of the West Indies, Mona Campus, Kingston 7, Jamaica

R. A. Taylor
University of the West Indies, St. Augustine Campus, St. Augustine, Trinidad and Tobago

N. F. Sauty, L. C. da Silva, M. D. Schulz, C. S. Few and K. B. Wagener
The George and Josephine Butler Polymer Research Laboratory, Department of Chemistry and Center for Macromolecular, Science and Engineering, University of Florida, Gainesville, FL 32611-7200, USA

N. F. Sauty, L. C. da Silva, M. D. Schulz, C. S. Few and K. B. Wagener
The George and Josephine Butler Polymer Research Laboratory, Department of Chemistry and Center for Macromolecular, Science and Engineering, University of Florida, Gainesville, FL 32611-7200, USA

P. Frank, A. Prasher and E. Berda
Department of Chemistry, University of New Hampshire, Durham, USA

B. Tuten and E. Berda
Materials Science Program, University of New Hampshire, Durham, USA

D. Chao
Alan G. MacDiarmid Institute, College of Chemistry, Jilin University, Changchun, People's Republic of China

A. A. Mohammed, S. E. K. Fateen, T. S. Ahmed and T. M. Moustafa
Chemical Engineering Department, Faculty of Engineering, Cairo University, Giza 12613, Egypt

H. Martinez, J. Zhang, S. Kobayashi, Y. Xu, L. M. Pitet, M. E. Matta and M. A. Hillmyer
Department of Chemistry, University of Minnesota, 207 Pleasant, St. SE, Minneapolis, MN 55455-0431, USA

J. S. J. Hargreaves
WestCHEM, School of Chemistry, University of Glasgow, Joseph Black Building, Glasgow G12 8QQ, UK

M. Takht Ravanchi and S. Sahebdelfar
Catalyst Research Group, Petrochemical Research and Technology Company, National Petrochemical Company, No. 27, Sarv Alley, Shirazi-south, Mollasadra, P.O. Box 14358-84711, Tehran, Iran

A. K. Wahab, S. Bashir, Y. Al-Salik and H. Idriss
SABIC-Corporate Research and Innovation (CRI) at KAUST, Thuwal, Saudi Arabia

M. H. Wilson, J. Groppo, A. Placido, S. Graham, E. Santillan-Jimenez, M. Crocker and R. Andrews
Center for Applied Energy Research, University of Kentucky, Lexington, KY 40511, USA

S. A. Morton III
Department of Engineering, James Madison University, Harrisonburg, VA 22807, USA

A. Shea and C. Crofcheck
Department of Biosystems and Agricultural Engineering, University of Kentucky, Lexington, KY 40506, USA

R. Andrews
Department of Chemical and Materials Engineering, University of Kentucky, Lexington, KY 40506, USA

A. A. Alsaygh, F. D. Alsewailem and I. M. Al-Najjar
Petroleum and Petrochemicals Research Institute, King Abdulaziz City for Science and Technology (KACST), P.O. Box 6086, Riyadh 11442, Saudi Arabia

J. Al-hamidi
Chemistry Department, College of Science, Princess Nora Bint Abdulrahman University, Riyadh, Saudi Arabia

V. L. Kuznetsov
Department of Chemistry, University of Oxford, South Parks Road, Oxford OX1 3QR, UK

J. Ali and S. D. Jackson
Department of Chemistry, Centre for Catalysis Research, WestCHEM, University of Glasgow, G12 8QQ Glasgow, Scotland, UK

A. M. Savage, X. Zhou, J. Huang and S. R. Turner
Department of Chemistry MC0212 and Macromolecules and Interfaces Institute MC0344, Virginia Tech, Blacksburg, VA 24061, USA

A. F. Lee
European Bioenergy Research Institute, Aston University, Aston Triangle, Birmingham B4 7ET, UK

F. M. Alotaibi, H. A. Al-Megren and M. C. Al-Kinany
Petrochemicals Research Institute, King Abdulaziz City for Science and Technology (KACST), P.O.Box 6086, Riyadh 11442, Saudi Arabia

R. H. Abudawood
Research and Development Center, Saudi Aramco, Dhahran 31311, Saudi Arabia

E. H. Jamea
SABIC Technology Center, Saudi Basic Industries Corporation, Riyadh 11551, Saudi Arabia

A. A. Garforth
School of Chemical Engineering and Analytical Science, The University of Manchester, Manchester M13 9PL, UK